AF576039

Advanced Materials and Technologies for Fuel Cells

Advanced Materials and Technologies for Fuel Cells

Editors

Massimo Viviani
Antonio Barbucci
Maria Paola Carpanese
Sabrina Presto

MDPI • Basel • Beijing • Wuhan • Barcelona • Belgrade • Manchester • Tokyo • Cluj • Tianjin

Editors

Massimo Viviani
National Research Council of Itay—CNR
Institute of Condensed Matter Chemistry and Energy Technologies
Italy

Antonio Barbucci
University of Genoa
Dept. of Civil, Chemical and Environmental Engineering—DICCA
Italy

Maria Paola Carpanese
University of Genoa
Dept. of Civil, Chemical and Environmental Engineering—DICCA
Italy

Sabrina Presto
National Research Council of Itay—CNR
Institute of Condensed Matter Chemistry and Energy Technologies
Italy

Editorial Office
MDPI
St. Alban-Anlage 66
4052 Basel, Switzerland

This is a reprint of articles from the Special Issue published online in the open access journal *Energies* (ISSN 1996-1073) (available at: https://www.mdpi.com/journal/energies/special_issues/Advanced_Materials_Technologies).

For citation purposes, cite each article independently as indicated on the article page online and as indicated below:

LastName, A.A.; LastName, B.B.; LastName, C.C. Article Title. *Journal Name* **Year**, *Volume Number*, Page Range.

ISBN 978-3-0365-0538-1 (Hbk)
ISBN 978-3-0365-0539-8 (PDF)

Cover image courtesy of Sabrina Presto.

Contents

About the Editors

Massimo Viviani received Physics and Ph.D. degrees from the University of Genova, Italy, in 1992 and 1998, respectively. Since 1999, he has been a Researcher at the National Research Council of Italy, alongside which he is currently the Director of Research at the Institute of Condensed Matter Chemistry and of Energy Technologies. His work is focused on the development of ceramic materials for energy applications. He is co-author of over 150 papers in multiple international journals and is a Section Board Member for the journal, *Energies*.

Antonio Barbucci received Chemical Engineering and Ph.D. degrees from the University of Genova, Italy, in 1991 and 1995, respectively. Since 1999, he has been a Researcher at the University of Genova, where he is currently a Professor in Material Science Engineering in the Dept. of Civil Chemical and Environmental Engineering. Since 2010, he has been an Associated Researcher at the National Research Council of Italy at the Institute of Condensed Matter Chemistry and Energy Technologies. His work is focused on electrochemical processes for energy applications. He is co-author of over 100 papers in multiple international journals and is a Section Board Member for the journal, *Energies*.

Maria Paola Carpanese received Chemical Engineering and Ph.D. degrees at the University of Genova, Italy, in 2003 and 2008, respectively. Since 2009, she has been a Researcher at the University of Genova, where she is currently Assistant Professor in Ceramic Materials and Electrochemical Energy Conversion and Storage Devices. Her work is addressed to solid state electrochemistry and solid oxide reversible cell production and characterization (notably, impedance spectroscopy). She is co-author of over 50 papers in multiple international journals and a Topics Board Member for the journal, *Energies*.

Sabrina Presto received her Physics degree from the University of Pisa, Italy, in 2000, and her Ph.D. degree from the University of Genova, Italy, in 2008. Since 2009, she has been a Researcher at the National Research Council of Italy, where she is currently a Researcher at the Institute of Condensed Matter Chemistry and Energy Technologies. Her work is focused on the synthesis and characterization of ceramic materials for energy applications. She is the co-author of over 50 papers in multiple international journals and a Topics Board Member for the journal, *Energies*.

Preface to "Advanced Materials and Technologies for Fuel Cells"

The urgent need for a transition toward a sustainable energy system is boosting research on clean and carbon net-zero processes, suitable for replacing the combustion of fossil fuels. Importantly, fuel cells (FCs) are a cornerstone that are able to sustain both the hydrogen and renewable-energy-based technologies. On one end, the power-to-fuel approach will be needed to fully exploit the intermittent and off-peak power availability of wind and solar energy, while on the other end the efficient conversion of H2 into electricity will be necessary, especially for non-stationary applications such as transportation. Fuel cells are available at different ranges of temperature and at different scales of power density; they operate with varying fuel types and also provide reversible operation as electrolyzers in order to obtain fuels with the highest efficiency. Not all types of fuel cells and electrolyzers share the same technology readiness level, and there is much available room still for new concepts and approaches. This Special Issue showcases compelling examples of the present developments in FC science and technology, with original and high-quality contributions from all over the world. Out of the 13 papers collected, eight are dedicated to different aspects of solid oxide fuel cells (SOFCs), from materials of single cells (electrolyte and electrodes) and stacks (glass sealing), to the simulation of CH4 internal reforming and to field tests in off-shore conditions. A detailed description of non-conventional operating modes in polymer electrolyte membrane (PEM) fuel cells is given in two brilliant papers, while hydrogen production by alkaline electrolyzers is the subject of the paper by Le Bideau et al. Furthermore, this Special Issue is enriched by two very interesting contributions from You et al. about microbial fuel cells and from Hernández Rivera et al. about soft drink–fueled paper cells. Editing this volume was an exciting professional and personal experience and, therefore, we are deeply grateful to all of the authors and reviewers that contributed to making it such a success. Finally, we would like to thank the *Energies* Editorial Office and especially the MDPI Managing Editor, Ms. Adele Min, for her constant support.

Massimo Viviani, Antonio Barbucci, Maria Paola Carpanese, Sabrina Presto

Editors

Article

$Ce_{0.9}Gd_{0.1}O_{2-x}$ for Intermediate Temperature Solid Oxide Fuel Cells: Influence of Cathode Thickness and Anode Functional Layer on Performance

Visweshwar Sivasankaran [1,2], Lionel Combemale [1], Mélanie François [1] and Gilles Caboche [1,*]

1 Laboratoire Interdisciplinaire Carnot de Bourgogne, ICB-UMR6303, FCLAB, CNRS—Université de Bourgogne, Franche-Comté, 9 Avenue Savary, CEDEX BP47870, 21078 Dijon, France; vsivasankaran@rgipt.ac.in (V.S.); lionel.combemale@u-bourgogne.fr (L.C.); melaniea.francois@gmail.com (M.F.)

2 Energy Institute Bangalore, A Centre of Rajiv Gandhi Institute of etroleum Technology—Bangalore Center, Nirman Bhavan, Third Floor Dr Rajkumar Road, Rajajinagar I Block, Bengaluru 560 010, India

* Correspondence: gilles.caboche@u-bourgogne.fr; Tel.: +33-380-39-6153

Received: 22 June 2020; Accepted: 25 August 2020; Published: 26 August 2020

Abstract: The performances of Intermediate Temperature Solid Oxide Fuel Cell (IT-SOFC) anode-supported planar cells with a 10 cm^2 active surface were studied versus the combination of cathode thickness and the presence of an Anode Functional Layer (AFL). The temperature range was 500 to 650 °C, and $Gd_{0.1}Ce_{0.9}O_{2-x}$ (GDC) was used as the electrolyte material, Ni-GDC as the anode material, and $La_{0.6}Sr_{0.4}Co_{0.2}Fe_{0.8}O_{3-d}$ (LSCF48) as the cathode material. The power density, conductivity, and activation energy of different samples were determined in order to investigate the influence of the cathode thickness and AFL on the performance. These results showed an improvement in the performances when the AFL was not present. The maximum power density reached 370 $mW{\cdot}cm^{-2}$ at 650 °C for a sample with a cathode thickness of 50 μm and an electrolyte layer that was 20 μm thick. Moreover, it was highlighted that a thinner cathode layer reduced the power density of the cell.

Keywords: tape casting process; open circuit voltage; activation energy; power density; IT-SOFC

---

1. Introduction

In recent years, there has been more need to develop Intermediate Temperature Solid Oxide Fuel Cells (IT-SOFCs), which work in the range 500–600 °C, compared to the commonly used SOFCs which work in the range 700–1000 °C [1]. A decrease in the SOFC operating temperature can lower the thermal stress on the SOFC stack and widen the types of materials used for structural components in the SOFC systems, which are supposed to provide a reduction in the operation costs and extension of the stack lifetime [2]. However, a reduction in operating temperature may be accompanied by a decrease in the electrochemical performance of each fuel cell system material [3].

Therefore, the choice of the cell materials is restricted to those electrochemically active at the chosen temperature range. Gadolinium-doped ceria ($Gd_{0.1}Ce_{0.9}O_{2-x}$: GDC) is a promising candidate for SOFC electrolyte at intermediate temperatures thanks to its high ionic conductivity [4], low activation energy, and chemical stability between the room temperature and its melting point.

For the cathode material, lanthanum strontium cobalt iron oxide with the specific composition $La_{0.6}Sr_{0.4}Co_{0.2}Fe_{0.8}O_{3\pm\delta}$ (LSCF48) is largely used because of its good chemical compatibility with GDC, its high electro-catalytic properties in the temperature range of 500 to 600 °C, and its thermal expansion coefficient in accordance with the GDC electrolyte [5]. For the anode material, a classical cermet (CERamic plus MEtal) Ni-GDC, showing an excellent electrochemical performance in the case of the use of pure H_2 as fuel [6], was used. Even though many studies deal with anode thickness and its porosity [7], few of them are interested in the thickness of the cathode and the impact of the Anode Functional Layer (AFL) presence.

This study presents the evolution of the electrical performance with respect to cathode morphology and the presence of an AFL. In order to explain in detail, the maximum power density, activation energies, and Area-Specific Resistance (ASR) measurements obtained at different temperatures were compared on the cells prepared by a single-step sintering process with multilayers.

2. Materials and Methods

2.1. Starting Materials

Commercial-grade $Gd_{0.1}Ce_{0.9}O_{2-x}$ (GDC-10 TC) powder provided by Neyco with a specific surface area of 12.1 $m^2 \cdot g^{-1}$ was chosen as the electrolyte material. NiO commercial nickel oxide from Sigma-Aldrich (99.99% purity with a specific surface area of 3 $m^2 \cdot g^{-1}$) was mixed with GDC powder thoroughly in a weight ratio of 65:35 by ball milling in ethanol for 4 h before bein dried for 13 h at 80 °C to obtain NiO-GDC anode material. For the cathode material, $La_{0.6}Sr_{0.4}Co_{0.2}Fe_{0.8}O_{3\pm\delta}$ (LSCF48) powder was prepared by solid state synthesis at Laboratoire Interdisciplinaire Carnot de Bourgogne (ICB Lab Dijon France).

The initial precursors brought from Sigma Aldrich—strontium carbonate ($SrCO_3$, 99.9%) and oxides of lanthanum (La_2O_3, 99.9%), cobalt (Co_2O_3, 99.0%), and iron (Fe_2O_3, 99.0%) powders—were mixed in a stoichiometric ratio and ball milled for 15 h in ethanol with zirconia balls before being dried for 24 h at 60 °C. Then, the dried powders were calcined at 1100 °C for 24 h with a heating and cooling ramp of 100 $°C \cdot h^{-1}$. The pure perovskite phase of LSCF48 was confirmed by X-ray diffraction (XRD, D5000, BRUKER Corporation) using Cu $K\alpha_1$ radiation (λ = 0.15406 nm), showing no impurities. The specific surface area was determined by the Brunauer–Emmett–Teller (BET) technique to be 0.64 $m^2 \cdot g^{-1}$.

2.2. Slurry Formation for Tape Casting Process

Starting powders have been used as active materials in slurry preparation. Compounds taken at the required weight percentage were mixed in a plastic container. For the preparation of the cathode slurry, LSCF48 was mixed with GDC to make a composite slurry to improve the cell performance by reducing the cathode polarization [8–10]. In this work, a mass ratio of 65:35 for LSCF48 to GDC was chosen, and carbon graphite (5% weight) was added as a pore former.

Concerning the anode material, two slurries were prepared, one with a pore former (5% graphite) for classical porous anode fabrication and the second one without pore former to understand the function of the AFL, between the electrolyte layer and the classical anode layer. The AFL was introduced to the cell to maximize the triple phase boundary (TPB) length and to restrain the activation polarization of the anode [11]. The electrolyte slurry was prepared using GDC powder and was mixed in a Turbula-T2F device with balls of zirconia with a diameter 10 mm for 24 h. This slurry was prepared in two steps, where in the first step GDC was mixed with ethanol and methyl ethyl ketone (MEK) as solvents and triethanolamine (TEA) as a dispersant. In the second step, the binder polyvinyl butyral (PVB) and the plasticizers polyethylene glycol (PEG) and benzylbutyl phthalate (BBP) were added and ball milled for 24 h.

2.3. Tape Casting and Sintering Process

The elaboration of the Membrane-Electrode Assemblies (MEA) composed of dense oxygen-ion-conducting electrolyte (GDC) sandwiched between cathode porous (LSCF-GDC) and anode porous (NiO-GDC) was performed from a multi-layer green tape of the different slurries [12,13]. The cathode slurry was first tape casted on a glass plate using an automatic tape caster (Elcometer) with a casting rate of 1 $cm \cdot s^{-1}$. The blade gap thickness was fixed, taking into account the sintering shrinkage.

Then, the cathode layer was dried in air at room temperature for 1 h and cut at the desired dimensions before the electrolyte layer was directly tape casted on the cathode layer. After 2 h of drying, the anode functional layer was casted above the electrolyte layer. In the last step, the anode layer was casted on AFL and dried, followed by sintering to obtain an anode-supported IT-SOFC. Figure 1 shows the diagrammatic representation of the process for the elaboration of the four anode-supported planar cells with a 10 cm^2 active surface. Long-term tests were carried out on these single cells manufactured using this process, and they showed a light performance degradation (~9%) after the first 100 h [13].

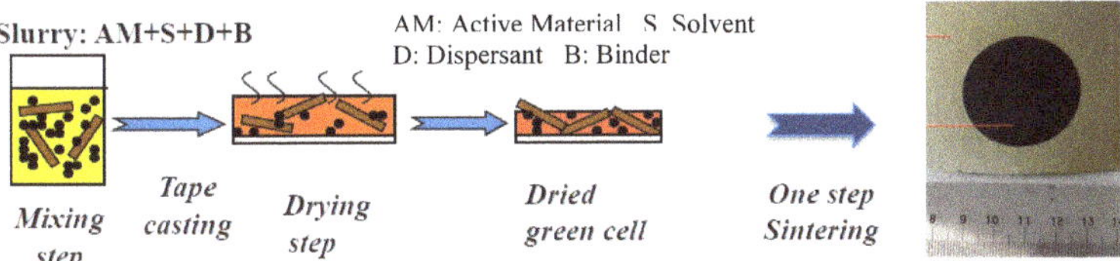

Figure 1. Diagrammatic representation of the elaboration process of the Membrane-Electrode Assemblies (MEAs).

The first MEAs were sintered at 1100, 1215, 1300, and 1400 °C for 5 h with one-hour isothermal at 280 °C, and a heating and cooling rate of 120 $K{\cdot}h^{-1}$ to determine the best sintering temperature. At 1100 °C, some pores are always present in the electrolyte layer, as shown in Figure 2. When the sintering temperature is higher than 1215 °C the electrolyte layer showed a better density. According to V. Sivasankaran et al. [13,14] and K. Raju et al. [15], the optimized sintering temperature was chosen to be 1215 °C.

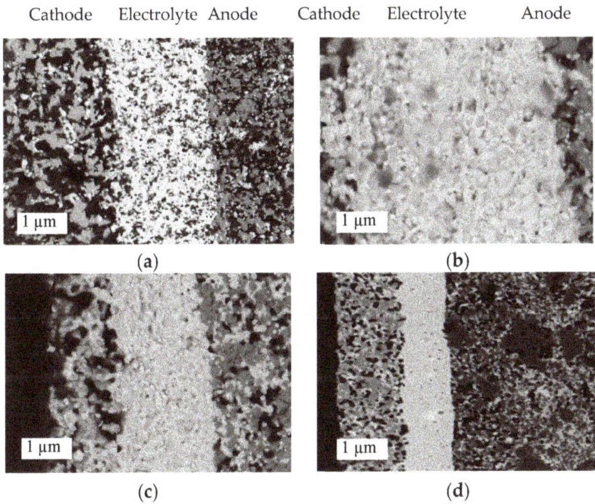

Figure 2. Cross-section of the different MEAs versus the sintering temperature obtained by SEM; the electrolyte layer in the middle shows a high density when the sintering temperature is equal to and over 1215 °C. (**a**) Sintered at 1100 °C; (**b**) sintered at 1215 °C; (**c**) sintered at 1300 °C; (**d**) sintered at 1400 °C.

The cross-sections of the MEAs samples 1 to 4 were observed by Scanning Electron Microscopy (SEM-HITACHI S4200) to determine the thicknesses of the different layers; see Figure 3a,b. In all cases, the electrolyte is very dense and the electrodes are porous. Table 1 gives the features of the different samples in this study. For samples 1 to 3, an AFL layer was deposited between the electrolyte layer and the classical porous anode-supported layer. For sample 4, the thicknesses of the different layers were optimized and no AFL was present.

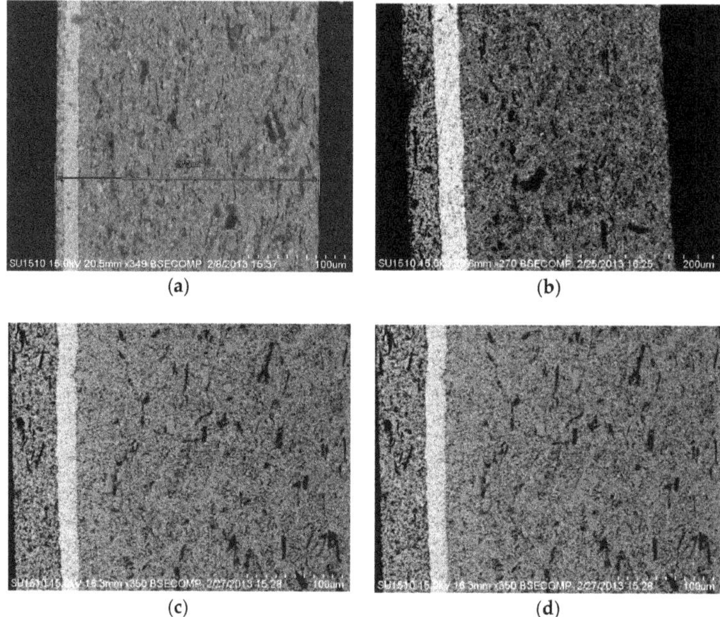

Figure 3. Evolution of the thicknesses of the cathode and electrolyte layers (values are reported in Table 1). (**a**) Cross section of sample 1; (**b**) cross section of sample 2; (**c**) cross section of sample 3; (**d**) cross section of sample 4.

Table 1. Features of the samples studied to improve the density power.

	Cathode Active Area (cm^2)	Thicknesses (μm)			
		Cathode	Electrolyte	Anode Functional Layer	Anode
Sample 1	10	9	15	18	242
Sample 2	10	43	37	33	245
Sample 3	10	48	21	29	249
Sample 4	10	50	20	-	270

3. Results and Discussion

Electrical measurements were conducted using a Fiaxell open flanges device. The cell consisted of a Ni-GDC anode substrate, a functional layer, a GDC electrolyte layer, and a LSCF-GDC cathode. The tests were carried out using a single cell with dimensions of 7.5 cm × 7.5 cm, with an active area of 10 cm^2 and a 3.6 cm diameter of the cathode part (Figure 1). All the samples were tested with gold mesh and nickel felt as current collectors at the cathode and anode sides, respectively. For better gas distribution, an alumina felt was also used at the anode and cathode sides. The samples were then heated at a heating rate of 150 $K \cdot h^{-1}$ in $H_2(1\%)/N_2$ to 750 °C and retained for 10 min, then the reduction process was continued in pure hydrogen with a flow rate of 500 $NmL \cdot min^{-1}$ at the anode side and air as the cathode gas with a flow rate of 600 $NmL \cdot min^{-1}$.

After reduction for 2 h, air and a mixture of $H_2/(3\%)$ H_2O were introduced into the cathode and the anode with a flow rate of 1600 and 800 $NmL \cdot min^{-1}$, respectively. Figure 4a shows *I-V* curves for sample 1 at different temperatures. An Open Circuit Voltage (OCV) of 920 mV at 500 °C was obtained, and it gradually decreases when the temperatures increase. The cell voltage decreases sharply at

500 °C when I increases due to a high value of internal resistance, as shown in Table 2. According to J. Li et al. [16], the internal resistance of a SOFC decreases when the temperature increases, leading to an increase in the power values.

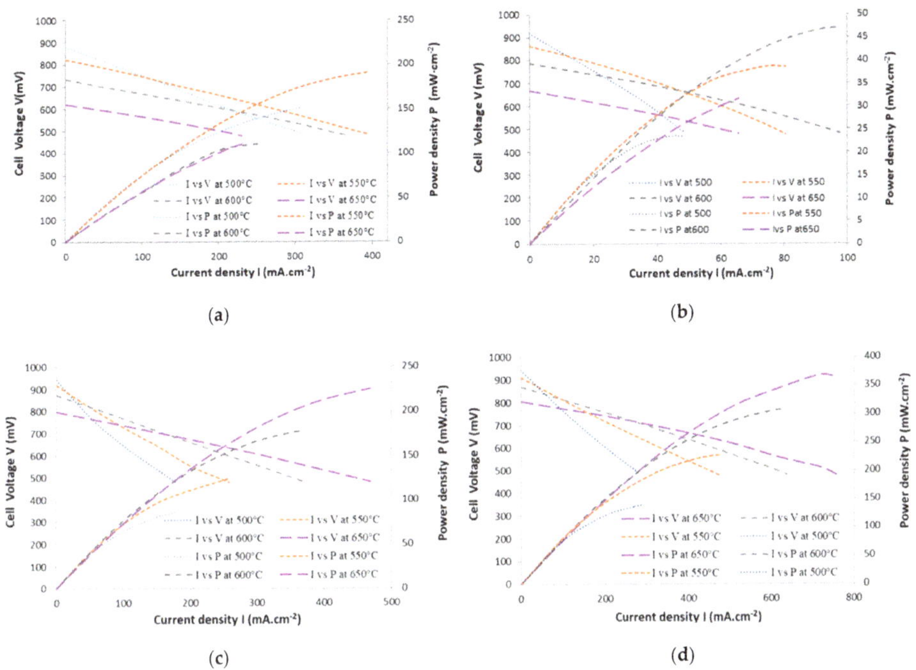

Figure 4. (**a**) *I-V* curves for sample 1 under operation temperatures of: 500, 550, 600, and 650 °C; (**b**) *I-V* curves for sample 2 under operation temperatures of: 500, 550, 600, and 650 °C; (**c**) *I-V* curves for sample 3 under operation temperatures of: 500, 550, 600, and 650 °C; (**d**) *I-V* curves for sample 4 under operation temperatures of: 500, 550, 600, and 650 °C.

The power density reaches 47 mW·cm^{-2}, the maximum value, at 600 °C. Figure 4b shows the *I-V* curves for sample 2 at different temperatures. The OCV is 880 mV at 500 °C and gradually decreases with the increase in the temperature; the power density reaches 190 mW·cm^{-2}, the maximum value, at 550 °C. Figure 4c shows the *I-V* curves for sample 3 at different temperatures. The OCV is 940 mV at 500 °C, and it strongly decreases at 500 °C when I increases. At 650 °C, the voltage is more stables versus I, and the power density reaches 225 mW·cm^{-2}, the maximum value. Sample 4 gives the best results; in Figure 4d, the OCV is 940 mV at 500 °C. At 650 °C, the voltage is more stable versus I, and the power density reaches 370 mW·cm^{-2}.

The OCV theoretical values are calculated using the thermodynamic laws applied to the H_2 oxidation and experimental conditions (Equations (1) and (2)). They are equal to 1.15 V at 500 °C and 1.13 V at 650 °C, in accordance with [3,17].

$$E^{OCV} = E^{\circ} - \frac{RT}{2F}\ln\left(\frac{P_{H_2O}}{P_{H_2}\,P_{O_2}^{0.5}}\right), \quad (1)$$

$$E^{\circ} = 1.253\text{–}2.4516 \times 10^{-4}\, T\,(\mathrm{K}). \quad (2)$$

Table 2. The conductivity of the individual layers for the $Gd_{0.1}Ce_{0.9}O_{2-x}$ (GDC)-based Solid Oxide Fuel Cell (SOFC).

Operating Temperature (°C)	σ_{el}, Ionic Conductivity of the Electrolyte ($S{\cdot}cm^{-1}$) [18]	σ_{cat}, Electrical Conductivity of the Cathode ($S{\cdot}cm^{-1}$) [19]	σ_{an}, Electrical Conductivity of the Anode ($S{\cdot}cm^{-1}$) [3]
500	1.32×10^{-3}	1.25	115
550	2.18×10^{-3}	1.37	119
600	3.39×10^{-3}	1.56	122
650	5.04×10^{-3}	1.75	124

The experimental OCVs versus temperatures are reported in Table 3. As usual, the temperature has a negative impact on the OCV value due to the Nernst law. For samples 1 and 2, the OCV values are lower and decrease more sharply than for the samples 3 and 4. The OCV values of samples 1 and 2 are lower than expected due to probable leakages in the open flange setup.

Table 3. Resistance of the individual layers, internal resistance, and ohmic overpotential at different operating temperatures for a specified current density of 200 $mA{\cdot}cm^{-2}$.

	Operating Temperature (°C)	$\frac{d_{el}}{\sigma_{el}}$ Resistance of the Electrolyte (Ω)	$\frac{d_{cat}}{\sigma_{cat}}$ Resistance of the Cathode (Ω)	$\frac{d_{an}}{\sigma_{an}}$ Resistance of the Anode (Ω)	R_{ohm} Internal Resistance (Ω)	η_{ohm} Ohmic Overpotential (V)
Sample 1	500	1.14	7.20×10^{-4}	2.35×10^{-4}	1.14	0.228
	550	6.88×10^{-1}	6.57×10^{-4}	2.27×10^{-4}	6.89×10^{-1}	0.138
	600	4.42×10^{-1}	5.77×10^{-4}	2.22×10^{-4}	4.43×10^{-1}	0.089
	650	2.98×10^{-1}	5.14×10^{-4}	2.18×10^{-4}	2.98×10^{-1}	0.060
Sample 2	500	2.80	3.44×10^{-3}	2.42×10^{-4}	2.81	n.d.
	550	1.70	3.14×10^{-3}	2.34×10^{-4}	1.70	n.d.
	600	1.09	2.76×10^{-3}	2.28×10^{-4}	1.09	n.d.
	650	7.34×10^{-1}	2.46×10^{-3}	2.24×10^{-4}	7.37×10^{-1}	n.d.
Sample 3	500	1.59	3.84×10^{-3}	2.42×10^{-4}	1.59	n.d.
	550	9.63×10^{-1}	3.50×10^{-3}	2.34×10^{-4}	9.67×10^{-1}	0.193
	600	6.19×10^{-1}	3.08×10^{-3}	2.28×10^{-4}	6.23×10^{-1}	0.125
	650	4.17×10^{-1}	2.74×10^{-3}	2.24×10^{-4}	4.20×10^{-1}	0.084
Sample 4	500	1.52	4.00×10^{-3}	2.35×10^{-4}	1.52	0.304
	550	9.17×10^{-1}	3.65×10^{-3}	2.27×10^{-4}	9.21×10^{-1}	0.184
	600	5.90×10^{-1}	3.21×10^{-3}	2.22×10^{-4}	5.93×10^{-1}	0.119
	650	3.97×10^{-1}	2.86×10^{-3}	2.18×10^{-4}	4.00×10^{-1}	0.080

By using the cell voltage versus the current density (Figure 4), different behavior, depending on the overpotential contributions, could be evidenced thanks to Equation (3).

$$V = E^{OCV} - \eta_{ohm} - \eta_{act,an} - \eta_{act,cat} - \eta_{conc,an} - \eta_{conc,cat}, \quad (3)$$

where η_{ohm} is the ohmic overpotential, $\eta_{act,an}$ is the activation overpotential due to the anode, $\eta_{act,cat}$ is the activation overpotential due to the cathode, $\eta_{conc,an}$ is the concentration overpotential due to the anode, and $\eta_{conc,cat}$ is the overpotential due to the cathode.

The internal resistance R_{ohm} is defined in Equation (4), and the ohmic overpotential is calculated by Equation (5).

$$R_{ohm} = \frac{d_{el}}{\sigma_{el}} + \frac{d_{cat}}{\sigma_{cat}} + \frac{d_{an}}{\sigma_{an}}, \quad (4)$$

$$\eta_{ohm} = I \times R_{ohm}, \quad (5)$$

where $\frac{d_{el}}{\sigma_{el}}$ corresponds to the resistance of the electrolyte, with d_{el} as the thickness and δ_{el} as the ionic conductivity; $\frac{d_{cat}}{\sigma_{cat}}$ corresponds to the resistance of the cathode, with d_{cat} as the thickness and δ_{cat} as the electrical conductivity; and $\frac{d_{an}}{\sigma_{an}}$ corresponds to the resistance of the anode, with d_{an} as the thickness and δ_{an} as the electrical conductivity.

The conductivity versus temperature of each layer were extracted from the references [3,18,19] and are reported in Table 2. The values of the electrolyte ionic conductivity were carefully selected to correspond to the electrolyte microstructure—i.e., the porosity and grain size of the cells sintered at 1215 °C. In Table 3, the calculated values of the resistance of each layer and internal resistance lead to the ohmic overpotential versus temperature of each sample for a specified current density of 200 mA·cm^{-2}.

The ohmic overpotential calculated in Table 3 is the major factor of the potential decrease reported in Equation (3). As expected, the influence of the electrolyte on the ohmic potential is largely dominant. The activation overpotentials are not strong enough to cause a decrease in potential for all *I-V* curves, although for samples 3 and 4 at 500 and 550 °C they are slightly noticeable. The concentration overpotential does not account for a significant portion of the potential; Bianchi and al. reported a proportion of ~3% [20]. These results explain the low performance of sample 2 regarding the thickness of the electrolyte layer and its importance in the potential decrease.

In a second step, the Area-Specific Resistance (ASR), which is an important performance parameter, was calculated using the following equation [21,22]:

$$ASR = \frac{OCV - 0.7}{I_{0.7}},$$

where $I_{0.7}$ is the discharging current density at a voltage of 0.7 V.

The results are summarized in Table 4. For samples 1 and 2, the ASR was not calculated at 650 °C because of the low value of OCV, which was inferior to 0.7 V. It is also apparent that the cell 3 and 4 show the lowest values of ASR at 650 °C, corresponding to their highest performance P_{max}. It is known that the resistance of the cell is greatly dependent on the operating temperature. This temperature dependence can be described using the Arrhenius empirical equation:

$$\ln(\sigma) = ln(A) - \frac{E_a}{RT},$$

where σ is the conductivity of the single cell equivalent to the reciprocal of ASR, E_a is the activation energy in eV, R is the gas constant, T is the absolute temperature of operation, and A is a constant independent of temperature.

Table 4. The electrochemical properties (including Pmax, Open Circuit Voltage (OCV), Area-Specific Resistance (ASR), lnσ, and calculated activation energy E_a) of SOFC-reduced cells operated at various temperatures (500 to 650 °C) under air and a mixture of H_2/(3%) H_2O (n.d. = non defined due to OCV < 0.7 V).

	Operating Temperature (°C)	P_{max} (mW·cm^{-2})	OCV (mV)	ASR (Ω·cm^2)	lnσ (S·cm^{-1})	Activation Energy E_a (eV)
Sample 1	500	23.5	920	8.13	−8.60	0.70
	550	38.5	862	3.89	−7.86	
	600	47	785	2.44	−7.39	
	650	31.5	667	n.d.	n.d.	
Sample 2	500	150	880	1.26	−5.83	0.43
	550	190	820	0.80	−5.38	
	600	110	725	0.61	−5.10	
	650	110	618	n.d.	n.d.	
Sample 3	500	82	950	2.72	−7.17	0.63
	550	121	914	1.78	−6.74	
	600	176	880	1.08	−6.24	
	650	225	800	0.66	−5.75	
Sample 4	500	145	950	1.73	−6.76	0.58
	550	226	905	0.92	−6.13	
	600	307	875	0.60	−5.70	
	650	366	800	0.36	−5.19	

The conductivity values of the four studied samples are in the range of other published results [23–25]. The conductivities of the cells are thus plotted logarithmically versus 1000/T, as shown in Figure 5, to calculate E_a, and the results are listed in Table 4. It is apparent that sample 2 obtains the lowest E_a of 0.43 eV in comparison with the other cells. Sample 1 exhibits the highest value of activation energy of 0.70 eV and the lowest conductivity. Samples 3 and 4 have "intermediate" values of 0.63 and 0.58 eV, respectively.

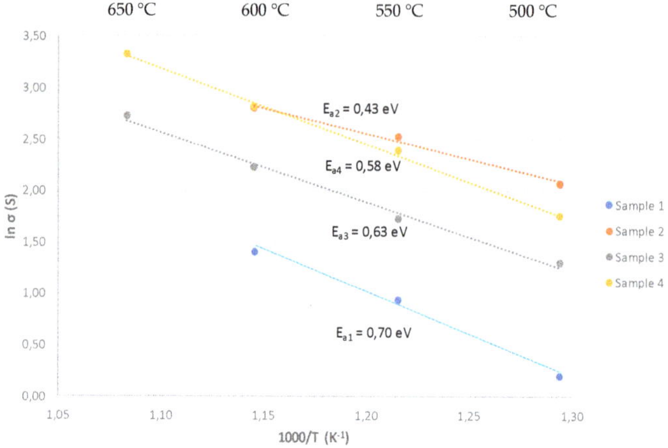

Figure 5. ln σ versus 1000/T for the 4 studied samples.

The anode support layer thicknesses with or without the Anode Functional Layer, varying from 260 to 278 μm, are practically the same for all the samples. Sample 1, exhibiting a low thickness of the cathode layer of 9 μm and electrolyte layer of 15 μm, gives the lowest performance: P_{max} = 47 mW·cm^{-2}. The oxygen reduction in the Mixed Ionic and Electronic Material (MIEC) material LSCF can be divided into several elemental steps [26–28]: (1) adsorption of O_2 onto the surface; (2) dissociation of the absorbed O_2 into oxygen atoms on the surface with charge transfer and oxygen incorporation at the MIEC interface; (3) surface diffusion of adsorbed oxygen on the MIEC surface; (4) bulk diffusion of adsorbed of O_2 in the bulk phase; (5) O_2 transfer at the interface between the MIEC and the electrolyte; (6) charge transfer and the incorporation of adsorbed oxygen at the TPB.

The oxygen reduction reaction rate can be influenced by any of these basic steps. As presented in a previous study [29], if the thickness of the cathode is too low, the efficiency of these previous elemental steps is not optimum. Consequently, sample 1 shows the lowest conductivity, the highest activation energy, and the worst power density performance. Sample 2 presents a thick electrolyte layer, with the power density of the cell reaching 150 mW·cm^{-2} at 500 °C. This high performance is in accordance with the lowest E_a value but cannot explain the low OCV value. Small cracks or pinholes were probably created during the fabrication process, and a direct contact between the H_2 and air was plausible. These cracks are not visible in the SEM. For this reason, the power density reaches a plateau, and the OCV values decrease dramatically with the increasing temperature. It is then impossible to give a conclusion on the electrolyte thickness effect. Finally, decreasing the thickness of the electrolyte to 20 μm while keeping the cathode thickness close to 50 μm, as in samples 3 and 4, allows to obtain the best results in terms of power density. Besides this, sample 4 exhibits a higher conductivity and also a lower activation energy than sample 3 leading to a better power density performance. The main AFL functions are to provide a flat surface for the electrolyte layer and to adjust the thermal expansion coefficient between the porous anode and the dense electrolyte. Even if the AFL, present in sample 3, plays its mechanical role, it slows down the chemical reactions, particularly the hydrogen diffusion at the TPB. However, sample 4 without an AFL layer exhibits the best performance in terms of power density. This being said a priori, this is linked to the greater efficiency of the hydrogen oxidation reaction on the anodic side.

4. Conclusions

Anode-supported planar cells Ni-GDC//GDC//LSCF-GDC with a 10 cm^2 active surface were shaped by a tape casting process before sintering at 1215 °C. The power density performances of four cells, differing by their cathode thickness and the presence of an AFL, were reported in the temperature range of 500 to 650 °C. The best performance, 368 $mW \cdot cm^{-2}$, was obtained for a cathode thickness of 50 μm and an electrolyte thickness of 20 μm. When the cathode thickness decreases, the power density decreases due to the limited oxygen integration. The ohmic potentials were calculated, and the results show that the electrolyte contribution is the preponderant factor in the potential decrease, while the activation and concentration overpotentials are almost negligible. The microstructure of the electrolyte leading to low values of conductivity seems to be the main reason. The activation energy has been concurrently determined, and the maximum value, 0.70 eV, corresponds to the sample with the lowest cathode/electrolyte thickness. The intermediate values of activation energies close to 0.60 eV are for samples 3 and 4, which respectively have a power density of 225 and 366 $mW \cdot cm^{-2}$. Assuming that the cathode and electrolyte layers have the same thicknesses for both samples, the power density difference is due to the presence of AFL in sample 3. This result, which shows a negative impact of the AFL on the power density performance, is surprising and is not in accordance with the common literature [30,31]. The absence of a TPB combined with the thickness and microstructure of the AFL could be a reason for the lower power density measured.

Author Contributions: Conceptualization, L.C. and G.C.; data curation, V.S. and M.F.; formal analysis, V.S., M.F., L.C. and G.C.; funding acquisition, L.C. and G.C.; investigation, V.S.; methodology, V.S., L.C., M.F. and G.C.; Project administration, L.C. and G.C.; resources, G.C. and L.C.; software, none; supervision, G.C.; visualization, V.S. and M.F.; writing—original draft preparation, V.S., M.F., L.C. and G.C.; writing—review and editing, G.C. and L.C. All authors have read and agreed to the published version of the manuscript.

Funding: This research was funded in part by the Regional Council of Bourgogne Franche-Comté and the French National Center for Scientific Research (CNRS) Graduate School EIPHI (Contract ANR-17-EURE-0002).

Acknowledgments: The authors thank the ESIREM Engineering college for the SEM experimentations.

Conflicts of Interest: The authors declare no conflict of interest.

References

1. Minh, N.Q. Ceramic Fuel Cells. *J. Am. Ceram. Soc.* **1993**, *76*, 563–588. [CrossRef]
2. Brett, D.J.L.; Atkinson, A.; Brandon, N.P.; Skinner, S.J. Intermediate temperature solid oxide fuel cells. *Chem. Soc. Rev.* **2008**, *37*, 1568–1578. [CrossRef]
3. Saeba, D.; Authayanum, S.; Patcharavorachot, Y.; Chatrattanawet, N.; Arpornwichanop, A. Electrochemical performance assessment of low-temperature solid oxide fuel cell with YSZ-based and SDC-based electrolytes. *Int. J. Hydrog. Energy* **2018**, *43*, 921–931. [CrossRef]
4. Steele, B.C.H.; Heinzel, A. Materials for fuel-cell technologies. *Nature* **2001**, *414*, 345–352. [CrossRef]
5. Kharton, V.V.; Figueiredo, F.M.; Navarro, L.; Naumovich, E.N.; Kovalevsky, A.V.; Yaremchenko, A.A.; Viskup, A.P.; Carneiro, A.; Marques, F.M.B.; Frade, J.R. Ceria-based materials for solid oxide fuel cells. *J. Mater. Sci.* **2001**, *36*, 1105–1117. [CrossRef]
6. Wandekar, R.V.; Basu, M.A.; Wani, B.N.; Bharadwaj, S.R. Physicochemical studies of NiO–GDC composites. *Mater. Chem. Phys.* **2006**, *99*, 289–294. [CrossRef]
7. Prokop, T.A.; Berent, K.; Iwai, H.; Szmyd, J.S.; Brus, G. A three-dimensional heterogeneity analysis of electrochemical energy conversion in SOFC anodes using electron nanotomography and mathematical modeling. *Int. J. Hydrog. Energy* **2018**, *43*, 10016–10030. [CrossRef]
8. Murray, E.P.; Barnett, S.A. (La,Sr)MnO_3–(Ce,Gd)O_{2-x} composite cathodes for solid oxide fuel cells. *Solid State Ion.* **2001**, *143*, 265–273. [CrossRef]
9. Lee, S.J.; Muralidharan, P.; Jo, S.H.; Kim, D.K. Composite cathode for IT-SOFC: Sr-doped lanthanum cuprate and Gd-doped ceria. *Electrochem. Comm.* **2010**, *12*, 808–811. [CrossRef]

10. Leng, Y.J.; Chan, S.H.; Khor, K.A.; Jiang, S.P. $(La_{0.8}Sr_{0.2})_{0.9}MnO_3$-$Gd_{0.2}Ce_{0.8}O_{1.9}$ composite cathodes prepared from $(Gd,Ce)(NO_3)_x$ -modified $(La_{0.8}Sr_{0.2})_{0.9}MnO_3$ for intermediate-temperature solid oxide fuel cells. *J. Solid State Electrochem.* **2006**, *10*, 339–347. [CrossRef]
11. Wang, Z.; Zhang, N.; Qiao, J.; Sun, K.; Xu, P. Improved SOFC performance with continuously graded anode functional layer. *Electrochem. Commun.* **2009**, *11*, 1120–1123. [CrossRef]
12. Sivasankaran, V.; Combemale, L.; Pera, M.-C.; Caboche, G. Initial Preparation and Characterization of Single Step Fabricated Intermediate Temperature Solid Oxide Fuel Cells (IT-SOFC). *Fuel Cells* **2014**, *14*, 533–536. [CrossRef]
13. Sivasankaran, V. Manufacturing and Characterization of Single Cell Intermediate Temperature Solid Oxide Fuel Cells for APU in Transportation Application. Ph.D. Thesis, University of Burgundy, Dijon, France, July 2014.
14. Sivasankaran, V.; Combemale, L.; Caboche, G. Method of Preparing a Fuel Cell. WO2014057218A2, 9 September 2013. PCT/FR2013/052408 World Patent.
15. Raju, K.; Kim, S.; Hyung, C.J.; Yu, J.H.; Seong, Y.-H.; Kim, S.-H.; Han, I.-S. Optimal sintering temperature for $Ce_{0.9}Gd_{0.1}O_{2-\delta}$–$La_{0.6}Sr_{0.4}Co_{0.2}Fe_{0.8}O_{3-\delta}$ composites evaluated through their microstructural, mechanical and elastic properties. *Ceram. Int.* **2019**, *45*, 1460–1463. [CrossRef]
16. Li, J.; Lv, T.; Dong, X.; Yu, J.; Yu, B.; Li, P.; Yao, X.; Zhao, Y.; Li, Y. Linear discharge model, power losses and overall efficiency of solid oxide fuel cell with thin film Sm doped ceria electrolyte. Part II: Power losses overall efficiency. *Int. J. Hydrog. Energy* **2017**, *42*, 17522–17527. [CrossRef]
17. Chiodelli, G.; Malavasi, L. Electrochemical open circuit voltage (OCV) characterization of SOFC materials. *Ionics* **2013**, *19*, 1135–1144. [CrossRef]
18. Sindiraç, C.; Büyükaksoy, A.; Akkurt, S. Electrical properties of gadolinia doped ceria electrolytes fabricated by infiltration aided sintering. *Solid State Ion.* **2019**, *340*, 115020–115028. [CrossRef]
19. Lubini, M.; Chinarro, E.; Moreno, B.; De Sousa, V.C.; Alves, A.K.; Bergmann, C.P. Electrical properties of $La_{0.6}Sr_{0.4}Co_{1-y}Fe_yO_3$ (y = 0.2–1.0) fibers obtained by electrospinning. *J. Phys. Chem.* **2016**, *120*, 64–69. [CrossRef]
20. Bianchi, F.R.; Bosio, B.; Baldinelli, A.; Barelli, L. Optimisation of a reference kinetic model for Solid Oxide Fuel Cells. *Catalysts* **2020**, *10*, 104. [CrossRef]
21. Hauch, A.; Mogensen, M.B. Testing of Electrodes, Cells, and Short Stacks. In *Advances in Medium and High Temperature Solid Oxide Fuel Cell Technology*; CISM International Centre for Mechanical Sciences: Udine, Italy, 2009; Volume 574, pp. 31–76. [CrossRef]
22. Li, T.; Wang, W.G.; Miao, H.; Chen, T.; Xu, C. Effect of reduction temperature on the electrochemical properties of a Ni/YSZ anode-supported solid oxide fuel cell. *J. Alloys Compd.* **2010**, *495*, 138–143. [CrossRef]
23. Zhen, Y.D.; Tok, A.I.Y.; Jiang, S.P.; Boey, F.Y.C. Fabrication and performance of gadolinia-doped ceria-based intermediate-temperature solid oxide fuel cells. *J. Power Sourc.* **2008**, *178*, 69–74. [CrossRef]
24. Yamaguchi, T.; Shimizu, S.; Suzuki, T.; Fujishiro, Y.; Awano, M. Fabrication and characterization of high performance cathode supported small-scale SOFC for intermediate temperature operation. *Electrochem. Commun.* **2008**, *10*, 1381–1383. [CrossRef]
25. Ding, C.; Lin, H.; Sato, K.; Hashida, T. A simple rapid spray method for preparing anode-supported solid oxide fuel cells with GDC electrolyte thin films. *J. Membr. Sci.* **2010**, *350*, 1–4. [CrossRef]
26. Gao, Z.; Liu, X.M.; Bergman, B.; Zhao, Z. Investigation of oxygen reduction reaction kinetics on $Sm_{0.5}Sr_{0.5}CoO_{3-\delta}$ cathode supported on $Ce_{0.85}Sm_{0.075}Nd_{0.075}O_{2-\delta}$ electrolyte. *J. Power Sourc.* **2011**, *196*, 9195–9203. [CrossRef]
27. Han, D.; Wu, H.; Li, J.L.; Wang, S.R.; Zhan, Z.L. Nanostructuring of $SmBa_{0.5}Sr_{0.5}Co_2O_{5+\delta}$ cathodes for reduced-temperature solid oxide fuel cells. *J. Power Sourc.* **2014**, *246*, 409–416. [CrossRef]
28. Horita, T.; Yamaji, K.; Sakai, N.; Yokokawa, H.; Weber, A.; Ivers-Tiffée, E. Oxygen reduction mechanism at porous $La_{1-x}Sr_xCoO_{3-d}$ cathodes/$La_{0.8}Sr_{0.2}Ga_{0.8}Mg_{0.2}O_{2.8}$ electrolyte interface for solid oxide fuel cells. *Electrochim. Acta* **2001**, *46*, 1837–1845. [CrossRef]
29. Li, W.; Shi, Y.; Luo, Y.; Cai, N. Theoretical modeling of air electrode operating in SOFC mode and SOEC mode: The effects of microstructure and thickness. *Int. J. Hydrog. Energy* **2014**, *39*, 13738–13750. [CrossRef]

30. Bi, L.; Fabbri, E.; Traversa, E. Effect of anode functional layer on the performance of proton-conducting solid oxide fuel cells (SOFCs). *Electrochem. Commun.* **2012**, *16*, 37–40. [CrossRef]
31. Suzuki, T.; Sugihara, S.; Yamaguchi, T.; Sumi, H.; Hamamoto, K.; Fujishiro, Y. Effect of anode functional layer on energy efficiency of solid oxide fuel cells. *Electrochem. Commun.* **2011**, *13*, 959–962. [CrossRef]

Article

In Situ High Pressure Structural Investigation of Sm-Doped Ceria

Cristina Artini [1,2,*], Sara Massardo [1], Maria Maddalena Carnasciali [1,3], Boby Joseph [4] and Marcella Pani [1,5]

1 Department of Chemistry and Industrial Chemistry, University of Genova, 16146 Genova, Italy; s.massardo93@gmail.com (S.M.); marilena@chimica.unige.it (M.M.C.); marcella@chimica.unige.it (M.P.)
2 Institute of Condensed Matter Chemistry and Technologies for Energy, National Research Council, CNR-ICMATE, 16149 Genova, Italy
3 INSTM, Genova Research Unit, Via Dodecaneso 31, 16146 Genova, Italy
4 GdR IISc-ICTP, Elettra-Sincrotrone Trieste S.C.p.A., ss 14, km 163.5, 34149 Trieste, Basovizza, Italy; boby.joseph@elettra.eu
5 CNR-SPIN, 16152 Genova, Italy
* Correspondence: artini@chimica.unige.it; Tel.: +39-010-353-6082

Received: 22 January 2020; Accepted: 23 March 2020; Published: 27 March 2020

Abstract: As a result of the lattice mismatch between the oxide itself and the substrate, the high-pressure structural properties of trivalent rare earth (RE)-doped ceria systems help to mimic the compressive/tensile strain in oxide thin films. The high-pressure structural features of Sm-doped ceria were studied by X-ray diffraction experiments performed on $Ce_{1-x}Sm_xO_{2-x/2}$ (x = 0.2, 0.3, 0.4, 0.5, 0.6) up to 7 GPa, and the cell volumes were fitted by the third order Vinet equation of state (EoS) at the different pressures obtained from Rietveld refinements. A linear decrease of the $\ln B_0$ vs. $\ln(2V_{at})$ trend occurred as expected, but the regression line was much steeper than predicted for oxides, most probably due to the effect of oxygen vacancies arising from charge compensation, which limits the increase of the mean atomic volume (V_{at}) vs. the Sm content. The presence of RE_2O_3-based cubic microdomains within the sample stiffens the whole structure, making it less compressible with increases in applied pressure. Results are discussed in comparison with ones previously obtained from Lu-doped ceria.

Keywords: solid oxides fuel cells; Sm-doped ceria; high pressure X-ray powder diffraction; diamond anvil cell; equation of state; Rietveld refinement

1. Introduction

Trivalent rare earth (RE)-doped ceria systems form a group of widely studied mixed oxides with technologically interesting values of ionic conductivity in the intermediate temperature range (673–973 K), which make them useful as solid electrolytes in solid oxide fuel cells (SOFCs). $Ce_{0.9}RE_{0.1}O_{1.95}$ oxides (RE = Gd, Sm, Nd) present values of ionic grain conductivity ranging between 5×10^{-2} and 7×10^{-2} S cm^{-1} at 773 K, with a remarkable increase occurring with increases of the RE ionic radius [1]. However, even a slight variation in RE content causes a non-negligible change in ionic conductivity; $Ce_{0.8}Gd_{0.2}O_{1.90}$, for example, is characterized by a total ionic conductivity of 1×10^{-3} S cm^{-1} at 773 K [2]. Ionic conductivity is in fact affected by many factors such as RE identity, composition, structure and microstructure of the oxide [3–7], and not least by the occurrence of the sample in the bulk or thin film form. With particular reference to this latter issue, thin films are essential when designing SOFC-based portable devices because of the need for fabricating fuel cells on chips [8,9]. Nevertheless, thermodynamic, structural and transport properties of the deposited electrolyte are not necessarily the same as in the bulk material; deposited Gd-doped ceria electrolytes, for instance, present ohmic losses up to 100 times lower than the corresponding oxides in bulk form [8], which is one of the main advantages to using doped ceria thin films. These issues make the investigation of such properties, and comparison with the ones of the bulk materials, of primary importance. In particular, it is known that the thinning of layer thickness often induces a significant ionic conductivity increase in doped ceria due to the oxide/substrate lattice mismatch [10,11], while at the same time causing a significant decrease in the same property due to a larger amount of defects [8]. The aim of this study is to separate these two competitive effects, as the high-pressure behavior of bulk samples can help to simulate the strain at the interface. The application of 5 GPa, for instance, roughly corresponds to a 1% compressive strain, which is not far from the one existing in a 250-nm-thick Gd-doped ceria film deposited on a $MgO/SrTiO_3$ layer [11]. Only a few papers in the literature have discussed the high-pressure structural properties of doped ceria [12,13]. Lattice mismatch is one of the main causes of microstrain at the interface [11,14], but for the sake of completeness it should also be noted that microstrain derives from thermal expansion coefficient mismatch or from properties resulting from processing issues (such as grain size) [8].

Together with Gd- and Nd-doped ceria, Sm-doped ceria is one of the most effective systems of ionic conductivity [1]. Ionic conductivity is strictly related to structural and microstructural issues, and this is the reason why so many papers in recent years have been devoted to the crystallographic properties of this material, even in the light of the non-negligible effects of extrinsic factors (e.g., thermal treatments) on the stability of the various oxide phases [15,16]. A particular attention has been paid to defect aggregations [3,17–21], since ionic conductivity in doped ceria occurs through the hopping of oxygen ions towards the vacancies ($V_{\ddot{O}}$) created for charge compensation when Ce^{4+} is partially substituted by RE^{3+}. The mechanism works until the fluorite-type cubic structure of CeO_2 (hereafter named *F*; space group: $Fm\bar{3}m$; Ce coordination number: 8) is retained (i.e., until the solid solution of RE^{3+} within the CeO_2 matrix is stable). Actually, maximum ionic conductivity for the solid F solution falls well within the stability range; for example, the compositional limit of F in $Ce_{1-x}Gd_xO_{2-x/2}$ is located at $x \sim 0.2$ [22], with its maximum conductivity at $x = 0.10–0.15$ [4,23]. This evidence is due to the creation of defect aggregates $RE'_{Ce} : V_{\ddot{O}}$ having the cubic structure of sesquioxides RE_2O_3 (hereafter named *C*; space group: $Ia\bar{3}$; RE coordination number: 6), which hinder the movement of oxygen ions. Due to their nanometric size, *C* aggregates are invisible to X-ray diffraction until the RE^{3+} amount reaches a threshold content that is defined by the RE's identity. On the contrary, *C* aggregates can also be revealed by Raman spectroscopy, due to the superior sensitivity of the latter technique toward oxygen displacement. Therefore, the position of the upper compositional limit of F as obtained from X-ray diffraction does not necessarily coincide with the limit deriving from Raman spectroscopy (i.e., $x \sim 0.3$ and ranging between $x = 0.2$ and 0.3 in $Ce_{1-x}Sm_xO_{2-x/2}$ for the former and latter techniques, respectively) [24]. For Sm-doped ceria beyond the F limit, this system presents an intermediate structure between *F* and *C*, as *F* and *C* are linked to each other through a structure/superstructure relation and—due to the close resemblance of the Ce^{4+} and Sm^{3+} ionic sizes (Ce^{4+}, CN:8, $r = 0.97$ Å; Sm^{3+}, CN:6, $r = 0.958$ Å [25])—peaks common to the two phases coincide, regardless of the sensitivity of the diffractometer employed [22]. This atomic arrangement, which can be interpreted as a solid solution with the $Sm'_{Ce} : V_{\ddot{O}}$ C-based domains acting as guests [24] was named H due to its hybrid character between F and C, and has even been described for other systems, such as Y- [26] and Gd-doped ceria [27]. Moreover, as a direct consequence of the structure/superstructure relation, the H cell parameter is roughly double that of *F*.

In the framework of a high-pressure structural study of doped ceria, it is very interesting to compare the properties of a system presenting the H phase (such as Sm-doped ceria) to that of a system that—due to ionic dimensional issues—behaves differently. In this respect, Lu-doped ceria ($Ce_{1-x}Lu_xO_{2-x/2}$) act as an ideal model: While the CeO_2-based solid solution with Lu^{3+} as a guest forms up to $x \sim 0.4$ [28], the large ionic size difference (Ce^{4+}, CN:8, $r = 0.97$ Å; Lu^{3+}, CN:6, $r = 0.861$ Å [25]) prevents the formation of the *H* phase and promotes the stability of a (*F* + *C*) two-phase field.

In this work the high-pressure structural properties of Sm-doped ceria were studied by analyzing X-ray diffraction patterns collected up to 7 GPa at the XPRESS diffraction beamline of the Elettra Synchrotron radiation facility [29]. The refined cell volumes were fitted to the Vinet equation of state (EoS) [30], and the obtained bulk moduli at zero pressure (B_0) were discussed in comparison to the ones obtained for Lu-doped ceria in a previous study [13]. A substantial insensitivity of B_0 to structural changes was revealed for both systems, while the compressibility of compositions containing C domains remarkably decreased starting from ~3 GPa. A significant effect of vacancies was also found for the trend of $\ln B_0$ vs. $\ln(2V_{at})$ (with $2V_{at}$ being the volume per atom pair).

2. Materials and Methods

2.1. Synthesis

Five compositions belonging to the $Ce_{1-x}Sm_xO_{2-x/2}$ system with nominal x = 0.2, 0.3, 0.4, 0.5 and 0.6 were synthesized by oxalate coprecipitation [31,32]. Stoichiometric amounts of Ce (Johnson Matthey ALPHA 99.99% wt.) and Sm_2O_3 were separately dissolved in HCl (13% vol.), then the two solutions were mixed together. Afterwards, the precipitation of the mixed Ce/Sm oxalate was accomplished by adding a solution of oxalic acid in large excess. Oxalates were filtered, washed and dried for 12 h then thermally treated at 1373 K in air for four days to obtain mixed oxides with a high degree of crystallinity.

2.2. Scanning Electron Microscopy-Energy-Dispersive System (SEM-EDS)

Scanning electron microscopy-energy-dispersive system (SEM-EDS, Oxford Instruments, model 7353 with Oxford-INCA software v. 4.07) was employed to check the overall rare earth contents of all the samples. Pellets of powders pressed and sintered at 1773 K were coated by a graphite layer and analyzed at a working distance of 15 mm, with an acceleration voltage of 20 kV. EDS analyses were carried out on at least six points for each sample.

2.3. High-Pressure Synchrotron X-ray Powder Diffraction (HP-XRPD)

Ambient and high-pressure X-ray powder diffraction acquisitions were done at the XPRESS diffraction beamline of the Elettra Synchrotron radiation facility located in Trieste (Italy) [29]. Data collection was performed by a monochromatic circular beam with a wavelength of 0.4957 Å and a diameter around 50 μm; pressure was applied from 0 to ~7 GPa using a gear-driven Boehler-Almax plate diamond anvil cell (plate DAC) with a large X-ray aperture containing diamonds with a culet size of 500 μm. In order to arrange the sample chamber, 200-μm-thick rhenium gaskets were pre-indented using the plate DAC, thus obtaining a thickness of about 110 μm; a through-hole with a diameter of 200 μm was subsequently drilled by spark erosion at the center of the pre-indented region. Fine powders of the sample were placed inside the sample chamber to obtain diffraction data with minimal preferred orientation effects. Pressure calibration was done either by Cu, making use of the Cu (111) diffraction peak (for samples Sm20 and Sm30), or by ruby, measuring the displacement in the position of its fluorescence lines (for samples Sm40, Sm50 and Sm60). The sample chamber was filled with silicon oil as pressure transmitting medium (PTM), and the diamonds of the anvil cell were subsequently brought into contact with the gasket. The PTM chosen provided good hydrostatic pressure conditions up to the maximum pressure applied in the present study. A schematic representation of the experimental set up is shown in Figure 1.

The experimental station was equipped with a MAR345 image plate detector, and images of the powder diffraction rings were converted into 2θ-intensity plots by means of the fit 2D software. The angular range spanned between 4° and 30°. Samples were named Sm20_7.54, Sm30_5.23 and so on, according to the nominal Sm atomic percent with respect to the total (Ce + Sm) content and applied pressure.

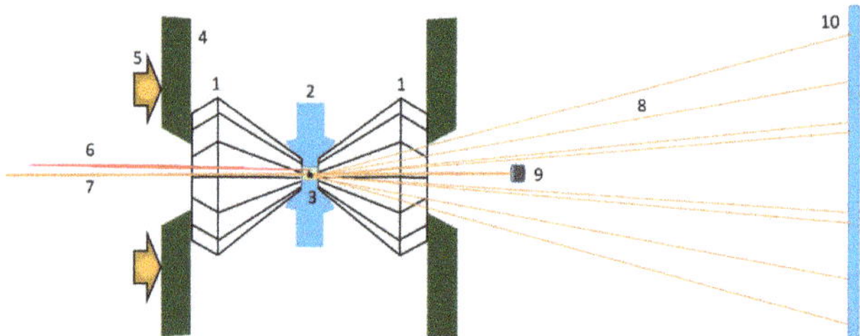

Figure 1. Scheme of the experimental set-up at the XPRESS beamline. (1) Pair of identical diamonds; (2) metal foil with a hole at the center (sample chamber); (3) sample chamber (sample, ruby, pressure transmitting medium (PTM)); (4) backing plates holding diamonds; (5) force pushing diamonds; (6) laser for ruby fluorescence excitation; (7) incident monochromatic X-ray microbeam; (8) diffracted X-rays; (9) beam stop blocking the direct beam; (10) image plate detector recording the position and intensity of the diffracted X-rays.

The FullProf suite [33] was used to refine structural models using the Rietveld method; the evolution of the lattice volume as a function of pressure was modeled through the third order Vinet EoS [30] by means of the EosFit7-Gui software [34]. High-pressure data are discussed in comparison to data collected using the same samples at ambient pressure at the MCX beamline, as described in [24] as well as to the samples of Lu-doped ceria collected at high pressure at the XRD1 beamline [12].

3. Results

3.1. Compositional Characterization

The experimental x of all the samples, as revealed by EDS, was found very close to the stoichiometric value, and results are reported in Table 1. Therefore, the occupancy factors of crystallographic sites hosting lanthanide ions were fixed at the nominal value and not refined during Rietveld cycles. No extra phases were detected, and backscattered images taken on the polished surfaces of all samples did not reveal the presence of any color inhomogeneity, as shown in Figure 2 with reference to sample Sm40.

Figure 2. Microphotograph taken by backscattered electrons on the polished surface of sample Sm40.

Table 1. Mean experimental x values as revealed by EDS analyses.

Sample	Experimental x in $Ce_{1-x}Sm_xO_{2-x/2}$
Sm20	0.17(1)
Sm30	0.30(1)
Sm40	0.38(1)
Sm50	0.48(1)
Sm60	0.60(1)

3.2. Structural Characterization

Peaks belonging to either the F or H phases were imposed to refine two different structural models according to their composition; namely, the F model for the Sm20 and Sm30 samples, and the H model for the Sm40, Sm50 and Sm60 samples. The F structure consisted of a cubic cell belonging to the $Fm\bar{3}m$ space group comprising four formula units per cell [35]; the two atomic positions were occupied by Ce/Sm (0,0,0) and O (1/4, 1/4, 1/4). The *H* phase was a solid solution of C domains within the CeO_2 matrix. The C structure consisted of a cubic cell belonging to the $Ia\bar{3}$ space group and comprising 32 formula units per cell. The atomic rearrangement with respect to F induced the position occupied by rare earths to split into two distinct crystallographic sites. Both the *F* and the *C* structural models are reported in the Supplementary Materials.

It is questionable whether the pyrochlore (P) structure even occurred in the studied samples at this stage. Pyrochlore is the name of the $(Na,Ca)_2Nb_2O_6(OH,F)$ mineral, which crystallizes in the $Fd\bar{3}m$ space group. It differs from the C structure just in regards to the position of oxygen vacancies, which are located in the next-nearest-neighbor sites with respect to RE^{3+}, while in the C structure they are placed in nearest-neighbor positions. In this sense, the P-type structure represents an alternative arrangement of defects in doped ceria. Due to the different extinction rules of the face-centered lattice of the P structure with respect to the body-centered one of the C, the reflections of the crystallographic planes (331) and (333) + (511) should be visible in the diffraction pattern in case of occurrence of the P phase, albeit with low intensity; since this condition is not fulfilled, we can exclude the presence of P in our samples. This conclusion is also supported by the literature, as is discussed hereafter. In principle, the presence of a $Ce_2RE_2O_7$ P-type phase in rare earth (RE)-doped ceria cannot be ruled out a priori; D.E.P. Vanpoucke et al. [36], for example, claimed there is a higher stability of P with respect to the F phase for $Ce_2La_2O_7$ based on density functional theory (DFT) calculations. Relying on the comprehensive theoretical study by L. Minervini and R.W. Grimes [37], $Ce_2La_2O_7$ has in fact been located at the P/F boundary, and its occurrence in the P phase cannot be excluded. Nevertheless, the stability range of A_2B_2O7 compounds (being A and B transition metals or lanthanide ions) is reported to depend on the A/B cationic ratio, and all the ceria-based systems doped with lanthanide ions smaller than La (i.e., from Nd to Lu) have been determined to be more stable in the *F* phase [37]. Even the exhaustive paper by W. Chen and A. Navrotsky [38], which makes use of the atomistic results by L. Minervini et al. [39], states that the formation enthalpies of $Ce_2RE_2O_7$ become thermodynamically unfavored when decreasing the RE^{3+} size, being −102 kJ/mol for $Ce_2La_2O_7$, −3.85 kJ/mol for $Ce_2Gd_2O_7$ and 175 kJ/mol for $Ce_2Y_2O_7$. For the sake of completeness, simulations performed by B. Wang et al. [40] showed that stable P defect clusters in Gd-doped ceria are expected to occur only when they form within ceria matrix domains in the sub-nanoscale range (i.e., at very low dopant amounts). On the basis of this experimental evidence, as well the other described considerations, we can thus exclude the occurrence of P phase in the current studied samples.

The location of the *F/H* boundary being close to $x \sim 0.3$ in $Ce_{1-x}Sm_xO_{2-x/2}$ agrees with previous studies performed using X-ray diffraction on the same system in room [24] and high-temperature [41] conditions. Very interestingly, as briefly mentioned in Section 1, these results differ slightly from the outcome of Raman spectroscopy, which has revealed the presence of a tiny amount of C-based domains even in Sm30 through the occurrence of a weak signal at ~370 cm^{-1}, which was caused by the $\left(A_g + F_g\right)$ Sm-O symmetrical stretching mode with Sm in a six-fold coordination [42]. The described discrepancy between data of X-ray diffraction and Raman spectroscopy is not surprising, as it was already observed for other similar systems, such as Tm-doped ceria [43]. This is due to the very high sensitivity of the latter technique toward oxygen displacement, which gives it the ability to clearly discriminate between the *F* and *C* structures. On the contrary, X-ray diffraction patterns of the two phases are essentially superimposable, with the only difference being the presence of the weak superstructure peaks in C. In Figure 3, the Rietveld refinement plot of sample Sm50_6.94 is reported as a representative example; the refined parameters and Rietveld agreement factors are collected in Table 2.

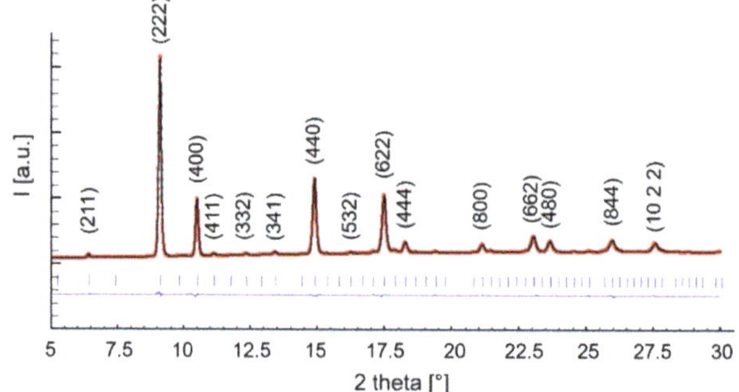

Figure 3. Rietveld refinement plot of sample Sm50_6.94. The dotted (red) and the continuous (black) lines are the experimental and the calculated diffractograms, respectively; the lower line is the difference curve; vertical bars indicate the calculated positions of Bragg peaks. Miller indexes related to the H structure (space group $Ia\bar{3}$) are reported too.

For each diffraction pattern the peak profile was refined by the pseudo-Voigt function, while the background was optimized by linear interpolation of a set of ~70 experimental points. For samples Sm20 and Sm30, the angular regions where peaks of Cu appear were excluded from refinements. In the last cycle, structural parameters (cell parameters; and for the *H* phase: the Ce/Sm1 x coordinate, the O1 x, y and z coordinates and both the occupancy factor and x coordinate of O2), atomic displacement parameters, the scale factor, five peak parameters and the background points were refined. For each composition, high-pressure data were optimized using atomic positions refined at ambient pressure as starting parameters.

Table 2. Applied pressure, refined cell parameter and Rietveld agreement factors for each studied sample.

Sample	P (GPa)	A Cell Parameter (Å)	R_B [a]	χ^2 [b]
Sm20	0	5.43531(5)	0.62	0.199
	2.75	5.41524(7)	0.48	0.384
	3.45	5.40982(8)	0.81	0.613
	4.20	5.40673(4)	0.50	0.829
	4.88	5.3966(1)	0.54	0.462
	7.54	5.3773(2)	0.64	0.569
Sm30	0	5.44092(5)	1.27	0.82
	1.15	5.43080(5)	0.90	0.66
	1.85	5.42734(6)	1.08	0.44
	1.95	5.42615(7)	0.57	0.45
	2.18	5.42500(9)	0.89	0.59
	2.63	5.4233(1)	1.12	0.84
	2.99	5.4212(1)	0.98	0.66
	4.18	5.4142(2)	0.93	0.60
	5.23	5.4081(2)	0.78	1.05

Table 2. *Cont.*

Sample	*P* (GPa)	*A* Cell Parameter (Å)	R_B [a]	χ^2 [b]
Sm40	0	10.89875(6)	3.67	2.47
	2.42	10.8646(1)	1.23	0.972
	3.00	10.8629(2)	0.89	0.620
	3.17	10.8551(2)	0.76	0.468
	3.73	10.8557(3)	0.77	0.549
	4.11	10.8548(4)	0.61	0.553
	4.67	10.8484(4)	0.60	0.44
Sm50	0	10.91507(5)	2.52	0.723
	0.20	10.91561(5)	1.85	0.451
	1.11	10.89095(5)	1.95	0.551
	1.30	10.89046(5)	2.29	0.652
	1.86	10.87945(6)	1.72	0.466
	2.60	10.86725(5)	1.62	0.540
	3.92	10.84993(8)	1.17	0.486
	5.05	10.8349(1)	0.95	0.381
	6.94	10.8159(1)	0.75	0.324
	7.32	10.8144(1)	0.68	0.331
Sm60	0.74	10.90674(4)	1.51	0.849
	1.58	10.89078(5)	1.52	0.814
	2.98	10.86970(7)	1.43	0.465
	4.11	10.85667(9)	1.14	0.359
	4.86	10.8490(1)	0.92	0.252
	5.43	10.8460(2)	0.71	0.22
	6.00	10.8389(2)	0.53	0.20
	6.56	10.8355(3)	0.48	0.19

Note: [a] $R_B = \frac{\sum_k |I_{k,obs} - I_{k,calc}|}{\sum_k |I_{k,obs}|}$; [b] $\chi^2 = \left(\frac{R_{wp}}{R_{exp}}\right)^2$.

Diffraction patterns collected from sample Sm50 at all the pressure values are shown in Figure 4 as representative examples of the behavior of all the compositions. No additional peaks can be observed by increasing the applied pressure with respect to the ones typical of the F or H structures; on the contrary, as evident from the inset, a shift toward higher diffraction angles occurred, indicating a lattice size reduction due to compression.

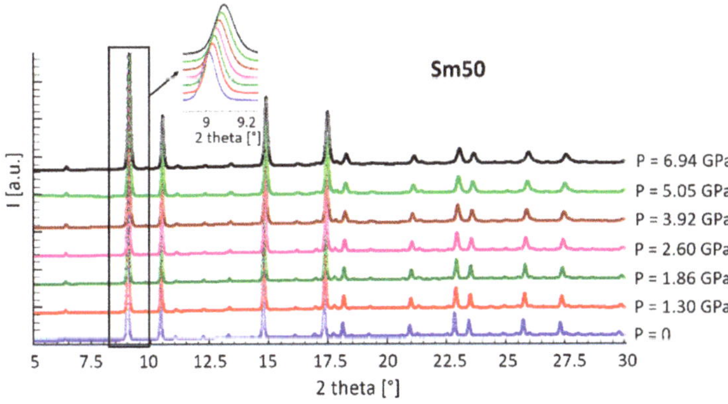

Figure 4. Diffraction patterns of sample Sm50 at all pressure values.

In Figure 5, the refined values of the lattice volume of each sample are reported as a function of the applied pressure. Within the considered pressure range, the cell volume decreased with pressure, and it can be observed that the trend was roughly linear for sample Sm20, while all the other compositions had a progressively more marked softening of the decrease starting from ~3 GPa, which is in agreement with the compressibility reduction that occurred with increases in applied pressure. No significant differences were observed with changing pressure among the refined values of the Sm1 x coordinate.

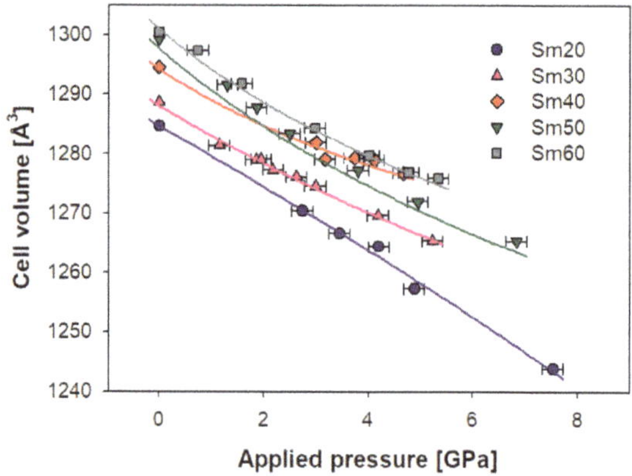

Figure 5. Evolution with pressure of the Sm-doped ceria cell volume for all the samples; experimental data are fitted by the third order Vinet EoS (continuous lines). Cell volumes of samples Sm20 and Sm30 are multiplied by 8 in order to make them comparable with those of Sm40, Sm50 and Sm60. Vertical error bars are hidden by data markers.

The bulk modulus that occurred at zero pressure (B_0) accounts for the resistance of a material to compression, and is defined as

$$B_0 = -V_0\left(\frac{\partial P}{\partial V}\right)_{P=0} \tag{1}$$

where V_0 is the cell volume at $P = 0$. B_0 can be estimated by fitting the refined lattice volumes (or cell parameters) obtained at different applied pressures to the proper equation of state. Several different EoSs were proposed in equations by Murnaghan [44], Birch-Murnaghan [45], Vinet [30] and Poirier-Tarantola [46], all of which were able to correctly predict the bulk modulus at ambient conditions [47]. All the aforementioned EoSs describe and model the pressure-volume correlation of a solid through the bulk modulus and its pressure derivatives. The Vinet EoS, for instance, is an exponential function with the following form:

$$P(V) = \frac{3B_0}{X^2}(1-X)exp[\eta_0(1-X)] \tag{2}$$

with

$$X = \left(\frac{V}{V_0}\right)^{1/3} \tag{3}$$

and

$$\eta_0 = \frac{3}{2}[B\prime_0 - 1] \tag{4}$$

In this work experimental data were fitted by the Vinet EoS, and fits of refined cell volumes vs. pressure are reported in Figure 5. The optimized B_0 as a function of Sm content appears in Figure 6 together with two B_0 values of CeO_2 chosen from the literature, which were obtained using the Hartree-Fock method [48] and through X-ray diffraction [49]. It is worth noting that B_0 values of CeO_2 available in the literature spanned a wide range, according to the employed technique [50]. A roughly linear decrease of B_0 with increasing Sm content can be observed in Figure 6.

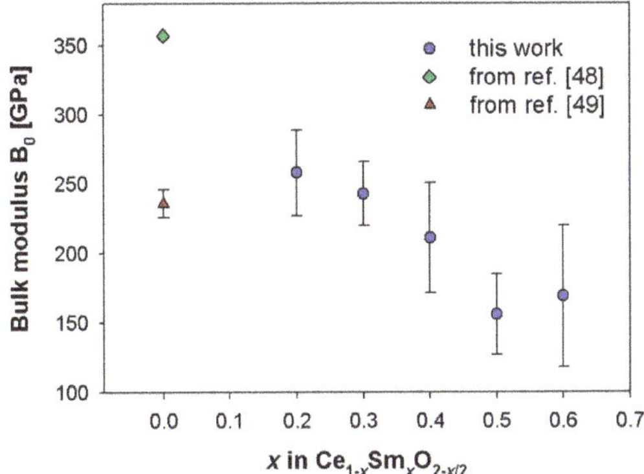

Figure 6. Zero-pressure bulk modulus (B_0) as a function of the Sm content alongside literature data for CeO_2 [48,49].

4. Discussion

The dependence of B_0 on Sm content can be analyzed and discussed in the light of the fundamental paper by Anderson and Nafe [51], which provides the theoretical background behind the behavior of B_0 in oxides. According to this work, a correlation exists between B_0 and the mean volume per ion pair (namely $2V_{at}$, calculated as the volume of a cell divided by the halved number of atoms therein contained). This correlation assumes different forms for ionic crystals and oxides. In particular, the expression

$$\ln B_0 = -\ln(2V_{at}) + \ln(Z_1 \times Z_2) + constant \tag{5}$$

with Z_1 and Z_2 as the cationic and anionic charge, respectively, applies to ionic crystals, while

$$\ln B_0 = -m \ln(2V_{at}) + constant \tag{6}$$

with m ranging between 3 and 4, is valid for oxides. A decreasing linear trend is observed in both cases, which accounts for the progressively increasing compressibility of a solid with an increasing mean atomic volume. Relying on Equations (5) and (6), it is thus clear that oxides are characterized by a much stronger dependence of B_0 on atomic volume.

In Figure 7, the diagram of $\ln B_0$ as a function of $\ln(2V_{at})$ for Sm-doped ceria is shown, with B_0 values obtained from the EoS fits, and $2V_{at}$ derived from the refined cell volumes. Firstly, it can be observed that the trend of data is decreasing and roughly linear.

It should be remembered at this point that, similar to all RE-doped ceria systems—where the total number of atoms changes with an increasing amount of RE—a linear correlation exists between the atomic mean volume and the doping ion amount, even in Sm-doped ceria, according to Zen's law [52]. This issue, widely discussed in [3,24], points to the existence of a solid solution along the whole compositional range, where the guests are Sm^{3+} ions within the F region and $Sm'_{Ce} : V_{O}^{\cdot\cdot}$ aggregates within the H. As a consequence, the linear decrease in $\ln B_0$ vs. $\ln(2V_{at})$ implies a corresponding linear trend in $\ln B_0$ vs. $\ln x$. Therefore, the absence of any discontinuity or slope change at $0.3 \leq x \leq 0.4$ means that B_0 is insensitive to the structural change occurring at the F/H boundary. For the sake of comparison, data deriving from Lu-doped ceria taken from [12] are reported in the inset to Figure 6. B_0 values of Lu-doped ceria were obtained by fitting experimental data to the third order Vinet EoS, in analogy with the data treatment performed on Sm-doped ceria. As mentioned above, the Lu-doped ceria system presents different structural regions depending on the Lu content: the F phase up to $x \sim 0.4$; the $(F + C)$ two-phase field for x ranging between ~0.4 and ~0.8; and a C-based solid solution containing F-structured CeO_2 domains at higher x. The occurrence of the two-phase field is due to the large size difference between Ce^{4+} (CN:8) and Lu^{3+} (CN:6), which hinders the formation of the H phase.

The regular decrease of ln B_0 vs. $\ln(2V_{at})$, observed in the inset to Figure 6, resembles the behavior of Sm-doped ceria, and suggests that even in this system B_0 is insensitive to structural changes.

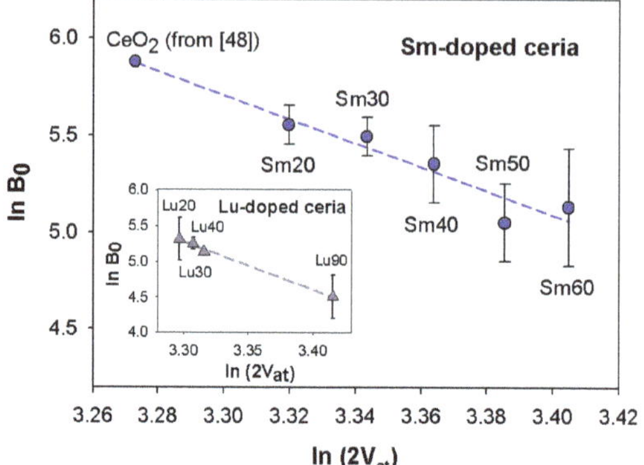

Figure 7. Trend of ln B_0 as a function of $\ln(2V_{at})$. B_0 and $2V_{at}$ values of Sm-doped ceria are obtained from EoS fits and Rietveld refinements, respectively. Data of CeO_2 are taken from [48]. Cell volumes of samples Sm20 and Sm30 are multiplied by 8 in order to make them comparable with those of Sm40, Sm50 ad Sm60. Inset: trend of ln B_0 vs. $\ln(2V_{at})$ for Lu-doped ceria: data are taken from [13]. In both diagrams the dashed line is the regression line interpolating data. Horizontal error bars are hidden by data markers.

ln B_0 vs. $\ln(2V_{at})$ data of both systems can be interpolated by a regression line that satisfactorily fits experimental points ($r^2 = 0.94$ and $r^2 = 0.99$ for Sm- and Lu-doped ceria, respectively), as highlighted in Figure 7. The striking result here is the closeness of the two slopes: $m = 6.1$ and $m = 6.6$ for the Sm- and Lu-doped ceria, respectively. The decreasing trend is thus much steeper than expected for an oxide, according to [51]. This evidence can be most probably be interpreted as a direct consequence of the Ce^{4+} replacement by Sm^{3+} or Lu^{3+}, the effects of which are two-fold: on the one hand, they cause an enlargement of the cell due to the incorporation of a larger ion (Sm^{3+}, CN:8, $r = 1.079$ Å; Lu^{3+}, CN:8, $r = 0.977$ Å [25]) in place of a smaller one (Ce^{4+}, CN:8, $r = 0.97$ Å [25]); on the other hand, the introduction of oxygen vacancies is generally considered responsible for lattice shrinkage [53–55]. In this respect, Raman spectroscopy offers a precious contribution to comprehension of the vacancies effect, since it is possible to apply a correction to the observed Raman shift of the CeO_2 F_{2g} symmetric vibration mode and separate the size from the vacancies effect [53]. Once subtracted from the size effect, the trend of the corrected Raman shifts increases along with increases of the amount of trivalent ions, which can be explained as a consequence of the lattice shrinkage brought about by the introduction of oxygen vacancies. The described result was obtained both for Sm- [24,41] and Gd-doped ceria [24] and can be conceivably predicted for all the RE-doped ceria systems within the F and H phases. From the previous discussion it can thus be expected that mean atomic volume is affected by the two competitive effects, namely, the dopant size and vacancies introduction. Therefore, the mean atomic volume increases with increases in Sm^{3+} or Lu^{3+} amounts, but less so than if vacancies were not created. Under these circumstances, a steeper decrease of ln B_0 vs. $\ln(2V_{at})$ is expected, and indeed was observed in our samples belonging to both systems with respect to the ones reported in [51].

A further remarkable issue concerning the effect of pressure on cell volume can be inferred from the data reported in Figure 5. Even at a first glance, Sm20 can be distinguished from other compositions due to roughly linear decreases in its cell volume vs. applied pressure, which indicate that the bulk modulus did not present any substantial change as a result of pressure within the considered range. All other samples, on the contrary, showed a progressively less steep decrease with increasing pressure, thus indicating an increase in the bulk modulus and a decrease in compressibility. Moreover, softening of the volume decrease vs. pressure decrease became more and more marked with increasing amounts of Sm. This substantial difference between samples with $x < \sim 0.3$ and $x \geq \sim 0.3$ seems to go hand in hand with the structural modifications occurring at $x \sim 0.3$, namely, the transition from F to H. Within this scenario, it can thus be deduced that the F phase is more tolerant toward compression with respect to C, or in other words that C-based microdomains tend to stiffen structures and make them less compressible. This conclusion is also corroborated by the general consideration that the fluorite structure of CeO_2 is indeed very prone to the replacement of Ce^{4+} by larger atoms up to a remarkable substitution degree [3]. An F-based solid solution preserving the atomic arrangement of CeO_2 is in fact stable in $Ce_{1-x}RE_xO_{2-x/2}$ systems (e.g., up to $x \sim 0.2$ or 0.25 for Gd^{3+} [28,56]; $x \sim 0.25$ for Y^{3+} [26]; $x \sim 0.3$ for Sm^{3+} [28] and Er^{3+} [57]; $x \sim 0.4$ for Tm^{3+} [54], Lu^{3+} [28] and Nd^{3+} [57]; $x \sim 0.5$ for Yb^{3+} [54] and an Nd^{3+}/Dy^{3+} mixture [58]; and $x \sim 0.6$ for La^{3+} [57]). On the contrary, a much narrower stability region occurs for RE_2O_3-based C-structured solid solutions where RE^{3+} is partially substituted by Ce^{4+}. For RE ≡ Lu, for example, a maximum 10% substitution by Ce^{4+} is in fact possible, while for RE ≡ Gd and Sm, the maximum substitution degree of trivalent ions reaches 20% [28].

The stiffening effect of the C phase is further confirmed by the behavior shown in Figure 8 related to the cell volume vs. applied pressure up to 7 GPa observed in Lu-doped ceria, which is based on the experimental results reported in [12]. It can be noticed that samples Lu20, Lu30 and Lu40, crystallizing in the F structure, are characterized by a roughly linear decrease within the pressure range considered; conversely, Lu90, crystallizing in the C structure, presents a softening of the volume decrease already revealed for C-containing compositions belonging to Sm-doped ceria, thus confirming the higher resistance to compression of this structure.

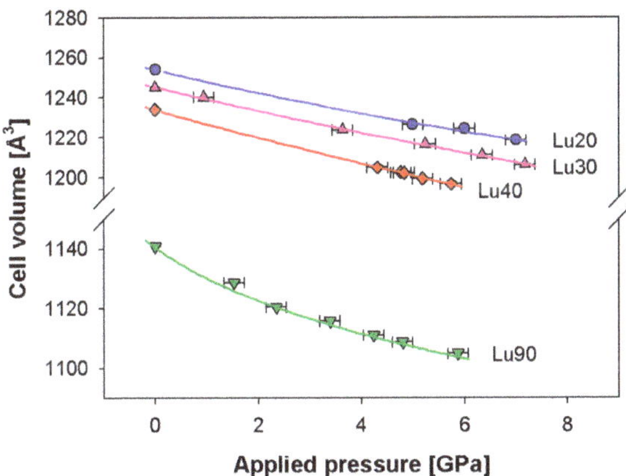

Figure 8. Dependence of the Lu-doped ceria cell volume vs. applied pressure; experimental data are taken from [13] and fitted by the third order Vinet EoS (continuous lines). Cell volumes of samples Lu20, Lu30 and Lu40 are multiplied by 8 in order to make them comparable with Sm90. Vertical error bars are hidden by data markers.

5. Conclusions

An in situ high-pressure structural investigation was performed by powder X-ray diffraction at the Elettra synchrotron facility on samples belonging to the $Ce_{1-x}Sm_xO_{2-x/2}$ system with nominal $x = 0.2$, 0.3, 0.4, 0.5 and 0.6. Rietveld refinements performed on the collected diffraction patterns provided optimized values of the cell volumes at the different pressures, which were then interpolated through the third order Vinet EoS. The fit allowed refined values of the bulk modulus to be obtained at zero pressure (B_0). Results were discussed in comparison to the ones previously obtained from Lu-doped ceria. Two main conclusions can be drawn from this study.

- The ln B_0 vs. $\ln(2V_{at})$ trend provided a linear decrease as expected, but the slope of the regression line was around −6 for both systems, instead of ranging between −3 and −4, which is at odds with predictions for oxides. This evidence can be interpreted considering that a linear correlation exists between the RE^{3+} content and the oxygen vacancies amount; the rise in the vacancies content occurring with increasing the RE^{3+} amount causes a lattice shrinkage and limits the increase of the mean atomic volume.
- While samples crystallizing in the F structure showed a roughly linear decrease of the cell volume vs. pressure, the ones containing C-based microdomains demonstrated a softening of cell volume decrease with increasing pressure. This evidence, emerging both from Sm- and Lu-doped ceria, suggests that the presence of C stiffens a structure, making it less and less compressible with increasing pressure.

Supplementary Materials: The following are available online at http://www.mdpi.com/1996-1073/13/7/1558/s1, Table S1: Hybrid (H) structural model compared to the F structure typical of CeO_2 and to the C structure typical of sesquioxides of heavy rare earths.

Author Contributions: Conceptualization, C.A., M.M.C. and M.P.; methodology, C.A.; formal analysis, C.A.; investigation, M.P. and B.J.; resources, M.P. and C.A.; data curation, S.M.; writing—original draft preparation, C.A.; writing—review and editing, C.A., S.M., M.M.C., B.J. and M.P.; visualization, S.M. All authors have read and agreed to the published version of the manuscript.

Funding: This research received no external funding.

Acknowledgments: The Elettra Synchrotron Radiation Facility is acknowledged for the provision of beam time. B.J. acknowledges IISc Bangalore and ICTP Trieste for the award of IISc-ICTP fellowship.

Conflicts of Interest: The authors declare no conflict of interest.

References

1. Omar, S.; Wachsman, E.D.; Jones, J.L.; Nino, J.C. Crystal structure-ionic conductivity relationships in doped ceria. *J. Am. Ceram. Soc.* **2009**, *92*, 2674–2681. [CrossRef]
2. Artini, C.; Pani, M.; Lausi, A.; Masini, R.; Costa, G.A. High temperature structural study of Gd-doped ceria by synchrotron X-ray diffraction (673 K ≤ T ≤ 1073 K). *Inorg. Chem.* **2014**, *53*, 10140–10149. [CrossRef] [PubMed]
3. Presto, S.; Artini, C.; Pani, M.; Carnasciali, M.M.; Massardo, S.; Viviani, M. Ionic conductivity and local structural features in $Ce_{1-x}Sm_xO_{2-x/2}$. *Phys. Chem. Chem. Phys.* **2018**, *20*, 28338–28345. [CrossRef]
4. Steele, B.C.H. Appraisal of $Ce_{1-x}Gd_yO_{2-y/2}$ electrolytes for IT-SOFC operation at 500 °C. *Solid State Ion.* **2000**, *129*, 95–110. [CrossRef]
5. Artini, C. RE-doped ceria systems and their performance as solid electrolytes: A puzzling tangle of structural issues at the average and local scale. *Inorg. Chem.* **2018**, *57*, 13047–13062. [CrossRef]
6. Coduri, M.; Checchia, S.; Longhi, M.; Ceresoli, D.; Scavini, M. Rare earth doped ceria: The complex connection between structure and properties. *Front. Chem.* **2018**, *6*, 526. [CrossRef]
7. Artini, C.; Presto, S.; Massardo, S.; Pani, M.; Carnasciali, M.M.; Viviani, M. Transport properties and high temperature Raman features of heavily Gd-doped ceria. *Energies* **2019**, *12*, 4148. [CrossRef]
8. Rupp, J.L.M.; Infortuna, A.; Gauckler, L.J. Thermodynamic stability of gadolinia-doped ceria thin film electrolytes for micro-solid oxide fuel cells. *J. Am. Ceram. Soc.* **2007**, *90*, 1792–1797. [CrossRef]
9. Taniguchi, I.; Van Landschoot, R.C.; Schoonman, J. Electrostatic spray deposition of $Gd_{0.1}Ce_{0.9}O_{1.95}$ and $La_{0.9}Sr_{0.1}Ga_{0.8}Mg_{0.2}O_{2.87}$ thin films. *Solid State Ion.* **2003**, *160*, 271–279. [CrossRef]
10. Burbano, M.; Marrocchelli, D.; Watson, G.W. Strain effects on the ionic conductivity of Y-doped ceria: A simulation study. *J. Electroceram.* **2014**, *32*, 28–36. [CrossRef]
11. Kant, K.M.; Esposito, V.; Pryds, N. Strain induced ionic conductivity enhancement in epitaxial $Ce_{0.9}Gd_{0.1}O_{2-\delta}$ thin films. *Appl. Phys. Lett.* **2012**, *100*, 033105. [CrossRef]
12. Rainwater, B.H.; Velisavljevic, N.; Park, C.; Sun, H.; Waller, G.H.; Tsoi, G.M.; Vohra, Y.K.; Liu, M. High pressure structural study of samarium doped CeO_2 oxygen vacancy conductor—Insight into the dopant concentration relationship to the strain effect in thin film ionic conductors. *Solid State Ion.* **2016**, *292*, 59–65. [CrossRef]

13. Artini, C.; Joseph, B.; Costa, G.A.; Pani, M. Crystallographic properties of Lu-doped ceria, $Ce_{1-x}Lu_xO_{2-x/2}$, at pressures up to 7 GPa. *Solid State Ion.* **2018**, *320*, 152–158. [CrossRef]
14. Lee, K.-R.; Ahn, K.; Chung, Y.-C.; Lee, J.-H.; Yoo, H.-I. Lattice distortion effect on electrical properties of GDC thin films: Experimental evidence and computational simulation. *Solid State Ion.* **2012**, *229*, 45–53. [CrossRef]
15. Spiridigliozzi, L.; Ferone, C.; Cioffi, R.; Accardo, G.; Frattini, D.; Dell'Agli, G. Entropy-stabilized oxides owning fluorite structure obtained by hydrothermal treatment. *Materials* **2020**, *13*, 558. [CrossRef]
16. Sarkar, A.; Loho, C.; Velasco, L.; Thomas, T.; Battacharya, S.S.; Hahn, H.; Djenadic, R. Multicomponent equiatomic rare earth oxides with a narrow band gap and associated praseodymium multivalenc. *Dalton Trans.* **2017**, *46*, 12167–12176. [CrossRef] [PubMed]
17. Coduri, M.; Masala, P.; Allieta, M.; Peral, I.; Brunelli, M.; Biffi, C.A.; Scavini, M. Phase transformations in the CeO_2-Sm_2O_3 system: A multiscale powder diffraction investigation. *Inorg. Chem.* **2018**, *57*, 879–891. [CrossRef]
18. Artini, C.; Nelli, I.; Pani, M.; Costa, G.A.; Caratto, V.; Locardi, F. Thermal decomposition of Ce-Sm and Ce-Lu mixed oxalates: Influence of the Sm- and Lu-doped ceria structure. *Thermochim. Acta* **2017**, *651*, 100–107. [CrossRef]
19. Omar, S.; Wachsman, E.; Nino, J. Higher conductivity Sm^{3+} and Nd^{3+} co-doped ceria-based electrolyte materials. *Solid State Ion.* **2008**, *178*, 1890–1897. [CrossRef]
20. Acharya, S.A.; Gaikwad, V.M.; D'Souza, S.W.; Barman, S.R. Gd/Sm dopant-modified oxidation state and defect generation in nano-ceria. *Solid State Ion.* **2014**, *260*, 21–29. [CrossRef]
21. Nitani, H.; Nakagawa, T.; Yamanouchi, M.; Osuki, T.; Yuya, M.; Yamamoto, T.A. XAFS and XRD study of ceria doped with Pr, Nd and Sm. *Mater. Lett.* **2004**, *58*, 2076–2081. [CrossRef]
22. Artini, C.; Costa, G.A.; Pani, M.; Lausi, A.; Plaisier, J. Structural characterization of the CeO_2/Gd_2O_3 mixed system by synchrotron X-ray diffraction. *J. Solid State Chem.* **2012**, *190*, 24–28. [CrossRef]
23. Tianshu, Z.; Hing, P.; Huang, H.; Kilner, J. Ionic conductivity in the CeO_2–Gd_2O_3 system ($0.05 \leq Gd/Ce \leq 0.4$) prepared by oxalate coprecipitation. *Solid State Ion.* **2002**, *148*, 567–573. [CrossRef]
24. Artini, C.; Pani, M.; Carnasciali, M.M.; Buscaglia, M.T.; Plaisier, J.; Costa, G.A. Structural features of Sm- and Gd-doped ceria studied by synchrotron X-ray diffraction and μ-Raman spectroscopy. *Inorg. Chem.* **2015**, *54*, 4126–4137. [CrossRef]
25. Shannon, R.D. Revised effective ionic radii and systematic studies of interatomic distances in halides and chalcogenides. *Acta Cryst. A* **1976**, *32*, 751–767. [CrossRef]
26. Coduri, M.; Scavini, M.; Allieta, M.; Brunelli, M.; Ferrero, C. Defect structure of Y-doped ceria on different length scales. *Chem. Mater.* **2013**, *25*, 4278–4289. [CrossRef]
27. Coduri, M.; Scavini, M.; Pani, M.; Carnasciali, M.M.; Klein, H.; Artini, C. From nano to microcrystals: Effect of different synthetic pathways on defects architecture in heavily Gd-doped ceria. *Phys. Chem. Chem. Phys.* **2017**, *19*, 11612–11630. [CrossRef]
28. Artini, C.; Pani, M.; Carnasciali, M.M.; Plaisier, J.R.; Costa, G.A. Lu-, Sm- and Gd-doped ceria: A comparative approach to their structural properties. *Inorg. Chem.* **2016**, *55*, 10567–10579. [CrossRef]
29. Lotti, P.; Milani, S.; Merlini, M.; Joseph, B.; Alabarse, F.; Lausi, A. Single-crystal diffraction at the high-pressure Indo-Italian beamline Xpress at Elettra, Trieste. *J. Synchrot. Radiat.* **2020**, *27*, 222–229. [CrossRef]
30. Vinet, P.; Smith, J.R.; Ferrante, J.; Rose, J.H. Temperature effects on the universal equation of state of solids. *Phys. Rev. B* **1987**, *35*, 1945–1953. [CrossRef]
31. Artini, C.; Costa, G.A.; Carnasciali, M.M.; Masini, R. Stability fields and structural properties of intra rare earths perovskites. *J. Alloy Compd.* **2010**, *494*, 336–339. [CrossRef]
32. Artini, C.; Costa, G.A.; Masini, R. Study of the formation temperature of mixed $LaREO_3$ (RE≡Dy, Ho, Er, Tm, Yb, Lu) and $NdGdO_3$ oxides. *J. Therm. Anal. Calorim.* **2011**, *103*, 17–21. [CrossRef]
33. Rodriguez-Carvajal, J. Recent advances in magnetic structure determination by neutron powder diffraction. *Phys. B Condens. Matter* **1993**, *192*, 55–69. [CrossRef]
34. Gonzalez-Platas, J.; Alvaro, M.; Nestola, F.; Angel, R. EosFit7-GUI: A new graphical user interface for equation of state calculations, analyses and teaching. *J. Appl. Cryst.* **2016**, *49*, 1377–1382. [CrossRef]
35. Eyring, L. The binary rare earth oxides. In *Handbook on the Physics and Chemistry of Rare Earths*; Gschneidner, K.A., Jr., Eyring, L., Eds.; North Holland: Amsterdam, The Netherlands, 1979; Volume 3, pp. 337–399.
36. Vanpoucke, D.E.P.; Bultinck, P.; Cottenier, S.; Van Speybroeck, V.; Van Driessche, I. Density functional theory study of $La_2Ce_2O_7$: Disordered fluorite versus pyrochlore structure. *Phys. Rev. B* **2011**, *84*, 054110. [CrossRef]

37. Minervini, L.; Grimes, R.W. Disorder in pyrochlore oxides. *J. Am. Ceram. Soc.* **2000**, *83*, 1873–1878. [CrossRef]
38. Chen, W.; Navrotsky, A. Thermochemical study of trivalent-doped ceria systems: CeO_2-$MO_{1.5}$ (M = La, Gd, and Y). *J. Mater. Res.* **2006**, *21*, 3242–3251. [CrossRef]
39. Minervini, L.; Zacate, M.O.; Grimes, R.W. Defect cluster formation in M_2O_3-doped CeO_2. *Solid State Ion.* **1999**, *116*, 339–349. [CrossRef]
40. Wang, B.; Lewis, R.J. Cormack, A.N. Computer simulations of large-scale defect clustering and nanodomain structure in gadolinia-doped ceria. *Acta Mater.* **2011**, *59*, 2035–2045. [CrossRef]
41. Artini, C.; Carnasciali, M.M.; Viviani, M.; Presto, S.; Plaisier, J.R.; Costa, G.A.; Pani, M. Structural properties of Sm-doped ceria electrolytes at the fuel cell operating temperatures. *Solid State Ion.* **2018**, *315*, 85–91. [CrossRef]
42. Ubaldini, A.; Carnasciali, M.M. Raman Characterisation of powder of cubic RE_2O_3 (RE=Nd, Gd, Dy, Tm, and Lu), Sc_2O_3 and Y_2O_3. *J. Alloy Compd.* **2008**, *454*, 374–378. [CrossRef]
43. Artini, C.; Carnasciali, M.M.; Costa, G.A.; Plaisier, J.R.; Pani, M. A novel method for the evaluation of the Rare Earth (RE) coordination number in RE-doped ceria through Raman spectroscopy. *Solid State Ion.* **2017**, *311*, 90–97. [CrossRef]
44. Murnaghan, F.D. The compressibility of media under extreme pressures. *Proc. Natl. Acad. Sci. USA* **1944**, *30*, 244–247. [CrossRef] [PubMed]
45. Birch, F. Finite elastic strain of cubic crystals. *Phys. Rev. B* **1947**, *71*, 809–824. [CrossRef]
46. Poirier, J.P.; Tarantola, A. A logarithmic equation of state. *Phys. Earth Planet Int.* **1998**, *109*, 1–8. [CrossRef]
47. Pavese, A. Pressure-volume-temperature equations of state: A comparative study based on numerical simulations. *Phys. Chem. Miner.* **2002**, *29*, 43–51. [CrossRef]
48. Hill, S.E.; Catlow, C.R.A. A Hartree-Fock periodic study of bulk ceria. *J. Phys. Chem. Solids* **1993**, *54*, 411–419. [CrossRef]
49. Gerward, L.; Staun Olsen, J. Powder diffraction analysis of cerium oxide at high pressure. *Powder Diffr.* **1993**, *8*, 127–130. [CrossRef]
50. Gerward, L.; Staun Olsen, J.; Petit, L.; Vaitheeswaran, G.; Kanchana, V.; Svane, A. Bulk modulus of CeO_2 and PrO_2 - An experimental and theoretical study. *J. Alloy Compd.* **2005**, *400*, 56–61. [CrossRef]
51. Anderson, O.L.; Nafe, J.E. The bulk modulus-volume relationship for oxides compounds and related geophysical problems. *J. Geophys. Res.* **1965**, *70*, 3951–3963. [CrossRef]
52. Zen, E. Validity of "Vegard's law". *Am. Mineral.* **1956**, *41*, 523–524.
53. McBride, J.R.; Hass, K.C.; Pointdexter, B.D.; Weber, W.H. Raman and X-ray studies of $Ce_{1-x}RE_xO_{2-y}$, where RE=La, Pr, Nd, Eu, Gd, and Tb. *J. Appl. Phys.* **1994**, *76*, 2435–2441. [CrossRef]
54. Mandal, B.P.; Grover, V.; Roy, M.; Tyagi, A.K. X-ray diffraction and Raman spectroscopic investigation on the phase relations in Yb_2O_3- and Tm_2O_3-substituted CeO_2. *J. Am. Ceram. Soc.* **2007**, *90*, 2961–2965. [CrossRef]
55. Kossoy, A.; Wang, Q.; Korobko, R.; Grover, V.; Feldman, Y.; Wachtel, E.; Tyagi, A.K.; Frenkel, A.I.; Lubomirsky, I. Evolution of the local structure at the phase transition in CeO_2-Gd_2O_3 solid solutions. *Phys. Rev. B* **2013**, *87*, 054101. [CrossRef]
56. Scavini, M.; Coduri, M.; Allieta, M.; Brunelli, M.; Ferrero, C. Probing complex disorder in $Ce_{1-x}Gd_xO_{2-x/2}$ using the pair distribution function analysis. *Chem. Mater.* **2012**, *24*, 1338–1345. [CrossRef]
57. Horlait, D.; Claparède, L.; Clavier, N.; Szencnekt, S.; Dacheux, N.; Ravaux, J.; Podor, R. Stability and structural evolution of $Ce^{IV}{}_{1-x}Ln^{III}{}_xO_{2-x/2}$ solid solutions: A coupled µ-Raman/XRD approach. *Inorg. Chem.* **2011**, *50*, 150–161. [CrossRef]
58. Artini, C.; Gigli, L.; Carnasciali, M.M.; Pani, M. Structural properties of the (Nd,Dy)-doped ceria system by synchrotron X-ray diffraction. *Inorganics* **2019**, *7*, 94. [CrossRef]

Article

Transport Properties and High Temperature Raman Features of Heavily Gd-Doped Ceria

Cristina Artini [1,2], Sabrina Presto [3,*], Sara Massardo [1], Marcella Pani [1,4], Maria Maddalena Carnasciali [1,5] and Massimo Viviani [3]

1 Department of Chemistry and Industrial Chemistry, University of Genova, Via Dodecaneso 31, 16146 Genova, Italy; artini@chimica.unige.it (C.A.); s.massardo93@gmail.com (S.M.); marcella@chimica.unige.it (M.P.); marilena@chimica.unige.it (M.M.C.)
2 Institute of Condensed Matter Chemistry and Technologies for Energy, National Research Council, CNR-ICMATE, Via De Marini 6, 16149 Genova, Italy
3 Institute of Condensed Matter Chemistry and Technologies for Energy, National Research Council, CNR-ICMATE, c/o DICCA-UNIGE, Via all'Opera Pia 15, 16145 Genova, Italy; massimo.viviani@ge.icmate.cnr.it
4 CNR-SPIN, Corso Perrone 24, 16152 Genova, Italy
5 INSTM, Genova Research Unit, Via Dodecaneso 31, 16146 Genova, Italy
* Correspondence: sabrina.presto@ge.icmate.cnr.it; Tel.: +39-010-353-6025

Received: 1 October 2019; Accepted: 29 October 2019; Published: 30 October 2019

Abstract: Transport and structural properties of heavily doped ceria can reveal subtle details of the interplay between conductivity and defects aggregation in this material, widely studied as solid electrolyte in solid oxide fuel cells. The ionic conductivity of heavily Gd-doped ceria samples ($Ce_{1-x}Gd_xO_{2-x/2}$ with x ranging between 0.31 and 0.49) was investigated by impedance spectroscopy in the 600–1000 K temperature range. A slope change was found in the Arrhenius plot at ~723 K for samples with x = 0.31 and 0.34, namely close to the compositional boundary of the CeO_2-based solid solution. The described discontinuity, giving rise to two different activation energies, points at the existence of a threshold temperature, below which oxygen vacancies are blocked, and above which they become free to move through the lattice. This conclusion is well supported by Raman spectroscopy, due to the discontinuity revealed in the Raman shift trend versus temperature of the signal related to defects aggregates which hinder the vacancies movement. This evidence, observable in samples with x = 0.31 and 0.34 above ~750 K, accounts for a weakening of Gd–O bonds within blocking microdomains, which is compatible with the existence of a lower activation energy above the threshold temperature.

Keywords: solid oxide fuel cells (SOFCs); ionic conductivity; raman spectroscopy; powder x-ray diffraction; doped ceria

1. Introduction

Rare earth (RE)-doped ceria systems form one of the most thoroughly studied families of RE oxides for their manifold properties, which make them interesting as catalysts for various processes of CO_2 valorization [1–3], and also as solid electrolytes in solid oxide fuel cells [4–6]. The latter application makes them a valid alternative to the commonly employed Y_2O_3-stabilized ZrO_2, as they present high ionic conductivity in the intermediate temperature range (673–973 K), which in principle allows them to lower the cell operating temperature. $Ce_{0.9}Gd_{0.1}O_{1.95}$, for instance, exhibits a remarkable value of 10^{-2} S cm^{-1} at 773 K [7].

The mechanism of ionic conductivity in doped ceria is quite well known: It occurs thanks to the presence of oxygen vacancies introduced into the structure by partially substituting Ce^{4+} with a trivalent ion, generally a lanthanide. At sufficiently high temperature, vacancies are provided with energy enough to overcome the activation barrier to diffusion, and are therefore free to move through the lattice. From the crystallographic point of view, the necessary condition for the occurrence of ionic conductivity is the preservation of the CeO_2 fluorite-based structure (hereafter called F), which can be accomplished up to the stability limit of the CeO_2-based solid solution, i.e., up to a threshold representing the maximum substitution of Ce^{4+} by RE^{3+} which can be tolerated by the CeO_2 structure [8]. Nevertheless, this condition is not sufficient, since the maximum ionic conductivity in doped ceria occurs within the stability region of the CeO_2-based solid solution at a substitution degree of Ce^{4+} strongly lower than the one corresponding to the aforementioned threshold. The maximum ionic conductivity is observed for instance at $x = 0.1$ [7] and $x = 0.2$ [9,10] in $Ce_{1-x}Gd_xO_{2-x/2}$, and at $x = 0.17$ [11] and $x \sim 0.2$ in $Ce_{1-x}Sm_xO_{2-x/2}$ [12], but the upper compositional limit of the F structure lies at x close to 0.3 in Gd-doped ceria [13,14] and at x ranging between 0.3 and 0.4 [15] or close to 0.3 [16] for Sm-doped ceria. The reason for the described non-coincidence of maximum ionic conductivity and maximum Ce^{4+} substitution is commonly attributed to the formation of $RE'_{Ce} : V_{O}^{\cdot\cdot}$ aggregates causing phase separation at the nanoscale, where vacancies are trapped at fixed positions, with their movement being consequently hindered [17–21]. The cited aggregates assume the cubic atomic arrangement (hereafter named C) typical of sesquioxides of the smallest rare earths, which is a superstructure of the F structure.

The presence of C nanodomains, responsible for the ionic conductivity reduction observed within the F region, is generally not detectable by average techniques, such as powder x-ray diffraction; on the contrary, they are clearly visible by local techniques, like pair distribution function (PDF) [14,16,22], Transmission Electron Microscopy (TEM) [23,24], and Extended X-ray Absorption Fine Structure (EXAFS) spectroscopy [25–27]. Among these experimental methods, Raman spectroscopy deserves a special mention: Its high sensitivity toward oxygen displacement and, thus, vacancies formation, and its superior ability to clearly distinguish between signals originating from F and C [28,29], make it the technique of choice to be coupled to x-ray diffraction in the experimental investigation of doped ceria [15,30–36].

The study of ionic conductivity in doped ceria is generally limited to lightly doped samples (i.e., up to $x \sim 0.20$–0.30, depending on the specific system); nevertheless, from the viewpoint of basic science, highly doped compositions are of great interest too, as they allow to investigate the effect of $RE'_{Ce} : V_{O}^{\cdot\cdot}$ aggregates on the reduction/suppression of vacancies migration, and the behavior of the activation barrier as a response to the presence of C domains. In this work, several samples belonging to the $Ce_{1-x}Gd_xO_{2-x/2}$ system with x ranging between 0.30 and 0.50 are studied by means of impedance and micro Raman spectroscopy, and results are compared to data deriving from lightly doped Gd-doped ceria [9] and Sm-doped ceria [12,30]. Compositions located close to the F compositional limit, i.e., with x close to 0.30, exhibit a two-fold behavior of total conductivity and two values of activation energy (E_a), depending on temperature, which suggest the existence of a threshold temperature separating a high from a low temperature conductivity mechanism. In this respect, Raman spectroscopy is of great help by providing information about the trend of Raman shift versus temperature of the F and C phases, and by revealing possible discontinuities which can unravel particular aspects of the conductivity mechanism.

2. Materials and Methods

2.1. Synthesis

Four samples belonging to the $Ce_{1-x}Gd_xO_{2-x/2}$ system having nominal x = 0.31, 0.34, 0.43, and 0.49 were prepared by co-precipitation of the corresponding mixed oxalates, as described in [37,38]. The co-precipitation technique was preferred to other synthetic methods, since it ensures a more intimate mixing of cations with respect for instance to solid state reactions, thus allowing a better homogeneity of the resulting samples [33]. Stoichiometric amounts of Ce (Johnson Matthey ALPHA

99.99% wt.) and Gd_2O_3 were separately dissolved in HCl (13% volume), and the two solutions were subsequently mixed. Oxalates were co-precipitated by adding an oxalic acid solution. Precipitates were filtered, washed, dried overnight, and then treated at 973 K in air for four days to obtain mixed oxides. Samples are named CGO31, CGO34, and so on, according to the Gd atomic percent with respect to the total lanthanide content.

2.2. X-ray Diffraction and Scanning Electron Microscopy—Energy-Dispersive System (SEM-EDS)

Powders of all the obtained mixed oxides were analyzed by x-ray diffraction using a Bragg-Brentano powder diffractometer (Philips PW1050/81, Fe-filtered Co Kα radiation, Amsterdam, The Netherlands); diffraction patterns were collected in the 20–80° angular range, with angular step 0.02°.

The powder morphology, as well as the Gd and Ce amount, and the feature of the sintered pellet surface, were analyzed by scanning electron microscopy coupled to energy-dispersive system (SEM-EDS, Phenom ProX, Phenom World, Eindhoven, The Netherlands).

2.3. Impedance Spectroscopy

Transport properties of all the samples were evaluated by impedance spectroscopy. Powders were pressed by a hydraulic press, and the obtained pellets were sintered in air at 1773 K for 1 h. Plane surfaces of sintered pellets were coated with metallic electrodes by brushing Pt ink (Metalor, Birmingham, UK), which were cured at 1273 K in air. Conductivity was measured by a Frequency Response Analyzer (Iviumstat.h, Ivium Technologies B.V., Eindhoven, The Netherlands) in the 600–1000 K temperature range, 10^{-2}–10^{6} Hz frequency range and under flowing pure O_2 atmosphere. The amplitude of the signal used for the measurements was 30 mV. Data was treated by the equivalent circuits method, as implemented in the Zview software (3.4), consisting in fitting data by a series of one or two (Ri//Qi), depending on the temperature, in series to a resistance R.

2.4. Micro Raman Spectroscopy

High temperature micro Raman analyses were carried out on all the mixed oxides samples employing a Renishaw System 2000 Raman imaging microscope. Specimens were pressed into pellets, and measurements were done at room temperature and at 673, 773, 873, 973, and 1073 K. Raman spectra were obtained by a 633 nm He-Ne laser in the 1000–100 cm^{-1} range as a result of one accumulation lasting 10 s. At least four different points of each sample were analyzed at room temperatures; afterwards, high temperature spectra were collected on a selected point during both heating and cooling, with a 20× magnification. The laser power was kept at around 2 mW. The wave number has an accuracy of 3 cm^{-1}.

3. Results

3.1. Structural and Morphological Analysis

Figure 1 shows the aspect of powders of samples CGO31 and CGO43 as revealed by scanning electron microscopy. Morphological analyses performed on powders of all samples suggest that the average particles size lies in the range 1–5 µm. Compositional analyses deriving from backscattered electrons reveal that the actual rare earth content is very close to the nominal one.

In Figure 2 SEM microphotographs taken (a) before and (b) after impedance spectroscopy measurements on the surface of sintered pellets of samples CGO43 are presented. Images of all samples show a good compaction degree of the pellet and no appreciable compositional inhomogeneities. Areas with grains having average sizes up to 5 µm are accompanied by much finer grains; this evidence could be most probably attributed to the presence of particles aggregates characterized by lower sinterability than other ones. No effect of impedance spectroscopy measurements was detected on such microstructural features. In Figure 2c the EDS spectrum obtained on a region of CGO43 sample is reported.

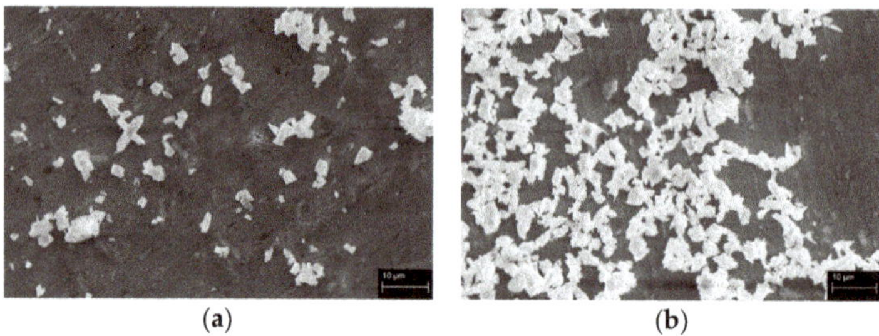

Figure 1. SEM microphotographs taken by secondary electrons on powders of samples (**a**) CGO31 and (**b**) CGO43.

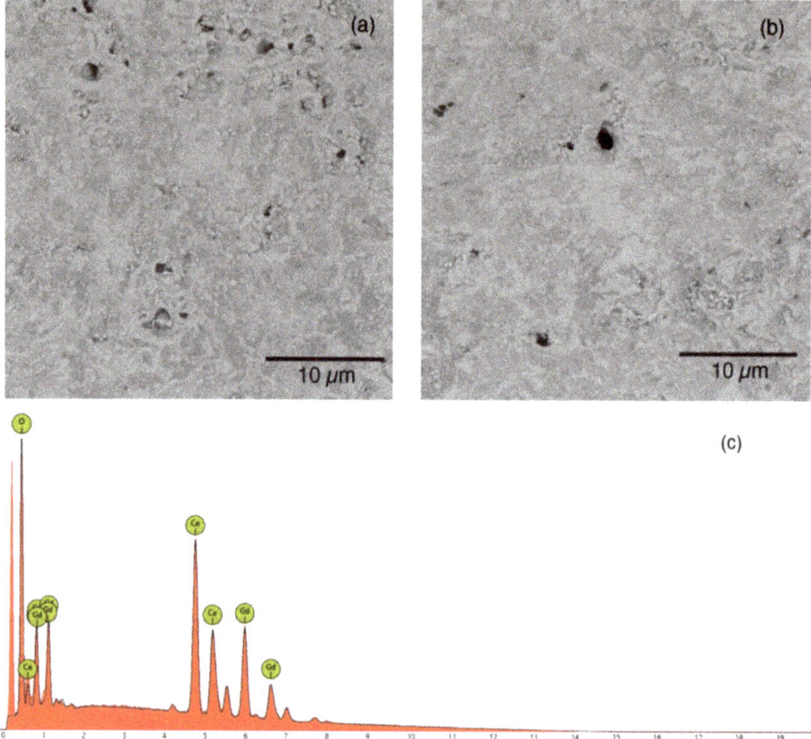

Figure 2. SEM microphotographs taken by back-scattered electrons on the surface of sintered pellets of CGO43 (**a**) before and (**b**) after impedance spectroscopy measurements. (**c**) energy-dispersive system (EDS) spectrum obtained on a region of CGO43 sample.

EDS spectra deriving from different regions reveal that for all samples the actual composition is very close to the nominal one.

All the collected diffraction patterns show only peaks attributable to doped ceria; no hints suggesting the presence of additional phases were detected. In Figure 3 diffractograms of all the samples are reported in a stack. It can be observed that, while sample CGO31 only presents peaks deriving from the F phase, in all the other patterns peaks originating from the C superstructure appear; the intensities of the latter increase with increasing the Gd content. It can also be noticed that peaks common to F and C are perfectly superimposed at each composition. This evidence was widely discussed in previous papers [13,15,32,39,40] and attributed to the onset of an atomic arrangement having an intermediate character between F and C. This F/C hybrid phase (hereafter called H) can be described as a solid solution where F behaves as a matrix hosting C-structured $Gd'_{Ce} : V_{O}^{\cdot\cdot}$ domains. C domains, in turn, grow as a finely interlaced phase within F, so that F and C cannot be distinguished at the average scale as two separate phases. Relying on the results of x-ray diffraction analyses, it can be thus concluded that CGO31 lies close to the F/H transition, thus confirming previous data reported in Reference [13].

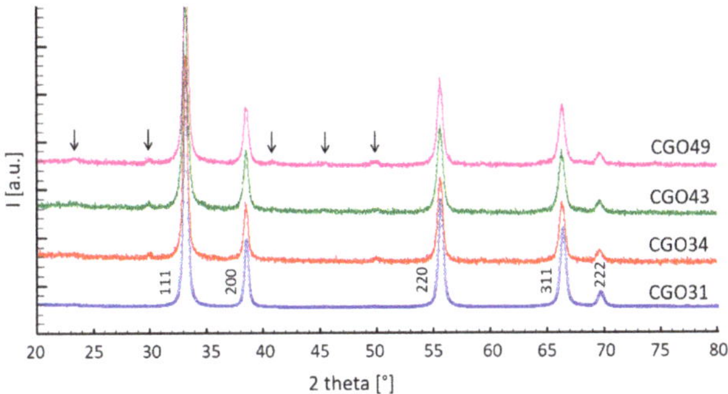

Figure 3. Diffraction patterns of samples CGO31, CGO34, CGO43, and CGO49. In CGO31 Miller indices of the F phase are reported; arrows indicate the position of peaks belonging to the C superstructure.

3.2. Ionic Conductivity

Figure 4 reports the Nyquist plots of impedance spectra obtained (a) at fixed temperature (998 K) for $x = 0.31$ and 0.34 and (b) increasing temperature from 947 K to 998 K for $x = 0.49$. Only one contribution can be recognized given by electrode-related phenomena. The presence of two partially overlapped arcs can be ascribed to some differences between the two Pt electrodes. The total resistance was estimated as the intercepts at high frequency with axis. In particular the total resistance increases with increasing Gd content at fixed temperature and decreases with increasing temperature at fixed Gd content.

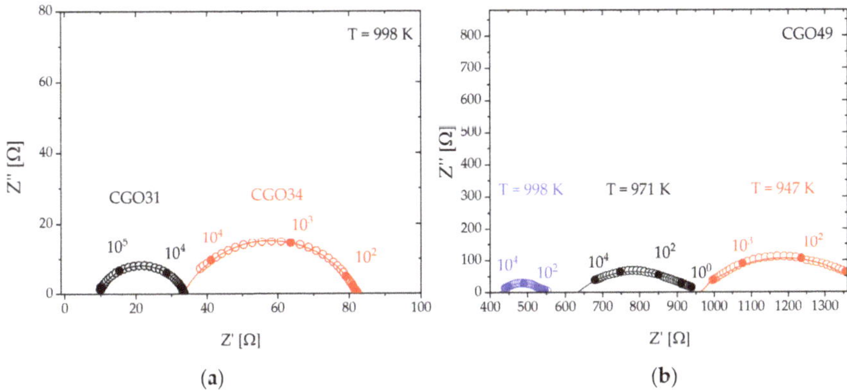

Figure 4. Nyquist plot of impedance spectra of (**a**) CGO31 and CGO34 samples measured at 998 K and (**b**) CGO49 sample measured at 998, 971, and 947 K, as reported within panels. Full markers point at frequency decades. Lines represent the best fit of data with model reported in the text.

From total resistance of samples, the conductivity values were calculated and used to evaluate E_a at different temperatures. Arrhenius plots are presented in Figure 5. In the whole temperature range, conductivity decreases with increasing Gd content, as expected in this compositional range. In addition, a slope change is present for sample CGO31 and, less pronounced, for sample CGO34. Experimental data of each of the two aforementioned samples were in fact fitted by two different regression lines, as evident from Figure 5. The described slope change is due to a variation of E_a taking place around 723–773 K. Such a change is not present in CGO43 and CGO49 samples.

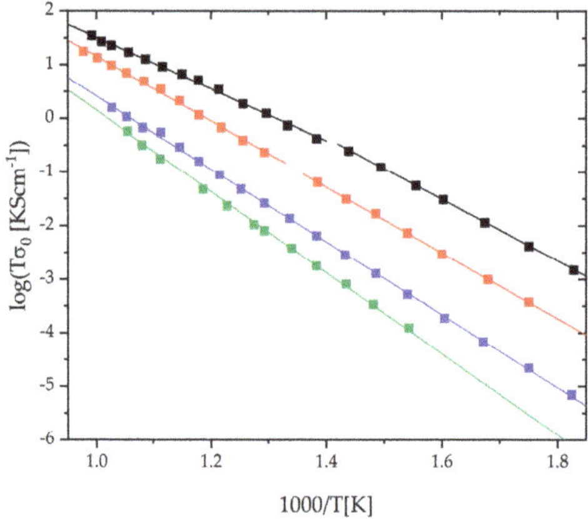

Figure 5. Arrhenius plots of total conductivity of samples CGO31–*black*, CGO34–*red*, CGO43–*blue*, and CGO49–*green*. Two different regression lines were used to fit data of samples CGO31 and CGO34.

The values of E_a calculated from the above-mentioned slopes are reported in Figure 6. In this plot, previously tested [9] samples with lower Gd content (10 at. % and 20 at. %), are also reported for comparison. Open circles represent unique E_a values, i.e., compositions which do not exhibit slope change in their Arrhenius plot. Solid markers indicate E_a values in the low-temperature (red) or high-temperature (black) regions, in case of slope change.

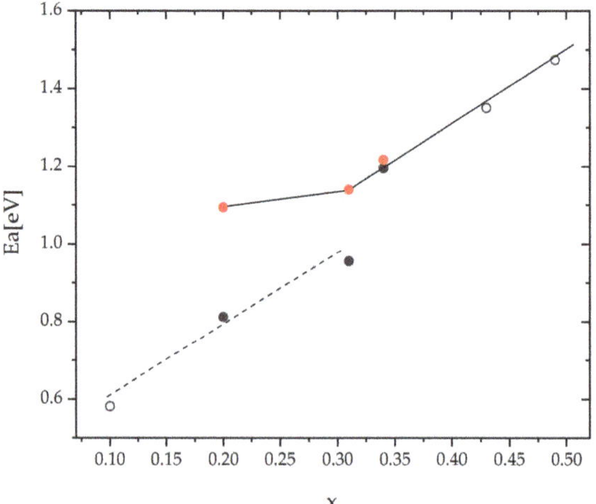

Figure 6. Activation energies of samples CGO31, CGO34, CGO43, and CGO49 as a function of x. Values found for two previously tested [9] samples with lower Gd content (10 at. % and 20 at. %), are also reported. Lines are guides for the eye indicating two series, namely the low-T (solid) and high-T (dashed).

E_a values displayed in Figure 6 can be conveniently divided into two series: The first one grouping low-T values and including samples with $x = 0.43$ and $x = 0.49$ (connected by the solid line); the second one grouping the high–T values and including the sample with $x = 0.10$ (connected by the dashed line).

3.3. Micro Raman Spectroscopy

Raman spectra collected on RE-doped ceria in the F and in the H region present signals attributable to three different structural features: (a) a signal at ~465 cm^{-1} due to the F_{2g} symmetric vibration mode of the Ce–O bond in eight-fold coordination [29,41], typical of CeO_2 and of CeO_2-based solid solutions; (b) signals caused by the introduction of the doping cation into F, respectively inducing the creation of vacancies within the Gd coordination sphere [28] (at ~540 cm^{-1} and ~250 cm^{-1}), and the formation of a REO_8-type complex not containing oxygen vacancies, due to the partial substitution of Ce by Gd [28] (at ~600 cm^{-1}); and (c) a broad band at ~370 cm^{-1}, to be ascribed to the $(A_g + F_g)$ Gd–O symmetrical stretching mode with Gd in six-fold coordination [42], namely the signature of the C structure. With increasing the dopant content, position, intensity, and width of all the cited bands change, as thoroughly discussed in [15,32]: The signal at ~465 cm^{-1} undergoes first a redshift and, at higher Gd content, a blueshift under the action of the Gd size and of vacancies, respectively; all the other signals increase in intensity and sharpen, since their occurrence is strictly related to the presence of the doping ion.

The effect of rising temperature on Raman spectra of Gd-doped ceria can be appreciated by observing Figure 7, where acquisitions performed on sample CGO34 from 298 K to 1073 K are reported. As generally expected, rising temperature leads to broaden the width and reduce the intensity of all the signals, as a consequence of the increased disorder caused by thermal motion; moreover, all the bands move toward lower Raman shifts, due to the decreasing energy necessary to excite vibrations along progressively looser bonds. The behavior of the main Raman signals of both the F and the C phase as a function of temperature, namely the ones at ~465 cm^{-1} and ~370 cm^{-1}, respectively, deserve a closer inspection. Figure 8 shows the trend of the aforementioned Raman shifts.

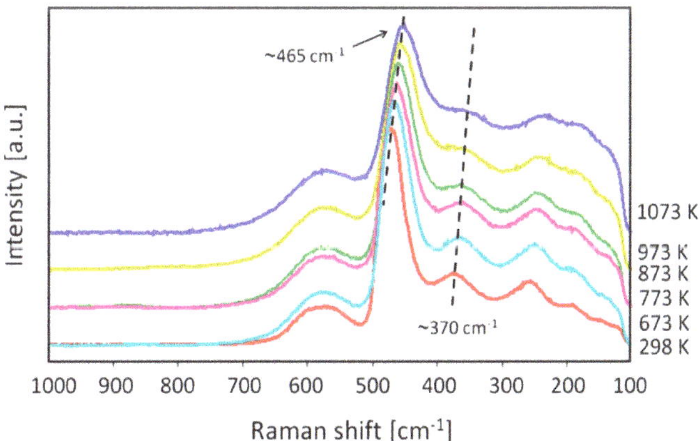

Figure 7. Raman spectra collected on sample CGO34 from 298 K to 1073 K. Dashed lines are a guide for the eye highlighting the movement of Raman shift of the main F (~465 cm^{-1}) and C (~370 cm^{-1}) signal toward lower values with increasing temperature.

It can be observed that the trends of both signals show a steeper decrease of the Raman shift above a certain threshold temperature, which is located between 673 K and 773 K, depending on the sample. The slope change occurs for all the samples except CGO49, and it most probably points at an abrupt lengthening and weakening of both the Ce–O and Gd–O bonds (with Ce and Gd in eight- and six-fold coordination, respectively), thus causing a sudden reduction of the energy needed to excite the vibration along the aforementioned bonds.

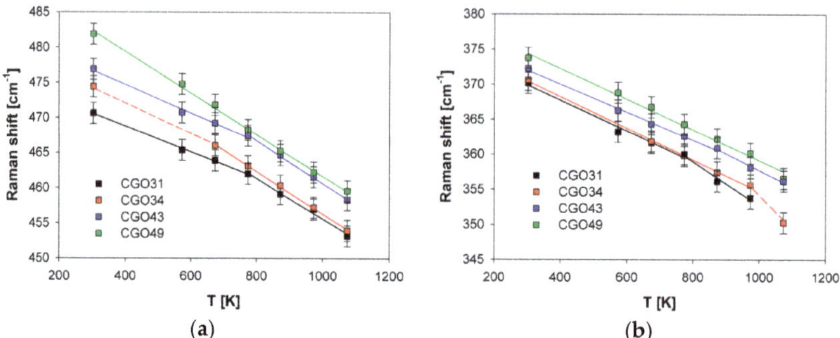

Figure 8. Trend of the Raman shift versus temperature of (**a**) the F_{2g} vibration mode of Ce–O and (**b**) the $(A_g + F_g)$ Gd–O stretching mode. Dashed lines are a guide for the eye, while solid lines are regression lines.

4. Discussion

Based on the results of impedance spectroscopy measurements, three different conductivity mechanisms can be recognized in the 0.1–0.5 x range.

(1) **0.10 ≤ x < 0.20**. In this range a unique E_a value of 0.6–0.8 eV is observed through the whole temperature span. Bulk E_a is considered to be the sum of two contributions, namely the migration enthalpy (H_m) and the association enthalpy (H_a), with the former being the energy needed to induce the hopping of vacancies through the lattice, and the latter the energy needed to force vacancies out of the C-based aggregates [43]. According to Inaba and Tagawa [44], while H_a increases with increasing the dopant amount due to the growing stability of progressively larger C domains, H_m remains essentially constant, assuming the value of around 0.60 eV. In the light of these considerations, it can be deduced that for x = 0.10 vacancies are randomly dispersed within the F matrix and essentially free to move through the lattice; this conclusion is in good agreement with the results of numerous studies, claiming that this is the region which provides the maximum ionic conductivity [7,9,10]. C nanodomains, existing in an embryonic form even for x = 0.125 as revealed by the pair distribution function technique [14], are not able to significantly affect ionic conductivity.

(2) **0.20 ≤ x ≤ 0.34**. In this compositional interval a two-fold behavior is observed: At temperature lower than ~723 K an E_a of ~1.1 eV is found, while above the value is approximately 0.8–0.9 eV. The cited temperature can be thus considered as a threshold: Below it, a low temperature behavior occurs, where vacancies, being trapped within C microdomains, are characterized by a reduced mobility, and therefore by a high E_a. This interpretation is fully compatible with the results of several local techniques, such as Raman spectroscopy [15,32], EXAFS [25,26], TEM [23,24], and PDF [14], which agree about the existence of C microdomains. The described scenario is also in good agreement with the drop in conductivity observed even within the F region at x > 0.1 [7] or 0.2 [9,10]. Above the threshold temperature, on the contrary, vacancies are mainly free to move through the lattice, and E_a is similar to the one observed in the 0.1–0.2 x range. It is noteworthy that for x = 0.34 the low temperature activation energies are still distinguishable but very close to the high temperature one, meaning that with increasing x even above the threshold temperature a significant part of C microdomains survives.

Raman spectra collected on samples CGO31 and CGO34 support the conclusions driven from the results of impedance spectroscopy, suggesting that above the threshold temperature both the release and the mobility of vacancies are enhanced. The slope change observable above ~750 K in the trend of Raman shift versus temperature of the C phase (see Figure 8b) can be ascribed to an abrupt lengthening and weakening of the Gd–O bond, responsible for facilitating the dissociation of the

C-structured aggregates and hence for promoting the release of oxygen vacancies. This interpretation is in good agreement with the strong reduction of H_a found above the threshold temperature. A similar discontinuity, also revealed in the Raman shift of the F structure (see Figure 8a), points at a lengthening even of the Ce–O bond. This phenomenon acts in the same direction as the lengthening of the Gd–O distance within C microdomains, as it weakens the bond and fosters its rupture, thus enhancing the mobility of vacancies through F.

(3) **0.43 ≤ *x* ≤ 0.49**. In this region ionic conductivity is again ruled by just one E_a, whose value is remarkably higher than in the 0.10–0.20 x region (1.3–1.5 eV with respect to 0.6–0.8 eV); moreover, data are well aligned with the higher value of samples, with x ranging between 0.20 and 0.34, so that all the cited experimental points can be conveniently fitted by a regression line. The described scenario suggests that on average vacancies are blocked within C microdomains. Moreover, the increasing trend of E_a with x is compatible with the corresponding increasing size of C microdomains [16,45], which makes them progressively more stable, consequently enhancing the energy amount needed to release a vacancy from the defect aggregate. Correspondingly, the trend of the Raman shifts related to the main signal of both C and F of CGO49 do not show any slope change as a function of temperature.

5. Conclusions

Ionic conductivity and high temperature Raman features of several samples of heavily Gd-doped ceria ($Ce_{1-x}Gd_xO_{2-x/2}$ with x = 0.31, 0.34, 0.43, and 0.49) were investigated and compared to data deriving from samples with x = 0.10 and 0.20. Aim of the study was to identify mechanisms lying behind the transport of the oxygen ion in the 600–1000 K temperature range. Three different behaviors were identified, namely: (a) in the $0.10 \leq x < 0.20$ range just one E_a appears. Its value (0.6–0.8 eV) is compatible with a random dispersion of vacancies within the CeO_2-based solid solution; consequently, vacancies are free to move through the lattice. (b) in the $0.20 \leq x \leq 0.34$ range two activation energies occur, giving rise to a threshold temperature (~723 K), below which vacancies are mainly trapped within defect aggregates, and above which they are free to move. This interpretation is fully confirmed by results of Raman spectroscopy, which demonstrate the existence of the aforementioned threshold temperature even in the trend of the Raman shift versus temperature of the main signals related both to the CeO_2-based solid solution and to the C-structured defect aggregates. (c) in the $0.43 \leq x \leq 0.49$ range a unique E_a, higher than in the $0.10 \leq x < 0.20$ interval, suggests that vacancies are mainly blocked within defect aggregates at each temperature considered.

Author Contributions: Conceptualization, C.A., S.P., S.M., M.P., M.M.C. and M.V.; methodology, C.A., S.P., M.P. and M.V.; validation, S.P. and S.M.; formal analysis, S.P. and S.M.; investigation, C.A., S.P., S.M., M.P., M.M.C. and M.V.; writing—original draft preparation, C.A.; writing—review and editing, C.A., S.P., S.M., M.P., M.M.C. and M.V.; visualization, S.P., S.M.; supervision, M.V.

Funding: This research received no external funding.

Conflicts of Interest: The authors declare no conflict of interest.

References

1. Boaro, M.; Colussi, S.; Trovarelli, A. Ceria-based materials in hydrogenation and reforming reactions for CO_2 valorization. *Front. Chem.* **2019**, *7*, 28. [CrossRef] [PubMed]
2. Le Saché, E.; Johnson, S.; Pastor-Pérez, L.; Horri, B.A.; Reina, T.R. Biogas upgrading via dry reforming over a Ni-Sn/CeO_2-Al_2O_3 catalyst: Influence of the biogas source. *Energies* **2019**, *12*, 1007. [CrossRef]
3. Smith, P.J.; Smith, L.; Dummer, N.F.; Douthwaite, M.; Willock, D.J.; Howard, M.; Knight, D.W.; Taylor, S.H.; Hutchings, G.J. Investigating the Influence of Reaction Conditions and the Properties of Ceria for the Valorisation of Glycerol. *Energies* **2019**, *12*, 1359. [CrossRef]
4. Skinner, S.J.; Kilner, J.A. Oxygen ion conductors. *Mater. Today* **2003**, *6*, 30–37. [CrossRef]
5. Jaiswal, N.; Tanwar, K.; Suman, R.; Kumar, D.; Upadhyay, S.; Parkash, O. A brief review on ceria based solid electrolytes for solid oxide fuel cells. *J. Alloys Compd.* **2019**, *781*, 984–1005. [CrossRef]

6. Presto, S.; Barbucci, A.; Viviani, M.; Ilhan, Z.; Ansar, A.; Soysal, D.; Thorel, A.S.; Abreu, J.; Chesnaud, A.; Politova, T.; et al. IDEAL-Cell, Innovative Dual mEmbrAne fueL-Cell: Fabrication and Electrochemical Testing of First Prototypes. *ECS Trans.* **2009**, *25*, 773–782.
7. Steele, B.C.H. Appraisal of Ce1-xGdyO2-y/2 electrolytes for IT-SOFC operation at 500 °C. *Solid State Ion.* **2000**, *129*, 95–110. [CrossRef]
8. Artini, C. RE-doped ceria systems and their performance as solid electrolytes: A puzzling tangle of structural issues at the average and local scale. *Inorg. Chem.* **2018**, *57*, 13047–13062. [CrossRef]
9. Artini, C.; Pani, M.; Lausi, A.; Masini, R.; Costa, G.A. High temperature structural study of Gd-doped ceria by synchrotron X-ray diffraction (673 K $\leq$ T $\leq$ 1073 K). *Inorg. Chem.* **2014**, *53*, 10140–10149. [CrossRef]
10. Tianshu, Z.; Hing, P.; Huang, H.; Kilner, J. Ionic conductivity in the CeO2–Gd2O3 system (0.05≤Gd/Ce≤0.4) prepared by oxalate coprecipitation. *Solid State Ion.* **2002**, *148*, 567–573. [CrossRef]
11. Aygün, B.; Őzdemr, H.; Faruk Őksuzömer, M.A. Structural, morphological and conductivity properties of samaria doped ceria (SmxCe1-xO2-x/2) electrolytes synthesized by electrospinning method. *Mater. Chem. Phys.* **2019**, *232*, 82–87. [CrossRef]
12. Artini, C.; Carnasciali, M.M.; Viviani, M.; Presto, S.; Plaisier, J.R.; Costa, G.A.; Pani, M. Structural properties of Sm-doped ceria electrolytes at the fuel cell operating temperatures. *Solid State Ion.* **2018**, *315*, 85–91. [CrossRef]
13. Artini, C.; Costa, G.A.; Pani, M.; Lausi, A.; Plaisier, J. Structural characterization of the CeO2/Gd2O3 mixed system by synchrotron X-ray diffraction. *J. Solid State Chem.* **2012**, *190*, 24–28. [CrossRef]
14. Scavini, M.; Coduri, M.; Allieta, M.; Brunelli, M.; Ferrero, C. Probing complex disorder in Ce1-xGdxO2-x/2 using the pair distribution function analysis. *Chem. Mater.* **2012**, *24*, 1338–1345. [CrossRef]
15. Artini, C.; Pani, M.; Carnasciali, M.M.; Buscaglia, M.T.; Plaisier, J.; Costa, G.A. Structural features of Sm- and Gd-doped ceria studied by synchrotron X-ray diffraction and μ-Raman spectroscopy. *Inorg. Chem.* **2015**, *54*, 4126–4137. [CrossRef]
16. Coduri, M.; Masala, P.; Allieta, M.; Peral, I.; Brunelli, M.; Biffi, C.A.; Scavini, M. Phase transformations in the CeO2-Sm2O3 system: A multiscale powder diffraction investigation. *Inorg. Chem.* **2018**, *57*, 879–891. [CrossRef]
17. Minervini, L.; Zacate, M.O.; Grimes, R.W. Defect cluster formation in M_2O_3-doped CeO_2. *Solid State Ion.* **1999**, *116*, 339–349. [CrossRef]
18. Acharya, S.A.; Gaikwad, V.M.; Sathe, V.; Kulkarni, S.K. Influence of gadolinium doping on the structure and defects of ceria under fuel cell operating temperature. *Appl. Phys. Lett.* **2014**, *104*, 113508. [CrossRef]
19. Inaba, H.; Sagawa, R.; Hayashi, H.; Kawamura, K. Molecular dynamics simulation of gadolinia-doped ceria. *Solid State Ion.* **1999**, *122*, 95–103. [CrossRef]
20. Ye, F.; Mori, T.; Ou, D.R.; Zou, J.; Auchterlonie, G.; Drennan, J. Compositional and structural characteristics of nano-sized domains in gadolinia-doped ceria. *Solid State Ion.* **2008**, *179*, 827–831. [CrossRef]
21. Wang, B.; Lewis, R.J.; Cormack, A.N. Computer simulations of large-scale defect clustering and nanodomain structure in gadolinia-doped ceria. *Acta Mater.* **2011**, *59*, 2035–2045. [CrossRef]
22. Coduri, M.; Scavini, M.; Pani, M.; Carnasciali, M.M.; Klein, H.; Artini, C. From nano to microcrystals: Effect of different synthetic pathways on defects architecture in heavily Gd-doped ceria. *Phys. Chem. Chem. Phys.* **2017**, *19*, 11612–11630. [CrossRef] [PubMed]
23. Ye, F.; Mori, T.; Ou, D.R.; Zou, J.; Drennan, J. A structure model of nano-sized domain in Gd-doped ceria. *Solid State Ion.* **2009**, *180*, 1414–1420. [CrossRef]
24. Ou, D.R.; Mori, T.; Ye, F.; Zou, J.; Auchterlonie, G.; Drennan, J. Oxygen-vacancy ordering in lanthanide-doped ceria: Dopant-type dependence and structure model. *Phys. Rev. B* **2008**, *77*, 024108. [CrossRef]
25. Deguchi, H.; Yoshida, H.; Inagaki, T.; Horiuchi, M. EXAFS study of doped ceria using multiple data set fit. *Solid State Ion.* **2005**, *176*, 1817–1825. [CrossRef]
26. Kossoy, A.; Wang, Q.; Korobko, R.; Grover, V.; Feldman, Y.; Wachtel, E.; Tyagi, A.K.; Frenkel, A.I.; Lubomirsky, I. Evolution of the local structure at the phase transition in CeO_2-Gd_2O_3 solid solutions. *Phys. Rev. B* **2013**, *87*, 054101. [CrossRef]
27. Nitani, H.; Nakagawa, T.; Yamanouchi, M.; Osuki, T.; Yuya, M.; Yamamoto, T.A. XAFS and XRD study of ceria doped with Pr, Nd or Sm. *Mater. Lett.* **2004**, *58*, 2076–2081. [CrossRef]
28. Nakajima, A.; Yoshihara, A.; Ishigame, M. Defect-induced Raman spectra in doped CeO2. *Phys. Rev. B* **1994**, *50*, 13297–13307. [CrossRef]

29. McBride, J.R.; Hass, K.C.; Pointdexter, B.D.; Weber, W.H. Raman and X-ray studies of Ce1-xRExO2-y, where RE=La, Pr, Nd, Eu, Gd, and Tb. *J. Appl. Phys.* **1994**, *76*, 2435–2441. [CrossRef]
30. Presto, S.; Artini, C.; Pani, M.; Carnasciali, M.M.; Massardo, S.; Viviani, M. Ionic conductivity and local structural features in Ce1-xSmxO2-x/2. *Phys. Chem. Chem. Phys.* **2018**, *20*, 28338–28345. [CrossRef]
31. Artini, C.; Carnasciali, M.M.; Costa, G.A.; Plaisier, J.R.; Pani, M. A novel method for the evaluation of the Rare Earth (RE) coordination number in RE-doped ceria through Raman spectroscopy. *Solid State Ion.* **2017**, *311*, 90–97. [CrossRef]
32. Artini, C.; Pani, M.; Carnasciali, M.M.; Plaisier, J.R.; Costa, G.A. Lu-, Sm- and Gd-doped ceria: A comparative approach to their structural properties. *Inorg. Chem.* **2016**, *55*, 10567–10579. [CrossRef] [PubMed]
33. Horlait, D.; Claparède, L.; Clavier, N.; Szenknekt, S.; Dacheux, N.; Ravaux, J.; Podor, R. Stability and structural evolution of CeIV1-xLnIIIxO2-x/2 solid solutions: A coupled μ-Raman/XRD approach. *Inorg. Chem.* **2011**, *50*, 150–161. [CrossRef] [PubMed]
34. Mandal, B.P.; Roy, M.; Grover, V.; Tyagi, A.K. X-ray diffraction, μ-Raman spectroscopic studies on CeO2-RE2O3 (RE=Ho, Er) systems: Observation of parasitic phases. *J. Appl. Phys.* **2008**, *103*, 033506. [CrossRef]
35. Mandal, B.P.; Grover, V.; Roy, M.; Tyagi, A.K. X-ray diffraction and Raman spectroscopic investigation on the phase relations in Yb_2O_3- and Tm_2O_3-substituted CeO_2. *J. Am. Ceram. Soc.* **2007**, *90*, 2961–2965. [CrossRef]
36. Grover, V.; Banerji, A.; Sengupta, P.; Tyagi, A.K. Raman, XRD and microscopic investigations on CeO_2-Lu_2O_3 and CeO_2-Sc_2O_3 systems: A sub-solidus phase evolution study. *J. Solid State Chem.* **2008**, *181*, 1930–1935. [CrossRef]
37. Artini, C.; Costa, G.A.; Carnasciali, M.M.; Masini, R. Stability fields and structural properties of intra rare earths perovskites. *J. Alloys Compd.* **2010**, *494*, 336–339. [CrossRef]
38. Artini, C.; Costa, G.A.; Masini, R. Study of the formation temperature of mixed $LaREO_3$ (RE≡Dy, Ho, Er, Tm, Yb, Lu) and NdGdO3 oxides. *J. Therm. Anal. Calorim.* **2011**, *103*, 17–21. [CrossRef]
39. Artini, C.; Gigli, L.; Carnasciali, M.M.; Pani, M. Structural properties of the (Nd,Dy)-doped ceria system by synchrotron X-ray diffraction. *Inorganics* **2019**, *7*, 94. [CrossRef]
40. Artini, C.; Nelli, I.; Pani, M.; Costa, G.A.; Caratto, V.; Locardi, F. Thermal decomposition of Ce-Sm and Ce-Lu mixed oxalates: Influence of the Sm- and Lu-doped ceria structure. *Thermochim. Acta* **2017**, *651*, 100–107. [CrossRef]
41. Weber, W.H.; Hass, K.C.; Mc Bride, J.R. Raman study of CeO_2: Second-order scattering, lattice dynamics, and particle-size effects. *Phys. Rev. B* **1993**, *48*, 178–185. [CrossRef] [PubMed]
42. Ubaldini, A.; Carnasciali, M.M. Raman characterisation of powder of cubic RE_2O_3 (RE = Nd, Gd, Dy, Tm, and Lu), Sc_2O_3 and Y_2O_3. *J. Alloys Compd.* **2008**, *454*, 374–378. [CrossRef]
43. Avila-Paredes, H.-J.; Shvareva, T.; Chen, W.; Navrotsky, A.; Kim, S. A correlation between the ionic conductivities and the formation enthalpies of trivalent-doped ceria at relatively low temperatures. *Phys. Chem. Chem. Phys.* **2009**, *11*, 8580–8585. [CrossRef] [PubMed]
44. Inaba, H.; Tagawa, H. Ceria-based solid electrolytes. *Solid State Ion.* **1996**, *83*, 1–16. [CrossRef]
45. Coduri, M.; Scavini, M.; Allieta, M.; Brunelli, M.; Ferrero, C. Defect structure of Y-doped ceria on different length scales. *Chem. Mater.* **2013**, *25*, 4278–4289. [CrossRef]

Article

Suitability of Sm^{3+}-Substituted $SrTiO_3$ as Anode Materials for Solid Oxide Fuel Cells: A Correlation between Structural and Electrical Properties

Saurabh Singh [1,2], Raghvendra Pandey [3], Sabrina Presto [4], Maria Paola Carpanese [4,5], Antonio Barbucci [4,5], Massimo Viviani [4] and Prabhakar Singh [1,*]

1 Department of Physics, Indian Institute of Technology (Banaras Hindu University), Varanasi 221005, India; saurabh.singhiitbhu@gmail.com
2 Department of Chemical Engineering, Indian Institute of Technology, Kanpur 208016, India
3 Department of Physics, A.R.S.D. College, University of Delhi, Dhaula Kuan, New Delhi 110021, India; raghvendra@arsd.du.AC.in
4 CNR-ICMATE, c/o DICCA-UNIGE, Via all'Opera Pia 15, 16145 Genova, Italy; sabrina.presto@cnr.it (S.P.); carpanese@unige.it (M.P.C.); barbucci@unige.it (A.B.); massimo.viviani@cnr.it (M.V.)
5 DICCA-UNIGE, Via all'Opera Pia 15, 16145 Genova, Italy
* Correspondence: psingh.app@iitbhu.AC.in; Tel.: +91-542-6701916

Received: 26 September 2019; Accepted: 22 October 2019; Published: 24 October 2019

Abstract: Perovskite anodes, nowadays, are used in any solid oxide fuel cell (SOFC) instead of conventional nickel/yttria-stabilized zirconia (Ni/YSZ) anodes due to their better redox and electrochemical stability. A few compositions of samarium-substituted strontium titanate perovskite, $Sm_xSr_{1-x}TiO_{3-\delta}$ (x = 0.00, 0.05, 0.10, 0.15, and 0.20), were synthesized via the citrate-nitrate auto-combustion route. The XRD patterns of these compositions confirm that the solid solubility limit of Sm in $SrTiO_3$ is x < 0.15. The X-ray Rietveld refinement for all samples indicated the perovskite cubic structure with a $Pm\bar{3}m$ space group at room temperature. The EDX mapping of the field emission scanning electron microscope (FESEM) micrographs of all compositions depicted a lower oxygen content in the specimens respect to the nominal value. This lower oxygen content in the samples were also confirmed via XPS study. The grain sizes of $Sm_xSr_{1-x}TiO_3$ samples were found to increase up to x = 0.10 and it decreases for the composition with x > 0.10. The AC conductivity spectra were fitted by Jonscher's power law in the temperature range of 500–700 °C and scaled with the help of the Ghosh and Summerfield scaling model taking ν_H and $\sigma_{dc}\,T$ as the scaling parameters. The scaling behaviour of the samples showed that the conduction mechanism depends on temperature at higher frequencies. Further, a study of the conduction mechanism unveiled that small polaron hopping occurred with the formation of electrons. The electrical conductivity, in the H_2 atmosphere, of the $Sm_{0.10}Sr_{0.90}TiO_3$ sample was found to be 2.7×10^{-1} S·cm^{-1} at 650 °C, which is the highest among the other compositions. Hence, the composition $Sm_{0.10}Sr_{0.90}TiO_3$ can be considered as a promising material for the application as the anode in SOFCs.

Keywords: solid oxide fuel cells (SOFCs); ionic conductivity; Raman spectroscopy; powder X-ray diffraction

1. Introduction

Solid oxide fuel cells (SOFC) and solid oxide electrolysis cells (SOEC) are of great interest for their high efficiency of the conversion between chemical energy and electric power without greenhouse gases emissions. Their net-zero environmental impact is becoming truly affordable, as demonstrated by the continuous development of the use of bio-fuels and by the production of hydrogen through SOEC technology [1–8]. SOFCs are more efficient in comparison to a conventional power plant and lower temperature polymer-based fuel cells [9,10]. Electrolyte and electrodes are the essential component of a solid oxide fuel cell. Even if anionic, protonic [11,12] and dual-membrane cells [13] are reported, the first are the most studied in the literature. The state of the art material for the electrolyte is fluorite-structured yttria-stabilized zirconia (YSZ) because of its wide range of stability in oxidizing and reducing media [14], although doped ceria [15] and perovskite Sr- and Mg-incorporated lanthanum gallate (LSGM) are considered as alternatives for intermediate operative temperatures. Lanthanum strontium manganite (LSM)-YSZ composite and Sr- and Fe-doped lanthanum cobaltite (LSCF) are extensively used for cathodes, while Ni-based cermets are conventionally used as anode materials for SOFCs [16,17], although this Ni-based cermet anode material should demonstrate volume instability upon redox cycling and low-tolerance to carbon deposition when exposed to hydrogen [18]. Therefore, nickel-free, alternate-structured anode materials are essential to overcome these issues. Recently, perovskite-structured $SrTiO_3$-based materials received much attention as an alternative anode materials for SOFCs [19]. Pure $SrTiO_3$ is not suitable as an anode material owing to its low electrical conductivity. Additionally, donor-substituted $SrTiO_3$ may be used owing to its better thermal and chemical stability, mixed ionic/electronic conductivity and also carbon or sulphur tolerance [20]. It is widely accepted that the mixed conductivity feature of $SrTiO_3$-based materials enables fuel oxidation to take place at the triple phase boundary region, also decreasing the associated polarization resistance [18,21]. However, combination of this electronic and ionic conductivity is not able to completely satisfy the requirement of electrode materials for SOFC. Therefore, many attempts have been made to enhance the conductivity of this existing anode materials. It was also reported that the doping, as an acceptor, has the possibility to increase the ionic conductivity [22,23], whereas the doping of a donor has been supposed to increase the electronic conductivity [18]. The radii of trivalent Sm^{3+} (1.24 Å) and divalent Sr^{2+} (1.44 Å) are nearly equivalent, which coordinated with O dodecahedrally (AO_{12}). Hence, Sm^{3+} may be one of the suitable candidates to be substituted at the Sr site as a donor dopant. It is also theoretically reported that Sm-doped $SrTiO_3$ may exhibit a high thermal expansion coefficient and conductivity [24].

In this work, Sm-substituted $SrTiO_3$ compositions were synthesized via the auto-combustion citrate-nitrate route and an attempt has been made to understand the influence of Sm^{3+} substitution on the electrical conductivities. The conductivity spectra of $Sm_xSr_{1-x}TiO_{3-\delta}$ systems with $0.0 \leq x \leq 0.2$ at measured temperatures have been fitted by using Jonscher's power law. The validity of the Ghosh and Summerfield scaling models have been studied to explicate the charge carrier dynamics in the system. We have also discussed the experimental authentication of polaron conduction mechanisms obtained from the temperature-dependent conductivity spectra and structural investigation of the studied samples.

2. Materials and Methods

A few compositions of $Sm_xSr_{1-x}TiO_{3-\delta}$ with x = 0.00, 0.05, 0.10, 0.15 and 0.20 were synthesized via the auto-combustion route using citric acid as a fuel agent. The starting materials Sm_2O_3 (99.9%), SrO (99.5%), $C_{12}H_{28}O_4Ti$ (97%) and citric acid were taken in stoichiometric amounts. The oxides were separately dissolved into nitric acid and then diluted with distilled water and mixed together to form a clear and homogeneous solution. The appropriate amount of citric acid to obtain a molar citrate-nitrate ratio (C/N) equal to 0.3 was also added to the solution. The required titanium isopropoxide ($C_{12}H_{28}O_4Ti$) solution with ethylene glycol was mixed at 100 °C to form a homogeneous solution. Finally, all the mixed solutions were dissolved in distilled water with a few drops of NH_4OH solution to keep the pH value close to 2. In the final solution, titanium hydroxide in the form of a white precipitate had formed. To dissolve the white precipitate, a few drops of diluted HNO_3 were added to keep the pH below 2. A translucent solution was formed which evaporated slowly with continuous stirring with magnetic stirrer on a hot plate at 250 °C. The resultant solution firstly converted into a brown coloured gel, then, after self-ignition, into a black coloured ash. This black coloured ash was pulverized by mortar and pestle to make the powder. Thereafter, the powder was calcined into the alumina crucible in air at 1000 °C for 10 h. The resultant calcined powders were pelletized via 12 mm diameter cylindrical die-set under a hydraulic press of pressure of 5 tons/m^3. After that, the resulting pellets were sintered in the air at 1200 °C for 12 h and then cooled to room temperature.

The phase composition of unreduced (as sintered) and reduced $Sm_xSr_{1-x}TiO_{3-\delta}$ samples (x = 0.00, 0.05, 0.10, 0.15 and 0.20) was characterised by using X-ray diffractometer (XRD, Rigaku Miniflex II desktop) with Cu-Kα_1 radiation (λ = 1.54098 Å) in 2θ range of 20–70° and a step size of Δ 2θ = 0.02° at room temperature. The lattice parameters, reliability fitting factors R_{exp}, R_B, R_F, etc., were also calculated using FullProf software (LLB, France) from the fitted XRD profiles. Archimedes' principle was employed to measure the relative density of the samples by using density measurement kit (Denver SI-234). The microstructural properties of the samples were examined by field emission scanning electron microscope (FESEM, model: NOVA NANOSEM 450) and local composition was checked by energy dispersive X-ray spectroscopy (EDX, Model: EDAX TEAM-Pegasus). The average grain size of the samples was calculated by using Image-J software through the linear intercept method. The XPS spectra of the samples were measured by using KRATOS (Model: Amicus) high-performance analytical instrument with Mg target under 1.0×10^{-6} Pa pressure. The XPS peaks fitted with XPSPEAK 4.1 software (The Chinese University of Hong Kong, Hong Kong) and further analysed. Moreover, the sintered pellets were polished with silver paste on both sides for the conductivity measurement. The resultant polished samples were matured by firing at 700 °C for 20 min. The conductivity measurement was performed in both O_2 and H_2 atmosphere. Initially, samples were subjected in O_2 atmosphere between room temperature and 700 °C and then reduced at 700 °C for 24 h in the pure H_2 atmosphere. Thereafter, electrical conductivity was measured again from high temperature to room temperature. A four-probe test station (Probostat, Norecs) was employed for measuring the electrical conductivity, and impedance was measured by a SOLARTRON 1255–1286, Schlumberger frequency response analyser in the frequency range of 1Hz-1MHz.

3. Results and Discussion

3.1. Structural Studies

The samples $Sm_xSr_{1-x}TiO_{3-\delta}$ are designated as SST and the compositions with x = 0.00, 0.05, 0.10, 0.15 and 0.20 are assigned as SST0, SST5, SST10, SST15 and SST20, respectively. In $SrTiO_3$, Sr^{2+} (1.44 Å) ion is coordinated with O dodecahedrally (AO12) whereas Ti forms octahedrally with O (0.605Å) (BO_6) [25,26]. The tolerance factor according to Goldschmidt relation [27,28] is nearly 1, which shows the formation of ideal perovskite structure. Additionally, the ionic radius of Sm^{3+} in coordination number 12 is 1.24 Å and with the Sm substitution, tolerance factor has been calculated which is observed to decrease from 1.0017 to 0.988 with an increase in Sm content from x = 0.0 to 0.20 as illustrated in Figure 1. Hence, Sm^{3+} may be one of the suitable substituent candidates to be substituted at the Sr site.

From XRD patterns, shown in Figure 2, it was found that all the samples show cubic phase with space group $Pm\bar{3}m$ (JCPDS card number: 86–0178). There were no impurity peaks for the samples with x ≤ 0.10, while secondary phase started to appear for samples x ≥ 0.15. This secondary phase was identified as $Sm_2Ti_2O_7$ (JCPDS card number: 73–1699) and the corresponding peak positions are also shown in Figure 2a. Thus, it is rational to state that the solid solubility limit of Sm in $SrTiO_3$ is less than 15 mol%. On the other hand, the diffraction peaks shifted towards higher angle with an increase in the concentration of samarium up to 15 mol% as shown in Figure 2b. All compositions were refined with Rietveld refinement using the FullProf Suite software package using Pseudo-Voigt wave function and 6th order polynomial as demonstrated in Figure 2a. The lattice parameters, cell volume, and reliability factors of Rietveld refinement are listed in Table 1.

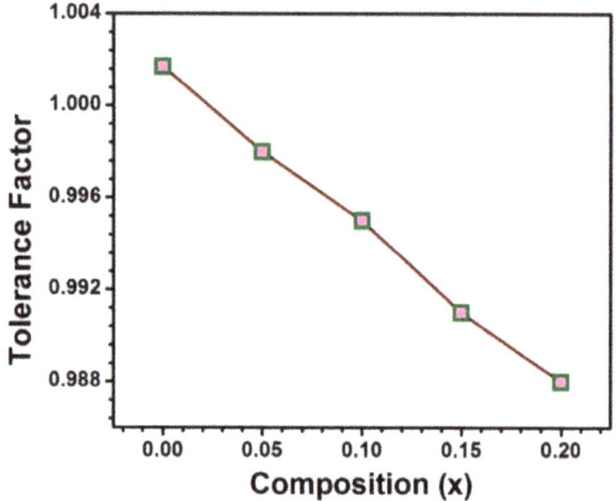

Figure 1. The tolerance factor with the compositions.

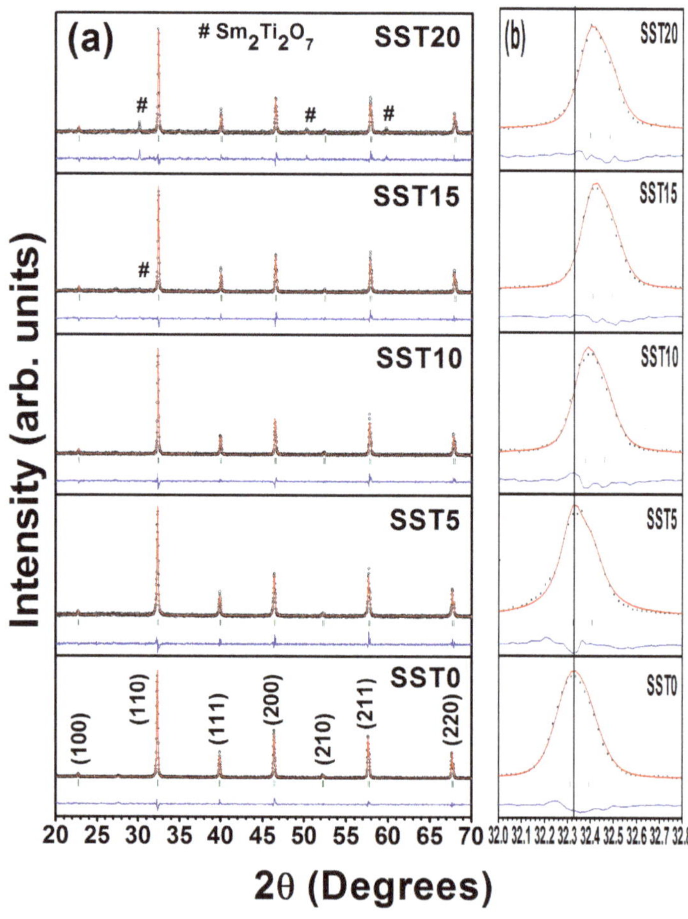

Figure 2. (**a**) The X-ray Rietveld refinement patterns of the sintered samples in air and (**b**) the shifting of XRD diffraction peaks towards higher angle with Sm concentration.

Table 1. The lattice parameter, cell volume, Rietveld refined parameters, and porosity of the studied samples.

Sample	Lattice Parameter (Å)	Cell Volume (Å^3)	χ^2	Re	RB	RF	Rp	Rwp	Porosity (%)
SST0	3.9151	60.011	5.81	9.21	4.61	2.93	23.0	22.2	3.5
SST5	3.9136	59.942	5.14	13.6	5.90	4.13	39.3	32.2	6.3
SST10	3.9069	59.634	5.45	13.1	6.46	4.97	41.9	31.3	19.9
SST15	3.9031	59.461	5.67	13.6	7.11	5.39	47.5	32.9	40.0
SST20	3.9045	59.525	5.92	15.9	5.86	5.70	59.1	39.4	38.6

Table 1 also shows that as the Sm concentration increases, the porosity also increases up to x = 0.15, and after that it decreases. It is also evident that, as Sm content increases in $SrTiO_3$, the unit cell volume decreases up to 15 mol%. This may be due to the replacement of the larger Sr^{2+} (1.44 Å, CN 12) by the smaller Sm^{3+} (1.24 Å, CN 12) [29,30] which also indicates that the corresponding structural parameters follow Vegard's rule [31–33].

Furthermore, Figure 3a shows that the XRD peak (110) broadens with composition (x) which ensures that the formation of polarons in the system [34], and observed that the width of the XRD peak (110) increases with the increase in Sm mol% in $SrTiO_3$. The inset of Figure 3a shows TiO_6 octahedra formation in pure $SrTiO_3$, i.e., without any distortion arising from Sm incorporation. Figure 3b shows the variation of lattice parameter and relative density with x and depicts that lattice parameter and relative density follow the same trend with x, i.e., the lattice parameter and density decrease with the increase in x up to $\leq$ 15. This can be attributed to smaller ionic radius of Sm^{3+} compared to Sr^{2+}. And slightly increases for $x > 15$, due to the solubility limit. According to the Williamson–Hall (W–H) model, the microstrain and crystallite size were studied from XRD patterns [34–36]. The W–H model as given by:

$$\beta cos\theta = 0.9\frac{\lambda}{t} + 4\epsilon sin, \quad (1)$$

where β is the full width at half maxima (FWHM) at Bragg's angle (2θ), λ is the X-ray wavelength of Cu-Kα (λ = 1.54098Å), t is the average crystallite size, and ϵ is the microstrain. The change in microstrain and crystallite size with the compositions (x) shown in Figure 3c. It was observed that the microstrain and crystallite size exhibit similar trends of variation with x, i.e., they increase with the increase in Sm content up to $x = 0.15$, while the microstrain and crystallite size decrease for $x > 0.15$ due to secondary phase formation, as shown in Figure 2a. For confirmation of polarons, the coherence length of polarons (L_{coh}) was calculated and is shown in Figure 3d along with the lattice parameters. It is well known that L_{coh} is less than Lattice parameters ($L_{coh} < a$) for the formation of small polarons while $L_{coh} > a$ indicates the formation of the large polarons [37]. From Figure 3d, it is observed that $L_{coh} < a$ shows the formation of small polarons, except for $x = 0.10$. Therefore, there may be the possibility of large polaron formation for the SST10 sample.

It is known that large polarons correspond to electrons free to move like in a conduction band, while small polarons are described as electrons hopping between neighbouring potential wells [37]. In some cases, the transition between the two regimes is associated to a step-like increase of conductivity. Therefore, the larger conductivity of SST10 (see below) could be associated with the formation of large polarons.

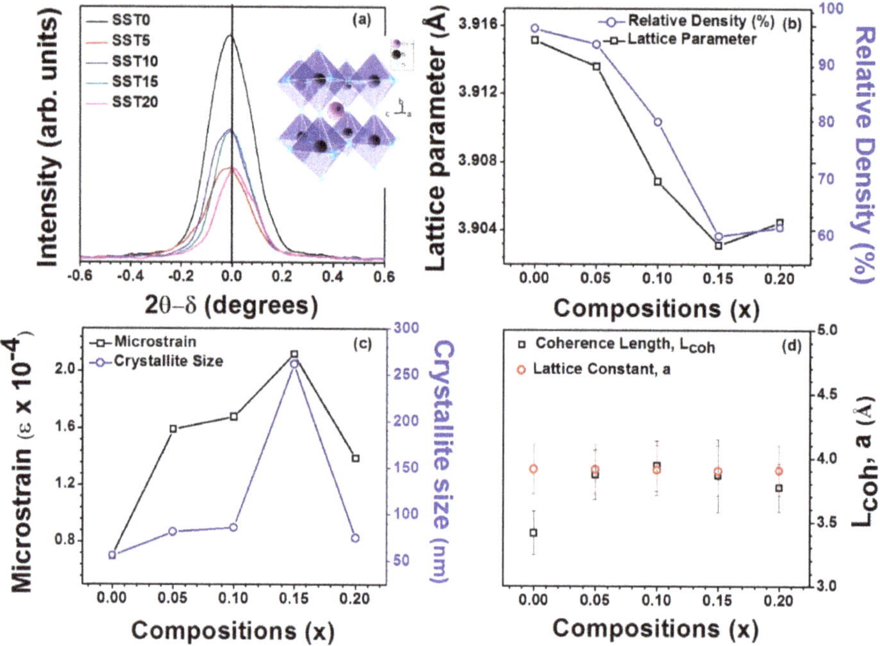

Figure 3. (**a**) Variation of intensity of XRD peak with 2θ-δ (°) where δ is the angle at which maxima occurs showing diffuseness of the XRD peak (110) with the compositions; (**b**) the variation of the lattice parameters and relative density with the compositions; (**c**) the variation of the microstrain and crystallite size with the compositions; and (**d**) the variation of the coherence length of the polarons (L_{coh}) with the lattice parameters.

3.2. Microstructural Analysis

Figure 4a–e show the FESEM micrographs of fractured samples sintered at 1200 °C in air and the grain size distribution (inset). The average grain sizes of all compositions were calculated by using the linear intercept method and were found to be in the range of 1.6–2.5 μm, with largest value associated with sample SST10. In particular, the grain size of $Sm_xSr_{1-x}TiO_{3-\delta}$ samples increases with Sm-doping concentration up to $x = 0.10$ and decreases for $x \geq 0.15$. This can be explained as an effect of Sm-doping activating grain growth during sintering [38] while the solubility limit is not exceeded. For heavy doping (≥ 15 mol), the formation of the pyrochlore secondary phase prevents the mass transportation during the sintering process, thus leading to higher porosity compared to pure $SrTiO_3$.

EDX data were collected in various positions for all specimens. Figure 4f shows the stoichiometric and average EDX at% for all elements in different compounds. An increasing amount of Sm was found along the series, although some deviations from stoichiometric values are visible. Such deviations are largely due to the intrinsic limitation of the technique for the detection of light elements, like oxygen. However, it can be noted that the oxygen concentrations in SST15 and SST20 are significantly higher than in SST5 and SST10. This can be partially attributed to the presence of the pyrochlore secondary phase, which has a higher molar oxygen concentration than the perovskite.

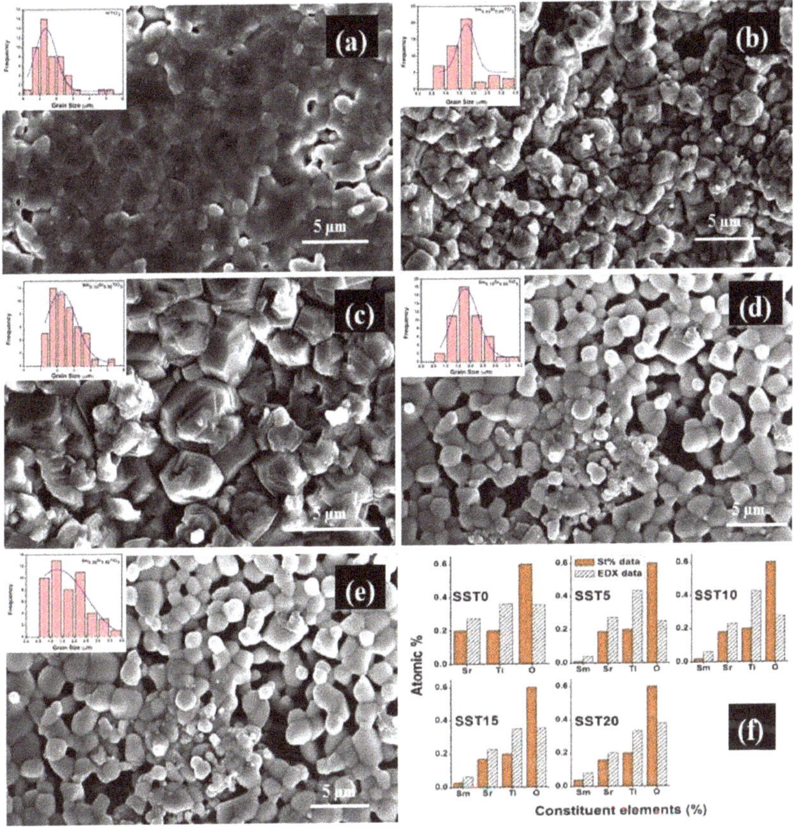

Figure 4. The FESEM micrographs of the fractured samples sintered at 1200 °C in air (**a**) SST0, (**b**) SST5, (**c**) SST10, (**d**) SST15, (**e**) SST20 and (**f**) The stoichiometric and EDX at% for the samples.

3.3. XPS Analysis

EDX analysis indicated that the samples are oxygen deficient. To verify the oxygen deficiency in the studied samples, X-ray photoelectron spectroscopy (XPS) measurements of each powdered composition were performed. Figure 5a shows the oxidation states of the constituent elements present in the SST system. The complete XPS spectra acquired under the range of 0–1150 eV which renders XPS binding energy (BE) peaks of Sm, Sr, C, Ti and O elements with satellite peaks of O (KLL). The XPS peaks of the constituent elements are assigned for different Sm substituted samples. The C 1s peak originated at ~286 eV due to external contamination of the samples. The binding energies of the constituent elements were estimated by considering C 1s reference peak at 284.6 eV. The XPS wide spectra of all samples revealed binding energy peaks of Sm 4d at ~130.8 eV, Sr 3p at ~269.4 eV, Ti 2p at ~ 459.4 eV and O 1s at ~ 530.6 eV. With increasing Sm^{3+} doping concentration at the Sr site, there is a slight shift in binding energy values to the higher energy side. The O (KLL) satellite peak is also noted at ~741.3 eV as shown in Figure 5a [39]. Furthermore, the O 1s spectral regions are used to obtain the information concerning the existence of oxygen deficiency present in the samples.

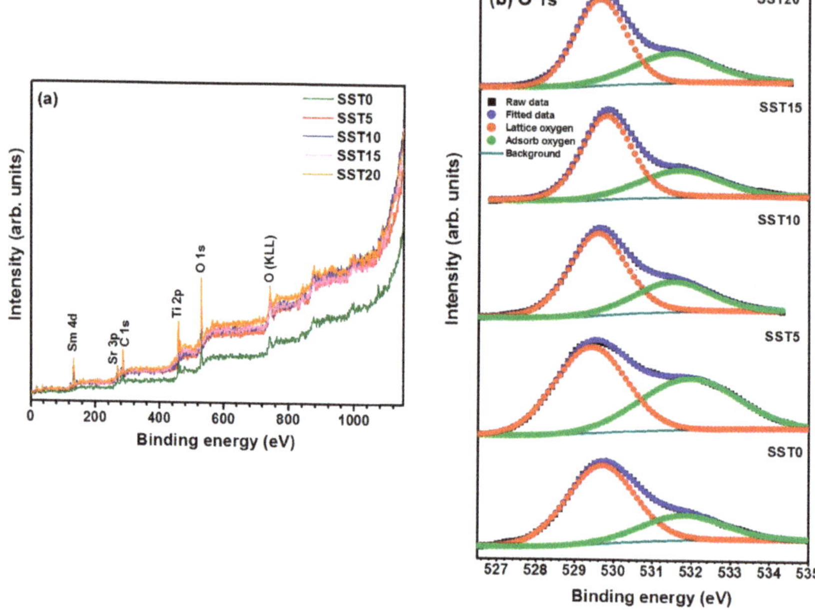

Figure 5. (**a**) The XPS wide spectra of Sm-substituted $SrTiO_3$ samples and (**b**) the illustration of the O 1s core level spectra of the samples.

Figure 5b shows the XPS spectra of the O 1s core-level. The peaks in these spectra are broad and asymmetric in nature for all the compositions. Therefore, the peak for all the samples can be split into two peaks, owing to two different types of oxygen species, i.e., adsorbed and lattice oxygen [40]. The amount of lattice oxygen is reported in Table 2, corresponding for each sample to the ratio between the area of the sub-peak centred at 529.7 eV and the total area of the O 1s peak. This quantity is a measure of the oxygen sites' occupancy or, equivalently, of oxygen vacancies.

Table 2. Activation energy and lattice oxygen content of all Sm-doped samples.

Sample	Ea in H_2 200–450 °C	Ea in H_2 450–700 °C	Lattice Oxygen (%)
SST5	0.16	0.33	55.8
SST10	0.15	0.22	66.6
SST15	0.12	0.27	64.7
SST20	0.13	0.17	62.4

The concentration of oxygen vacancies increases for $x \geq 0.1$ as expected considering an increasing incorporation of Sm in the perovskite lattice. On the other hand, the lowest occupancy found in the sample SST5 can be explained as an indirect effect of porosity. When materials are fired at 1200 °C for a few hours some of the lattice oxygen is eliminated in the whole body of the ceramic. During the cooling phase, oxygen re-equilibration may be incomplete, especially if the material reaches high density. In fact, SST5 showed much higher density than other samples, which can explain the higher oxygen non-stoichiometry.

3.4. Electrical Conductivity and Chemical Stability

The impedance analysis of Sm-doped $SrTiO_3$ samples was performed by using a frequency response analyser under the frequency range of 1–10^6 Hz. Several models have been suggested to understand the ion dynamics and conduction mechanism [41]. However, in the present system, Jonscher's power law (JPL) has been used to study the ion dynamics of the specimens. The conductivity spectra of any polycrystalline materials consist of two components viz. DC and AC conductivities. Thus, the electrical conductivity can be described by Jonscher's power law using following equation [42,43]:

$$\sigma = \sigma_{dc} + A\nu^n, \tag{2}$$

where σ_{dc} is the DC conductivity, A is the pre exponential factor which depends upon temperature, ν is the frequency and n is the frequency exponent factor varies in between 0 and 1.

Figure 6a shows the conductivity spectra fitted with Jonscher's power law (JPL) using equation. 2 for the sample SST5 in measured temperature range. The electrical conductivity under low frequency region is found to be independent of frequency which corresponds to the DC or bulk conductivity caused by the random motion of charge carriers. However, the conductivity at high frequency region is frequency dependent and successive dispersion occurs representing hopping motion of charge carriers due to relaxation processes and this also corresponds the AC conductivity [44,45]. Figure 6a clearly indicates that the conductivity spectra of SST5 found to follow Jonscher's Power Law [46]. It was also observed that all the other samples follow the JPL at the measured temperatures. The variables σ_{dc}, ν_H and n have been obtained by fitting of data points at all the measured temperatures. The conductivity spectra were also examined to obey the scaling models. The scaling behaviour of conductivity spectra was checked by using Ghosh and Summerfield scaling laws. Ghosh scaling function using scaling parameters ν_H can be expressed with the following equation [47]:

$$\sigma/\sigma_{dc} = F(\nu/\nu_H), \tag{3}$$

where F is the scaling function, independent of temperature and the scaling parameters σ_{dc} and ν_H are the DC conductivity and hopping frequency, respectively. It was observed that the conductivity spectra at measured temperatures do not merge into a single master curve in the high frequency range as shown in Figure 6b. A similar kind of pattern was also observed with the Summerfield scaling model in which ν_H introduced as $\sigma_{dc}{\cdot}T$ is shown in inset Figure 6b. Additonally, all the samples do not follow the time temperature superposition principle in the high frequency range. This indicates that the conduction mechanism changes along with the number of charge carriers with temperature in the high-frequency regime.

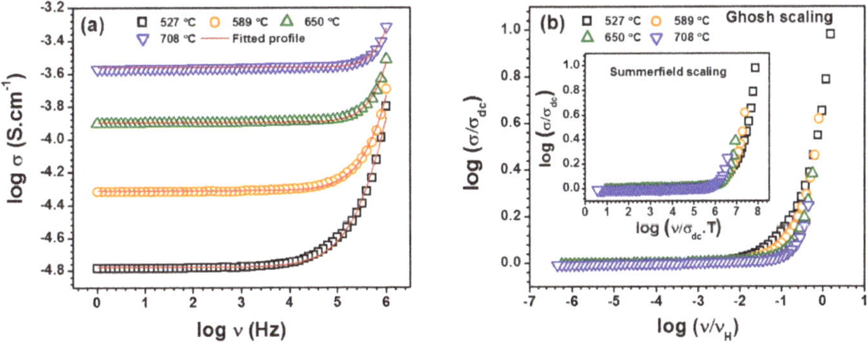

Figure 6. (**a**) The typical electrical conductivity spectra of SST5 composition fitted with Jonscher's power law and (**b**) scaling of conductivity spectra at measured temperatures for the composition SST5.

Furthermore, the electrical behaviour of all the compositions was investigated via complex impedance spectroscopy techniques. In Figure 7, the Nyquist plot shows the bulk relaxation for measured temperatures in air for the sample SST5. It was observed that the sample shows the lowest impedance at a higher temperature. The complex impedance plane plots for all other studied samples also show similar behaviour. In the studied samples, only one semi-circular arc has been observed at measured temperatures due to the grain/bulk contribution, and the frequency range of these arcs shift to the higher frequency side with increasing temperature [48,49]. From the impedance plots, the electrical conductivity of the Sm-doped $SrTiO_3$ system was extracted by the following formula:

$$\sigma = \frac{1}{R_t} \times \frac{l}{S}, \tag{4}$$

where σ, R_t, l and S are the electrical conductivity, total resistance, thickness, and surface area of the pelletized sample, respectively.

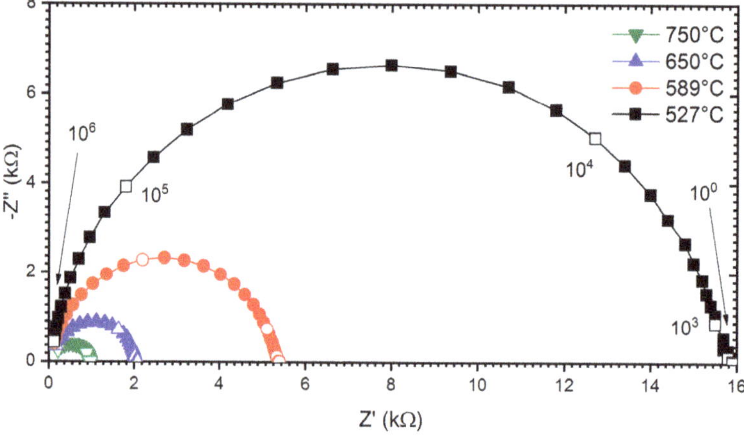

Figure 7. Nyquist plots of impedance measured on the SST5 sample under O_2 and at different temperatures. Open markers point at frequency decades (Hz) on all plots.

The DC electrical conductivity of all compositions has also been calculated by Jonscher's power law in air and hydrogen atmospheres. Thereafter, the activation energy of conduction for all studied compositions was calculated by using the Arrhenius relation [50,51]:

$$\sigma = \sigma_0 / T \cdot e^{-E_a/KT}, \tag{5}$$

where σ_0, E_a, K and T are the pre-exponential factor, activation energy for conduction, Boltzmann constant, and absolute temperature, respectively. The Arrhenius plot between log (σT) and 1000/T for Sm-doped samples in a H_2 atmosphere is shown in Figure 8a. The Arrhenius plot exhibits two slopes in the two temperature ranges (i.e., 200–450 °C, 450–700 °C) for the studied samples [52], as shown in Figure 8a,b. The maximum activation energy (Ea) is found to be 0.325 eV for the system in a H_2 atmosphere, which indicates high electronic contribution as reported in Table 2. It was observed that SST10 possesses the highest conductivity (i.e., 2.6×10^{-1} S.cm^{-1} at 700 °C) with 0.218 eV activation energy.

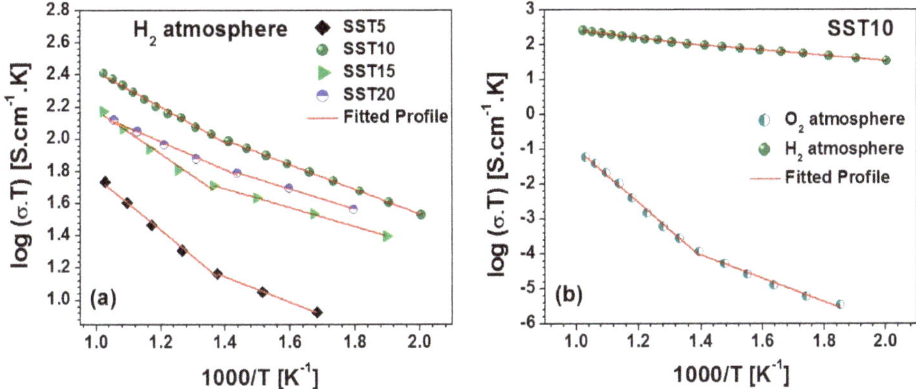

Figure 8. (**a**) The Arrhenius plot (log (σ_{dc}T) vs. 1000/T) for the all studied samples and (**b**) the comparative Arrhenius plot for the sample SST10 in O_2 and H_2 atmospheres.

It is observed that SST10 possesses higher conductivity in both atmospheres. An activation energy lower than 0.5 eV indicates that the samples contribute to more electronic conductivity than ionic conductivity [53], as shown in Table 2. Additionally, the grain size of the SST10 sample is larger in comparison to others, resulting in a lower concentration of grain boundaries which frequently reduce the total conductivity.

The conductivity of SST10 is also considerable when compared to other doped $SrTiO_3$ compositions. For comparison's sake, the conductivity at 650 °C of SST10 and other reference materials are reported in Table 3. The reducing conditions, i.e., the temperature and time of treatment under a reducing atmosphere, clearly influence the total conductivity. However, SST10 shows a higher conductivity among compositions treated at temperatures ≤ 850 °C.

Table 3. Comparison of conductivity at 650 °C for various doped $SrTiO_3$ compositions after reduction. Reducing conditions and literature references are also listed.

Sample	σ (Scm^{-1}) @ 650°C	Reduction T (°C)/Time (h)	Reference
$Sm_{0.1}Sr_{0.9}TiO_{3-\sigma}$	2.7 x10^{-1}	700/24	this work
$Dy_{0.08}Sr_{0.92}TiO_{3-\sigma}$	1.4 x10^{-1}	700/24	[54]
$Y_{0.08}Sr_{0.92}TiO_{3-\sigma}$	6.8 x10^{-2}	700/24	[55]
$Y_{0.08}Sr_{0.88}TiO_{3-\sigma}$	1.0 x10^{-1}	850/24	[56]
$Y_{0.07}Sr_{0.93}TiO_{3-\sigma}$	5.0 x10^{0}	1400/5	[30]
$Dy_{0.1}Sr_{0.9}TiO_{3-\sigma}$	7.1 x10^{1}	1450/4	[57]
$La_{0.1}Sr_{0.9}TiO_{3-\sigma}$	7.0 x10^{1}	1450/12	[58]
$Y_{0.08}Sr_{0.92}TiO_{3-\sigma}$	5.5 x10^{1}	1500/10	[52]

Additionally, with the substitution of Sm, there is a slight decrease in the lattice parameters, whereas, for 10 mol% Sm-substituted $SrTiO_3$, no appreciable change in the lattice parameters in comparison to pure $SrTiO_3$ is observed. Further, these samples are found to be oxygen-deficient samples as confirmed through XPS or EDX analysis. The increased conductivity after reduction could also be understood by defect chemistry.

The lattice oxygen might be lost under reducing atmosphere and thus oxygen vacancies will be generated according to the following defect equation:

$$2TiO_2 \rightarrow 2Ti'_{Ti} + V_O^{\cdot\cdot} + 3O_O^x + \frac{1}{2}O_2, \tag{6}$$

and the following defect equation must be satisfied for balancing the electro-neutrality:

$$\left[Ti'_{Ti}\right] = \left[Sm^{\cdot}_{Sr}\right] + 2\left[V^{\cdot\cdot}_{O}\right], \tag{7}$$

Therefore, both extrinsic defects, caused by the Sm doping at the Sr-site and intrinsic oxygen vacancies [59], formed under the reducing conditions, subsidize the formation of electrons and the increase of electrical conductivity. The investigation of chemical stability for the reduced compositions was performed by XRD measurements and microstructural studies. From the results as reported in Figure 9, it was found that the XRD patterns of the samples SST5, SST10, SST15, and SST20 show similarity with earlier samples, and no signature of structural and chemical degradations was observed [60,61].

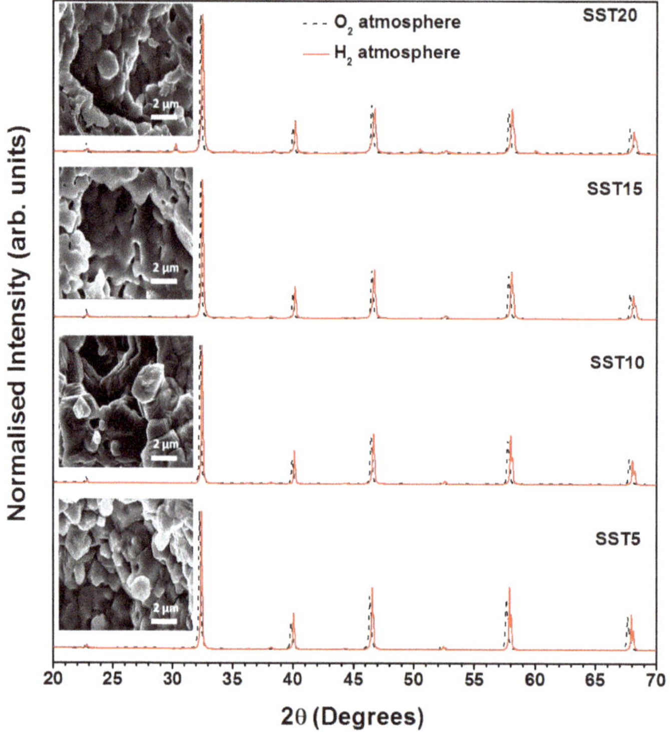

Figure 9. XRD patterns of unreduced and reduced samples and inset FESEM micrographs of reduced compositions.

Additionally, FESEM images of corresponding reduced compositions (inset in Figure 9) also confirm that the microstructure was not modified by reduction. Moreover, the XRD peaks of all reduced samples have been slightly shifted towards the right which illustrated lattice contraction. In a H_2 atmosphere, the lattice contraction may be attributed to increasing the defects along with oxygen vacancies. Amongst all the explored samples, the most capable anodic material for SOFCs was $Sm_{0.10}Sr_{0.90}TiO_{3-\delta}$, due to its excellent chemical stability without involvement of a secondary phase coupled with high electrical conductivity under reducing conditions [62].

4. Conclusions

The structural and electrical properties of Sm-substituted $SrTiO_3$ (SST) were studied to investigate its suitability as an anode material for SOFCs. SST compositions exhibit high electrical conductivity with excellent chemical stability in reducing atmospheres. It was observed that Sm-substitution at the Sr site was compensated by the formation of electrons. Therefore, there is a decrease in the lattice parameters observed up to the solubility limit. Ghosh and Summerfield scaling laws also predict the change in the conduction mechanism at a higher frequency range and it also depends upon the temperature. The $Sr_{0.90}Sm_{0.10}TiO_{3-\delta}$ (SST10) sample shows higher electronic conductivity rather than ionic conductivity without involvement of a secondary phase. Hence, the $Sr_{0.90}Sm_{0.10}TiO_{3-\delta}$ (SST10) can be proposed as a suitable anode material for SOFC applications.

Author Contributions: Conceptualization: S.S., R.P., S.P., M.P.C., A.B., M.V. and P.S.; methodology: S.S., R.P. and P.S.; formal analysis: S.S. and S.P.; investigation: S.S., R.P., S.P., M.P.C. and A.B.; writing—original draft preparation: S.S.; writing—review and editing: R.P., S.P., M.V. and P.S.; supervision: M.V. and P.S.

Funding: This research was funded by the BRNS, India through the project grant no. 34/14/15/2015/BRNS.

Acknowledgments: Authors are thankful to A. S. K. Sinha for providing the XPS facility.

Conflicts of Interest: The authors declare no conflict of interest.

References

1. De Lorenzo, G.; Fragiacomo, P. Electrical and thermal analysis of an intermediate temperature IIR-SOFC system fed by biogas. *Energy Sci. Eng.* **2018**, *6*, 60–72. [CrossRef]
2. Fragiacomo, P.; Corigliano, O.; De Lorenzo, G.; Mirandola, F.A. Experimental Activity on a 100-W IT-SOFC Test Bench Fed by Simulated Syngas. *J. Energy Eng.* **2018**, *144*, 04018006. [CrossRef]
3. Lo Faro, M.; Trocino, S.; Zignani, S.C.; Italiano, C.; Vita, A.; Aricò, A.S. Study of a solid oxide fuel cell fed with n-dodecane reformate. Part II: Effect of the reformate composition. *Int. J. Hydrog. Energy* **2017**, *42*, 1751–1757. [CrossRef]
4. Lo Faro, M.; Trocino, S.; Campagna Zignani, S.; Machado Reis, R.; Monforte, G.; Ticianelli, E.A.; Arico, A.S. Ni-based Alloys as Protective Layer for a Conventional Solid Oxide Fuel Cell Fed with Biofuels. *ECS Trans.* **2015**, *68*, 2653–2658. [CrossRef]
5. Yang, C.; Ren, C.; Yu, L.; Jin, C. High performance intermediate temperature micro-tubular SOFCs with $Ba_{0.9}Co_{0.7}Fe_{0.2}Nb_{0.1}O_3{-}\delta$ as cathode. *Int. J. Hydrog. Energy* **2013**, *38*, 15348–15353. [CrossRef]
6. Miyake, M.; Matsumoto, S.; Iwami, M.; Nishimoto, S.; Kameshima, Y. Electrochemical performances of Ni1−xCux/SDC cermet anodes for intermediate-temperature SOFCs using syngas fuel. *Int. J. Hydrog. Energy* **2016**, *41*, 13625–13631. [CrossRef]
7. Fragiacomo, P.; De Lorenzo, G.; Corigliano, O. Performance Analysis of an Intermediate Temperature Solid Oxide Electrolyzer Test Bench under a CO_2-H_2O Feed Stream. *Energies* **2018**, *11*, 2276. [CrossRef]
8. Milewski, J.; Wołowicz, M.; Lewandowski, J. Comparison of SOE/SOFC system configurations for a peak hydrogen power plant. *Int. J. Hydrog. Energy* **2017**, *42*, 3498–3509. [CrossRef]
9. Steele, B.C.H.; Heinzel, A. Materials for fuel-cell technologies. *Nature* **2001**, *414*, 345–352. [CrossRef]
10. Zhang, X.; Liu, L.; Zhao, Z.; Tu, B.; Ou, D.; Cui, D.; Wei, X.; Chen, X.; Cheng, M. Enhanced Oxygen Reduction Activity and Solid Oxide Fuel Cell Performance with a Nanoparticles-Loaded Cathode. *Nano Lett.* **2015**, *15*, 1703–1709. [CrossRef]
11. Da Silva, F.S.; de Souza, T.M. Novel materials for solid oxide fuel cell technologies: A literature review. *Int. J. Hydrog. Energy* **2017**, *42*, 26020–26036. [CrossRef]
12. Rashid, N.; Samat, A.; Jais, A.; Somalu, M.; Muchtar, A.; Baharuddin, N., Isahak, W. Review on zirconate-cerate-based electrolytes for proton-conducting solid oxide fuel cell. *Ceram. Int.* **2019**, *45*, 6605–6615. [CrossRef]
13. Presto, S.; Barbucci, A.; Viviani, M.; Ilhan, Z.; Ansar, S.A.; Soysal, D.; Thorel, A.; Abreu, J.; Chesnaud, A.; Politova, T.; et al. IDEAL-cell, an innovative dual membrane fuel-cell: Fabrication and electrochemical testing of first prototypes. *ECS Trans.* **2009**, *25*, 773–782.

14. Singh, R.K.; Singh, P. Electrical conductivity of LSGM–YSZ composite materials synthesized via coprecipitation route. *J. Mater. Sci.* **2014**, *49*, 5571–5578.
15. Artini, C.; Carnasciali, M.M.; Viviani, M.; Presto, S.; Plaisier, J.R.; Costa, G.A.; Pani, M. Structural properties of Sm-doped ceria electrolytes at the fuel cell operating temperatures. *Solid State Ion.* **2018**, *315*, 85–91. [CrossRef]
16. Brunaccini, G.; Lo Faro, M.; La Rosa, D.; Antonucci, V.; Aricò, A.S. Investigation of composite Ni-doped perovskite anode catalyst for electrooxidation of hydrogen in solid oxide fuel cell. *Int. J. Hydrog. Energy* **2008**, *33*, 3150–3152. [CrossRef]
17. Presto, S.; Barbucci, A.; Carpanese, M.P.; Viviani, M.; Marazza, R. Electrochemical performance of Ni-based anodes for solid oxide fuel cells. *J. Appl. Electrochem.* **2009**, *39*, 2257–2264. [CrossRef]
18. Li, X.; Zhao, H.; Zhou, X.; Xu, N.; Xie, Z.; Chen, N. Electrical conductivity and structural stability of La-doped $SrTiO_3$ with A-site deficiency as anode materials for solid oxide fuel cells. *Int. J. Hydrog. Energy* **2010**, *35*, 7913–7918. [CrossRef]
19. Verbraeken, M.C.; Ramos, T.; Agersted, K.; Ma, Q.; Savaniu, C.D.; Sudireddy, B.R.; Irvine, J.T.S.; Holthappels, P.; Tietz, F. Modified strontium titanates: From defect chemistry to SOFC anodes. *RSC Adv.* **2015**, *5*, 1168–1180. [CrossRef]
20. Moos, R.; Härdtl, K.H. Electronic transport properties of $Sr_{1-x}La_xTiO_3$ ceramics. *J. Appl. Phys.* **1996**, *80*, 393–400. [CrossRef]
21. Slater, P.R.; Fagg, D.P.; Irvine, J.T.S. Synthesis and electrical characterisation of doped perovskite titanates as potential anode materials for solid oxide fuel cells. *J. Mater. Chem.* **1997**, *7*, 2495–2498. [CrossRef]
22. Hui, S.; Petric, A. Electrical Properties of Yttrium-Doped Strontium Titanate under Reducing Conditions. *J. Electrochem. Soc.* **2002**, *149*, J1–J10. [CrossRef]
23. Li, X.; Zhao, H.; Gao, F.; Zhu, Z.; Chen, N.; Shen, W. Synthesis and electrical properties of Co-doped $Y_{0.08}Sr_{0.92}TiO_{3-\delta}$ as a potential SOFC anode. *Solid State Ion.* **2008**, *179*, 1588–1592. [CrossRef]
24. Shannon, R.D. Revised effective ionic radii and systematic studies of interatomic distances in halides and chalcogenides. *Acta Crystallogr. Sect. A* **1976**, *32*, 751–767. [CrossRef]
25. Wang, Y.; Lee, K.H.; Hyuga, H.; Kita, H.; Inaba, K.; Ohta, H.; Koumoto, K. Enhancement of Seebeck coefficient for $SrO(SrTiO_3)_2$ by Sm substitution: Crystal symmetry restoration of distorted TiO_6 octahedra. *Appl. Phys. Lett.* **2007**, *91*, 242102. [CrossRef]
26. Panahi, S.N.; Samadi, H.; Nemati, A. Effect of Samarium Oxide on the Electrical Conductivity of Plasma-Sprayed SOFC Anodes. *JOM* **2016**, *68*, 2569–2573. [CrossRef]
27. Goldschmidt, V.M. Die Gesetze der Krystallochemie. *Die Nat.* **1926**, *21*, 477–485. [CrossRef]
28. Goldschmidt, V.M. Crystal structure and chemical composition. *Ber. Dtsch. Chem. Ges.* **1927**, *60*, 1263–1268. [CrossRef]
29. Maletic, S.; Popovic, D.; Dojcilovic, J. Dielectric measurements, Raman scattering and surface studies of Sm-doped $SrTiO_3$ single crystal. *J. Alloy. Compd.* **2010**, *496*, 388–392. [CrossRef]
30. Ma, Q.; Tietz, F.; Stöver, D. Nonstoichiometric Y-substituted $SrTiO_3$ materials as anodes for solid oxide fuel cells. *Solid State Ion.* **2011**, *192*, 535–539. [CrossRef]
31. Johnson, D.W.; Cross, L.E.; Hummel, F.A. Dielectric Relaxation in Strontium Titanates Containing Rare–Earth Ions. *J. Appl. Phys.* **1970**, *41*, 2828–2833. [CrossRef]
32. Eror, N.G.; Balachandran, U. Self-compensation in lanthanum-doped strontium titanate. *J. Solid State Chem.* **1981**, *40*, 85–91. [CrossRef]
33. Bamberger, C.E. Homogeneity Ranges of Phases $Sr_{4-x}Ln_{2x/3}Ti_4O_{12}$ (Ln = Sm to Lu) and $Sr_{1-y}Eu_y||TiO_3$. *J. Am. Ceram. Soc.* **2005**, *80*, 1024–1026. [CrossRef]
34. Shimomura, S.; Wakabayashi, N.; Kuwahara, H.; Tokura, Y. X-Ray Diffuse Scattering due to Polarons in a Colossal Magnetoresistive Manganite. *Phys. Rev. Lett.* **1999**, *83*, 4389–4392. [CrossRef]
35. Ohon, N.; Stepchuk, R.; Blazhivskyi, K.; Vasylechko, L. Structural Behaviour of Solid Solutions in the $NdAlO_3$-$SrTiO_3$ System. *Nanoscale Res. Lett.* **2017**, *12*, 148. [CrossRef]
36. Prabhu, Y.T.; Rao, K.V.; Kumar, V.S.S.; Kumari, B.S. X-ray Analysis by Williamson-Hall and Size-Strain Plot Methods of ZnO Nanoparticles with Fuel Variation. *World. J. Nano Sci. Eng.* **2014**, *4*, 21–28. [CrossRef]
37. Mannella, N.; Yang, W.L.; Tanaka, K.; Zhou, X.J.; Zheng, H.; Mitchell, J.F.; Zaanen, J.; Devereaux, T.P.; Nagaosa, N.; Hussain, Z.; et al. Polaron coherence condensation as the mechanism for colossal magnetoresistance in layered manganites. *Phys. Rev. B* **2007**, *76*, 233102. [CrossRef]

38. Nasar, R.S.; Cerqueira, M.; Longo, E.; Varela, J.A. Sintering mechanisms of ZrO_2·MgO with addition of TiO_2 and CuO. *Ceram. Int.* **2004**, *30*, 571–577. [CrossRef]
39. Singh, P. Influence of Bi_2O_3 additive on the electrical conductivity of calcia stabilized zirconia solid electrolyte. *J. Eur. Ceram. Soc.* **2015**, *35*, 1485–1493.
40. Kumar, P.; Presto, S.; Sinha, A.S.K.; Varma, S.; Viviani, M.; Singh, P. Effect of samarium (Sm^{3+}) doping on structure and electrical conductivity of double perovskite Sr_2NiMoO_6 as anode material for SOFC. *J. Alloys Compd.* **2017**, *725*, 1123–1129. [CrossRef]
41. Aziz, S.B.; Woo, T.J.; Kadir, M.F.Z.; Ahmed, H.M. A conceptual review on polymer electrolytes and ion transport models. *J. Sci. Adv. Mater. Devices* **2018**, *3*, 1–17. [CrossRef]
42. Praharaj, S.; Rout, D. Study of electrical properties of $Pb(Yb_{0.5}Ta_{0.5})O_3$ by impedance spectroscopy. *IOP Conf. Ser. Mater. Sci. Eng.* **2016**, *149*, 12181. [CrossRef]
43. Dhahri, A.; Dhahri, E.; Hlil, E.K. Electrical conductivity and dielectric behaviour of nanocrystalline $La_{0.6}Gd_{0.1}Sr_{0.3}Mn_{0.75}Si_{0.25}O_3$. *RSC. Adv.* **2018**, *8*, 9103–9111. [CrossRef]
44. Jonscher, A.K. Dielectric relaxation in solids. *J. Phys. D Appl. Phys.* **1999**, *32*, R57–R70. [CrossRef]
45. Dwivedi, R.K.; Kumar, D.; Parkash, O. Valence compensated perovskite oxide system $Ca_{1-x}La_xTi_{1-x}Cr_xO_3$ Part I Structure and dielectric behaviour. *J. Mater. Sci.* **2001**, *36*, 3641–3648. [CrossRef]
46. Peláiz-Barranco, A.; Guerra, J.D.S.; López-Noda, R.; Araújo, E.B. Ionized oxygen vacancy-related electrical conductivity in $(Pb_{1-x}La_x)(Zr_{0.90}Ti_{0.10})_{1-x/4}O_3$ ceramics. *J. Phys. D. Appl. Phys.* **2008**, *41*, 215503. [CrossRef]
47. Ghosh, A.; Pan, A. Scaling of the Conductivity Spectra in Ionic Glasses: Dependence on the Structure. *Phys. Rev. Lett.* **2000**, *84*, 2188–2190. [CrossRef]
48. Sen, S.; Choudhary, R.N.; Pramanik, P. Structural and electrical properties of Ca^{2+}-modified PZT electroceramics. *Phys. B Condens. Matter* **2007**, *387*, 56–62. [CrossRef]
49. Katiliute, R.M.; Seibutas, P.; Ivanov, M.; Grigalaitis, R.; Stanulis, A.; Banys, J.; Kareiva, A. Dielectric and Impedance Spectroscopy of $BaSnO_3$ and Ba_2 SnO_4. *Ferroelectrics* **2014**, *464*, 49–58. [CrossRef]
50. Liu, J.; Wang, C.L.; Li, Y.; Su, W.B.; Zhu, Y.H.; Li, J.C.; Mei, L.M. Influence of rare earth doping on thermoelectric properties of $SrTiO_3$ ceramics. *J. Appl. Phys.* **2013**, *114*, 223714. [CrossRef]
51. Li, G.; Liu, H.; Wang, Z.; Hao, H.; Yao, Z.; Cao, M.; Yu, Z. Dielectric properties and relaxation behaviour of Sm substituted $SrTiO_3$ ceramics. *J. Mater. Sci. Mater. Electron.* **2014**, *25*, 4418–4424. [CrossRef]
52. Gao, F.; Zhao, H.; Li, X.; Cheng, Y.; Zhou, X.; Cui, F. Preparation and electrical properties of yttrium-doped strontium titanate with B-site deficiency. *J. Power Sources* **2008**, *185*, 26–31. [CrossRef]
53. Lai, Y.W.; Wei, W.C. Synthesis and Study on Ionic Conductive $(Bi_{1-x},V_x)O_{1.5-\delta}$ Materials with a Dual-Phase Microstructure. *Materials* **2016**, *9*, 863. [CrossRef]
54. Singh, S.; Singh, P.; Viviani, M.; Presto, S. Dy doped SrTiO3: A promising anodic material in solid oxide fuel cells. *Int. J. Hydrog. Energy* **2018**, *43*, 19242–19249. [CrossRef]
55. Singh, S.; Jha, P.A.; Presto, S.; Viviani, M.; Sinha, A.S.K.; Varma, S.; Singh, P. Structural and electrical conduction behaviour of yttrium doped strontium titanate: Anode material for SOFC application. *J. Alloy. Compd.* **2018**, *748*, 637–644. [CrossRef]
56. Kolodiazhnyi, T.; Petric, A. The Applicability of Sr-deficient n-type $SrTiO_3$ for SOFC Anodes. *J. Electroceramics* **2005**, *15*, 5–11. [CrossRef]
57. Liu, J.; Wang, C.L.; Peng, H.; Su, W.B.; Wang, H.C.; Li, J.C.; Zhang, J.L.; Mei, L.M. Thermoelectric Properties of Dy-Doped $SrTiO_3$ Ceramics. *J. Electron. Mater.* **2012**, *41*, 3073–3076. [CrossRef]
58. Presto, S.; Barbucci, A.; Carpanese, M.; Han, F.; Costa, R.; Viviani, M. Application of La-Doped SrTiO3 in Advanced Metal-Supported Solid Oxide Fuel Cells. *Crystals* **2018**, *8*, 134. [CrossRef]
59. Buscaglia, M.T.; Viviani, M.; Buscaglia, V.; Bottino, C.; Nanni, P. Incorporation of Er^{3+} into $BaTiO_3$. *J. Am. Ceram. Soc.* **2002**, *85*, 1569–1575. [CrossRef]
60. Beltrán Rodríguez, R.; Landínez Téllez, D.; Roa-Rojas, J. Chemical Stability and Crystallographic Analysis of the the Sr_2HoNbO_6 Cubic Perovskite as Potential Substrate for $YBa_2Cu_3O_{7-\delta}$ Superconducting Films. *Mater. Res.* **2016**, *19*, 877–888. [CrossRef]

61. Bernuy-Lopez, C.; Høydalsvik, K.; Einarsrud, M.A.; Grande, T. Effect of A-Site Cation Ordering on Chemical Stability, Oxygen Stoichiometry and Electrical Conductivity in Layered $LaBaCo_2O_{5+\delta}$ Double Perovskite. *Materials* **2016**, *9*, 154. [CrossRef]
62. Barison, S.; Battagliarin, M.; Cavallin, T.; Doubova, L.; Fabrizio, M.; Mortalò, C.; Boldrini, S.; Malavasi, L.; Gerbasi, R. High conductivity and chemical stability of $BaCe_{1-x-y}Zr_xY_yO_{3-\delta}$ proton conductors prepared by a sol–gel method. *J. Mater. Chem.* **2008**, *18*, 5120. [CrossRef]

Article

Infiltrated $Ba_{0.5}Sr_{0.5}Co_{0.8}Fe_{0.2}O_{3-\delta}$-Based Electrodes as Anodes in Solid Oxide Electrolysis Cells

Xavier Majnoni d'Intignano [1], Davide Cademartori [2], Davide Clematis [2], Sabrina Presto [3], Massimo Viviani [3], Rodolfo Botter [2], Antonio Barbucci [2,3], Giacomo Cerisola [2], Gilles Caboche [1] and M. Paola Carpanese [2,3,*]

1 ICB-UMR6303, FCLAB, CNRS–Université de Bourgogne Franche-Comté, 9 Avenue Savary, BP47870 21078 Dijon CEDEX, France; Xavier_Majnoni-D-Intignano@etu.u-bourgogne.fr (X.M.d.); gilles.caboche@u-bourgogne.fr (G.C.)

2 DICCA-UNIGE, Via all'Opera Pia 15, 16145 Genova, Italy; davidecade95@gmail.com (D.C.); davide.clematis@edu.unige.it (D.C.); rodolfo.botter@unige.it (R.B.); barbucci@unige.it (A.B.); giacomo.cerisola@unige.it (G.C.)

3 CNR-ICMATE, c/o DICCA-UNIGE, Via all'Opera Pia 15, 16145 Genova, Italy; sabrina.presto@cnr.it (S.P.); massimo.viviani@ge.icmate.cnr.it (M.V.)

* Correspondence: carpanese@unige.it; Tel.: +39-010-335-6020

Received: 9 June 2020; Accepted: 13 July 2020; Published: 15 July 2020

Abstract: In the last decades, several works have been carried out on solid oxide fuel cell (SOFC) and solid oxide electrolysis cell (SOEC) technologies, as they are powerful and efficient devices for energy conversion and electrochemical storage. By increasing use of renewable sources, a discontinuous amount of electricity is indeed released, and reliable storage systems represent the key feature in such a future energy scenario. In this context, systems based on reversible solid oxide cells (rSOCs) are gaining increasing attention. An rSOC is an electrochemical device that can operate sequentially between discharging (SOFC mode) and charging (SOEC mode); then, it is essential the electrodes are able to guarantee high catalytic activity, both in oxidation and reduction conditions. $Ba_{0.5}Sr_{0.5}Co_{0.8}Fe_{0.2}O_{3-\delta}$ (BSCF) has been widely recognized as one of the most promising electrode catalysts for the oxygen reduction reaction (ORR) in SOFC technology because of its astonishing content of oxygen vacancies, even at room temperature. The purpose of this study is the development of BSCF to be used as anode material in electrolysis mode, maintaining enhanced energy and power density. Impregnation with a $La_{0.8}Sr_{0.2}MnO_3$ (LSM) discrete nanolayer is applied to pursue structural stability, resulting in a long lifetime reliability. Impedance spectroscopy measurements under anodic overpotential conditions are run to test BSCF and LSM-BSCF activity as the electrode in oxidation mode. The observed results suggest that BSCF is a very promising candidate as an oxygen electrode in rSOC systems.

Keywords: BSCF; SOEC; SOFC; rSOC; anodic overpotential; impedance spectroscopy

---

1. Introduction

The increasing penetration of renewable energy sources in the power market, guided by new energy policies to address the climate change, poses new challenges that need to be tackled [1]. The intermittent nature of wind and solar power requires the development of large-scale energy storage as a key to improve the flexibility of the electric grid. Electrical energy storage (EES) is envisioned as the key factor to boost the development of advanced grid-energy management systems [2,3]. In this context, systems based on reversible solid oxide cells (rSOCs) are gaining increasing attention and interest. An rSOC is an electrochemical device that can operate in both power-producing (solid oxide fuel cell SOFC) and energy storage (solid oxide electrolysis cell SOEC) modes. The system operates sequentially between discharging (SOFC) and charging (SOEC) modes [4–9]. Electrochemical reactions can be based on either H-O or H-O-C elemental systems. In H-O systems, only hydrogen, water and oxygen are involved, while in H-O-C ones, hydrocarbons also participate in reactions. SOFC converts hydrogen-rich fuels into electricity and heat [10]. On the other side, the electricity supplied to SOEC leads to the conversion of H_2O and CO_2 into a syngas usually containing H_2 and CO [11]. Water electrolysis can also be performed by either alkaline or proton exchange membrane electrolyzers. The most commercial electrolyzers to date belong to the alkaline series with sizes ranging from 0.6 to 125 MW of produced H_2 [5,12]. However, the operating voltage for splitting an H_2O molecule can be significantly reduced at high temperature. For this reason, SOEC can represent an attractive and effective solution, with an operative temperature in the intermediate range (500–650 °C), which appears as the optimum trade-off between durability and efficiency [13].

A typical rSOC is constituted by a solid electrolyte sandwiched between two porous electrodes. Among the most commonly used materials for the electrolyte, fluorite-structured reference electrode (RE)-doped ceria is considered the best candidate to operate SOC at the targeted operative temperatures [14–16]. Various oxides with perovskite [17–20], double perovskite [21–23] and Ruddlesden–Popper (RP) [24–26] structures showing suitable mixed ionic-electronic conductivity and/or electrocatalytic activity have been proposed for SOFC and SOEC electrode layers.

Depending on the operating mode, each electrode can be the location where oxidation or reduction takes place. Global losses in such systems are related to the losses due to each component, and it is widely proved that losses due to electrode processes can be lowered by an optimization of their microstructure, in terms of capability of gas exchange with gas phase as well as of ion migration. Among different strategies that have been pursued to improve the long-lasting time operation (durability), the infiltration of porous electrodes by discrete or continuous thin layers have been shown to be one of the most efficient. The main advantages of infiltration are (i) the electrode is fabricated by a two-step deposition process, i.e., firstly, a supporting porous backbone is deposited and sintered to get strong adhesion with the electrolyte, then the catalytic layer can be deposited and sintered at a lower temperature to keep the optimal microstructure; (ii) different couplings of backbone/catalyst can be used, since some fundamental parameters such as the thermal expansion coefficient (TEC) mismatch or chemical interactions can be minimized [27–29].

In the literature, it was already affirmed that BSCF material displays better performance under anodic polarization, suggesting that this material could be successfully used in reversible SOFC–SOEC systems [30]. Moreover, the authors have already experienced the positive effect of LSM-infiltrated, nano-sized layers on porous backbones, used in oxygen reduction conditions, finding that infiltration resulted in improved activity performance as well as durability [31].

In this study, a similar approach is carried out, to investigate the influence of LSM-infiltration on BSCF porous electrodes, to test its catalytic activity in water reduction conditions and evaluate BSCF-based material as a possible electrode at the air side of rSOC.

2. Materials and Methods

The shape of the electrolyte support must respect appropriate geometric criteria. Namely, the distance between the reference electrode (RE) and working electrode (WE) should be at least three-times the electrolyte thickness in order to avoid artefact formation in impedance experimental spectra [17,32]. According to the relative shrinkage value obtained from previous sintering experience (22%), a weight of $Ce_{0.8}Sm_{0.2}O_{2-\delta}$ (SDC20) electrolyte powder equal to 2.4 g was uniaxially pressed at 37 MPa to obtain electrolyte supports. The green pellets were sintered at 1500 °C for 5 h, obtaining sintered discs of 20 mm in diameter and 1.1 mm in thickness. Before the electrode deposition, the SDC discs were sanded down (P320 SiC paper) for obtaining a rough surface reliable for an easy adhesion of the BSCF electrode to the electrolyte layer.

A mixture of graphite (KS6, TIMREX®, TIMCAL Graphite & Carbon, Bodio, Switzerland) and BSCF (Treibacher Industry AG, Althofen, Austria) powders, according to the 60/40 v/v% composition, was ball-milled for 40 h at 40 rpm in distilled water at room temperature (R.T.), and zirconia balls (Tosoh) were employed as mixing bodies. This volume concentration was chosen in order to have a proper porosity resulting in an easy evacuation of the oxygen gas phase. After the mixing, a freeze-drying procedure was applied (24 h, at −52 °C and 22 Pa). Finally, after sieving, the BSCF powder was ready to be deposited. Alpha-terpineol (Sigma-Aldrich, >96%) was added to the powder (BSCF graphite) in a mortar to obtain a mixture suitable for deposition. By applying an appropriate tape mask, a WE and a counter electrode (CE) were slurry-coated on the sides of the SDC pellets. An RE, used for the three-electrode impedance measurements, was applied around the WE. The geometry is described in detail elsewhere [33]. The electrodes were co-sintered at 1100 °C for 1 h. After the sintering process, the geometric area for both the WE and the CE was 0.28 cm^2.

For the infiltration of a porous BSCF electrode backbone, an aqueous solution of hydrated nitrates, namely $La(NO_3)\cdot 2H_2O$, $Sr(NO_3)_2$ and $Mn(NO_3)_3\cdot xH_2O$, with x = 4 or 6, was prepared. In order to estimate the actual cation concentration in precursors, gravimetric titration of the above-mentioned nitrates was carried out. About 5 g of each precursor was weighted out in a clean and dry Pt crucible and heated up to 900 °C for 2 h, in order to allow for the formation of oxides, i.e., La_2O_3, SrO and Mn_3O_4.

The impregnating solution was prepared by adding stoichiometric amounts of hydrated nitrates, reported in Table 1, to 50 mL of water in order to obtain a concentration of 0.6 mol L^{-1}. Glycine and polyvinylpyrrolidone (PVP) were added, respectively, as a chelating agent and as a surfactant (see Table 1), and the solution was heated at 200 °C under stirring for 10 min to completely dissolve the components. Impregnation was carried out by using different amounts of the nitrate solution, in the range 1.5–6 µL, either as-prepared or diluted. The samples tested in this work were impregnated with 1.5 µL of a solution with a concentration of 0.06 mol L^{-1}. In order to eliminate air in the pores, infiltrated samples were placed under a vacuum at 94 Pa for 2 min. The infiltrated cells (both in the WE and the CE) were finally heated at 800 °C for 3 h, with a 1 h dwelling step at 300 °C.

LSM-BSCF/SDC/LSM-BSCF half-cells (with an RE) were tested inside a ProboStat^{TM} (NorECs Norwegian Electro Ceramics AS, Oslo, Norway) setup system. Two Pt nets were placed on the surface of each electrode as current collectors. Electrochemical impedance spectroscopy (EIS) measurements were carried out through a Solartron Analytical potentiostat (SI 1286) coupled with a frequency response analyzer (SI 1255) and the ZPLOT software (version 14.1.3, 45485, Scribner Associates Inc., Southern Pines, NC, USA) Impedance tests were run between 475 and 650 °C in a frequency range 10^4–10^{-2} Hz. An amplitude varying between 5 and 20 mV was applied, according to the system response. Measurements were performed at open circuit voltage (OCV) conditions as well as under anodic overpotential load, up to +0.2 V. An N_2/O_2 = 80/20 v/v% mixture was fed at both sides of the cell, corresponding to a flow of 40 NmL min^{-1} and 10 NmL min^{-1} of N_2 and O_2, respectively. To test the system durability, a 48 h long-lasting test was run under an anodic potential of +0.150 V, collecting an OCV impedance test every 6 h. Experimental impedance spectra were analyzed by ZView-Impedance Software (version 14.1.3, 45485, Scribner Associates Inc., Southern Pines, NC, USA).

To investigate the morphology and the adhesion between the layers, observations were taken by scanning electron microscopy (SEM, Phenom Pro-X) equipped with Energy-dispersive X-ray spectroscopy (EDXS, Bruker) on the as-sintered and tested samples in order to check the distribution of LSM on the top and cross the section of the BSCF electrode.

Table 1. Composition of $La_{0.8}Sr_{0.2}MnO_{3-\delta}$-infiltrated layer.

$La(NO_3)_2\cdot 2H_2O$	$Sr(NO_3)_2$	$Mn(NO_3)_3\cdot xH_2O$	Glycine	PVP
10.477 g	0.622 g	7.289 g	2.468 g [1]	1.158 g [2]

[1] 1/4 of the total molar amount of starting nitrates; [2] 0.05 wt.% of $La_{0.8}Sr_{0.2}MnO_3$ (LSM) amount.

3. Results and Discussion

3.1. Microstructural

In Figure 1, the surface SEM images of the pristine BSCF and of the 1.5 μL, 0.06 mol L^{-1} LSM-infiltrated electrode are reported, before and after the electrochemical characterization. In the following, the LSM-infiltrated BSCF will be named just LSM-BSCF electrode.

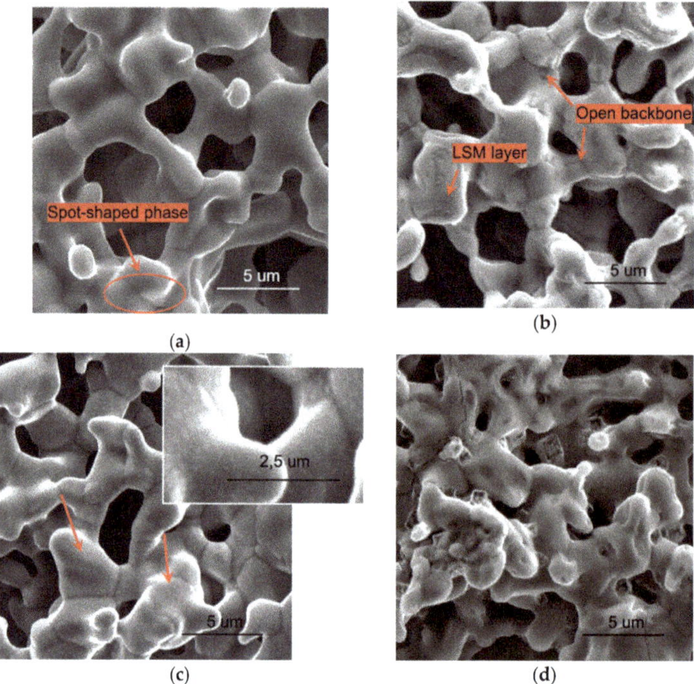

(a) (b) (c) (d)

Figure 1. Surface SEM images of $Ba_{0.5}Sr_{0.5}Co_{0.8}Fe_{0.2}O_{3-\delta}$ (BSCF) (**a**,**c**) and of LSM-BSCF (**b**,**d**) electrodes before (**a**,**b**) and after (**c**,**d**) the electrochemical testing. Inset in (**c**) shows an improved view of the spot-shaped phase. BSCF backbones were sintered at 1100 °C for 1 h.

After the sintering process at 1100 °C, the electrodes showed a homogeneous structure, with open porosity and an appreciable coarsening of the grains, resulting in the formation of a good interconnected ceramic network (see Figure 1).

The comparison of images obtained before (Figure 1a,b) and after the electrochemical testing (Figure 1c,d) shows that the porosity of structures was not affected by testing. By contrast, transformations were evident at the surface of both electrodes after the electrochemical investigation. On the surface of the blank BSCF electrode, a dot-shaped minor phase appeared (Figure 1c); likewise, in previous findings [31], it was supposed that this phase could result from the aggregation, during testing time between 475 and 650 °C, of the distributed phase already present on the BSCF starting electrode surface (Figure 1a, red arrow and oval). Concerning the infiltrated electrode, the LSM was distributed as a continuous and homogeneous layer at the beginning (Figure 1b), while, after the testing, several extended defects with sharp edges and micropores appeared (Figure 1d).

Figure 2 shows an SEM surface image of the as-prepared LSM-infiltrated layer and the corresponding atomic concentration obtained through a map elemental analysis. Further EDXS observations performed through the cross-section of the infiltrated electrode confirmed the penetration of the LSM layer up to the electrode/electrolyte interface.

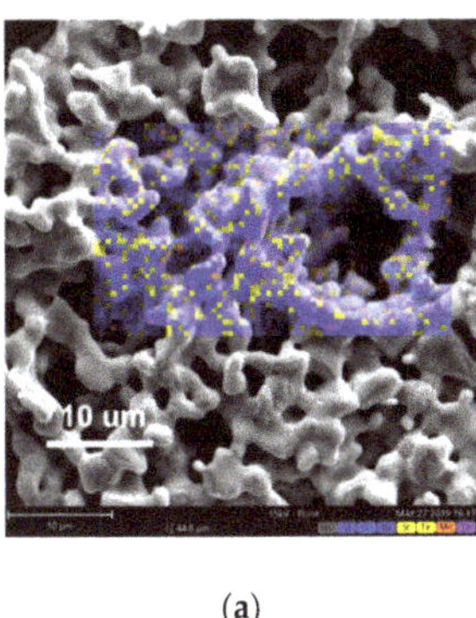

Element Number	Element Symbol	Element Name	Atomic Conc.	Weight Conc.
8	O	Oxygen	45.59	12.91
27	Co	Cobalt	19.23	20.06
56	Ba	Barium	16.89	41.05
38	Sr	Strontium	10.19	15.81
26	Fe	Iron	4.74	4.68
25	Mn	Manganese	1.88	1.83
57	La	Lanthanum	1.48	3.65

(a) (b)

Figure 2. SEM-EDXS of LSM-BSCF infiltrated electrodes (before testing). Surface SEM image (**a**) with corresponding atomic concentration obtained through a map elemental analysis (**b**).

3.2. Electrochemical

Figure 3 reports an optical image (Figure 3a) of the partial side of SDC electrolyte with the slurry-coated WE (in the center) and RE (ring-shaped), while the geometry features of the three-electrode configuration are reported in Figure 3b, following the indications of [32,33], as aforementioned. The measurements were performed using an RE in order to have accurate control of the WE potential, especially when an overpotential was applied.

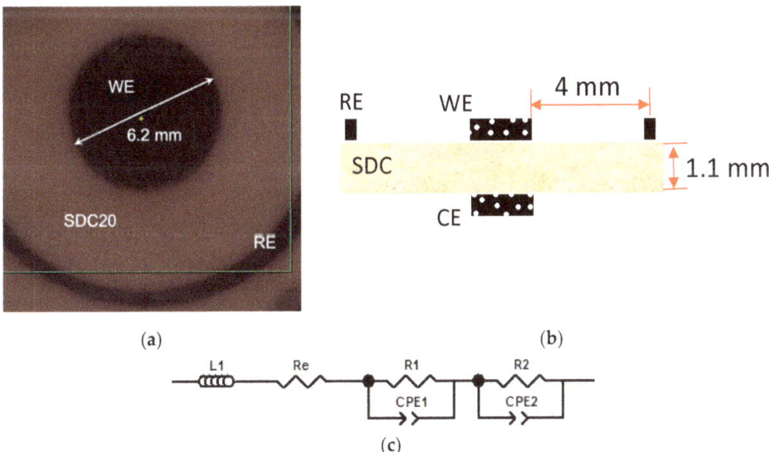

Figure 3. Image and scheme of the cell used for electrochemical investigation. Working electrode (WE) and reference electrode (RE) on the SDC20 electrolyte (**a**); drawing of the three-electrode cell configuration (**b**). In (**c**) the equivalent circuit used to model impedance spectra is shown: inductance, L1; electrolyte resistance, R_e; resistance/constant phase element, R1/CPE1 and R2/CPE2, at high and low frequency, respectively, of the electrode-related arch.

Impedance spectra were modeled through an equivalent circuit in order to obtain the different parameters, such as polarization resistance. The equivalent circuit consisted of a resistance (R_e) in series with an inductance (L1) and two resistance/constant phase elements (R1/CPE1 and R2/CPE2) in series as shown by Figure 3c. The constant phase elements were chosen in view of their flexibility to model the circuits obtained in the different conditions of operations.

First, impedance measurements were carried out at OCV, then tests were run by applying an anodic overpotential, up to +0.2 V.

In Figure 4, impedance spectra of blank BSCF (Figure 4a) and LSM-BSCF (Figure 4b) are shown. It is evident that the values of polarization resistance (R_p), extracted as the difference between the low and high frequency intercepts with the real axis, are larger for the infiltrated electrodes than for pristine BSCF. At OCV and 650 °C, the value of the area specific resistance (ASR), as obtained by modeling, was 0.121 Ω cm^2 for the blank electrode and 0.160 Ω cm^2 for the infiltrated one. These results agree with other works, dealing with both blank and LSM-infiltrated electrodes [34–36]. The polarization resistance values obtained for the blank BSCF and LSM-BSCF infiltrated electrode were in contrast with previous findings obtained by the authors on LSM-infiltrated BSCF electrodes. In that case [31], the LSM infiltration had a positive effect in terms of R_p at OCV. The reason for this discrepancy is likely due to the different morphology of the LSM-infiltrated layer. In the previous work, we dealt with a nano-sized discrete layer, resulting in an increased electrode–gas surface exchange area and consequent improved electrode activity; in this study, a continuous LSM layer was obtained, whose exchange properties with the gas phase are very different from those of the nano-distributed one.

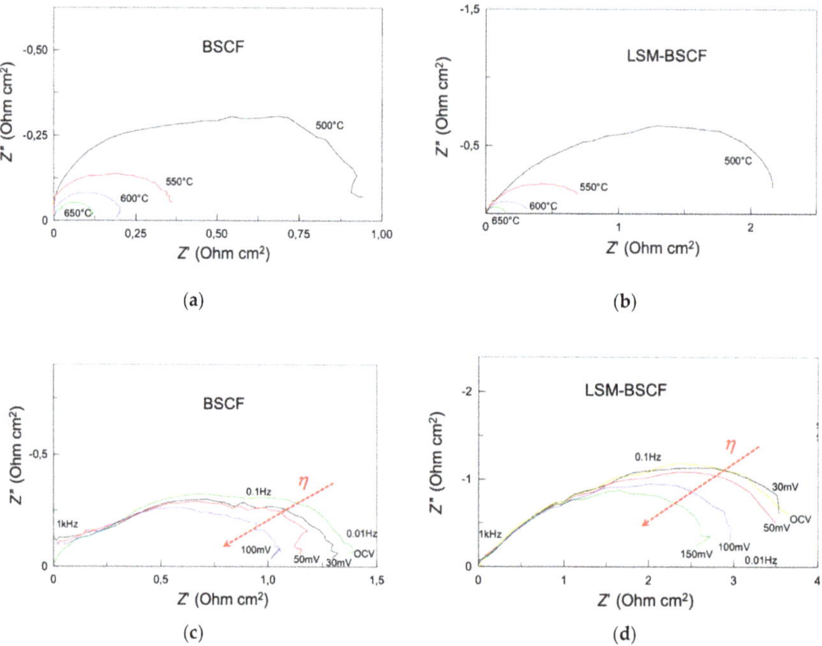

Figure 4. Impedance spectra at open circuit voltage (OCV) and under anodic overpotential (η up to +0.150 V). (**a**) OCV for blank BSCF; (**b**) OCV for infiltrated LSM-BSCF electrode; (**c**) with η for blank BSCF; (**d**) with η for LSM-BSCF infiltrated electrodes. Temperature conditions for (**a**,**b**): between 500 and 650 °C; for (**c**,**d**): 475 °C. In every case, SDC20 constitutes the electrolyte support. Measurements were performed in three-electrode configuration, using the ring-shaped RE.

Considering the effect of anodic overpotential (Figure 4c,d), it had a positive effect in terms of R_p values, both in case of blank and infiltrated samples. It was already found in previous study that cathodic overpotential had, on the contrary, a negative influence on the electrocatalytic performance of

BSCF-based electrodes [17]. The positive effect of anodic overpotential was further confirmed through a sweep voltammetry measurement. Although not reported in the paper, the current density measured under anodic overpotential was slightly higher than that under cathodic overpotential, indicating a higher activity in anodic conditions (namely, when the electrode is involved as anode in the electrolysis mode of operation).

The presence of LSM did not improve the trend of impedance when η was increased. At the temperature of 475 °C, R_p at OCV was 4.43 Ω cm^2 for LSM-BSCF, while it decreased until 1.72 Ω cm^2 for the blank electrode. For the latter, an ASR decrease of 32% was observed under increasing anodic overpotential; the LSM-BSCF displayed a 24% decrease between OCV and +0.100 V, becoming 34.5% between OCV and +0.150 V. Unfortunately, beyond +0.1 and +0.15 V, the overpotential applied to blank and infiltrated electrodes, respectively, was destroyed, with the infiltrated one appearing more resistant.

In Figure 5, the inverse of polarisation resistance ($1/R_p$) is reported as a function of temperature. An Arrhenius-like behaviour allowed for obtaining the values of apparent activation energy (E_a) for both the systems: 85.3 and 110.6 kJ mol^{-1} for blank BSCF and LSM-BSCF, respectively. These activation energy values concern the activity during the dynamic equilibrium state at OCV. It has been proven in the literature that LSM activity drops under anodic conditions [37], and, actually, the impedance spectra extracted under anodic overpotential conditions confirm that the presence of the LSM hinders the activity of the electrode (see Figure 4c,d). Despite this detrimental effct due to LSM, the increase in the catalytic activity of the BSCF was confirmed when a net anodic current flowed through the cell. As observable in Figure 4d, the LSM hindering effect was not able to counterbalance the enhanced activity of BSCF. Moreover, the LSM-infiltrated layer resulted in a positive effect on the long-lasting performance of the SDC/LSM-BSCF system. In previous work, it was found that LSM contributed to maintaining the polarization stability of the BSCF electrode, likely because of a positive influence on the electrochemical potential gradient at the interface LSM/BSCF, with this stabilizing effect indeed also observed for LSCF-based and other electrodes [14,17].

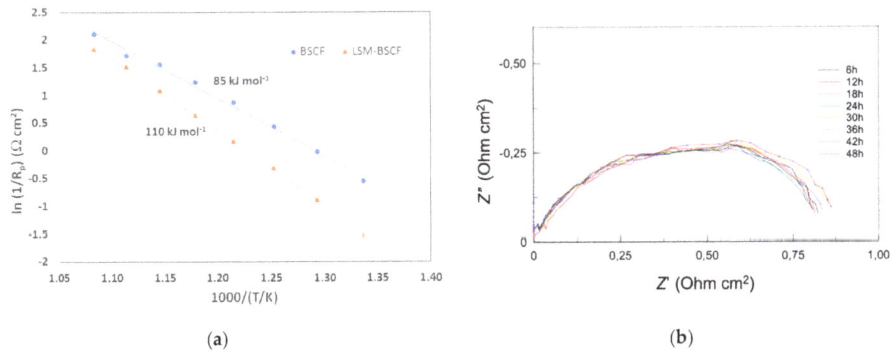

Figure 5. (**a**) Trend of the inverse of polarization resistance (R_p) vs temperature, for blank BSCF and LSM-infiltrated BSCF. The values shown are the apparent activation energy from the linear regression; (**b**) impedance spectra obtained for the LSM-BSCF electrode every 6 h, at OCV, during the anodic current load of +0.150 V for 48 h at 600 °C.

An anodic load of 0.150 V was run for 48 h at 600 °C, and an impedance test was performed every 6 h. The result is shown in Figure 5. It is observed that the activity behavior, in terms of polarization resistance, was stable for the duration of the test. During the voltage load, a current of 14 mA cm^2 was observed at 600 °C. This value should certainly be improved, but optimization of the electrode morphology and performance under different gas partial pressures will be the subject of further investigations.

4. Conclusions

In this study, blank BSCF and LSM-infiltrated BSCF electrodes are analyzed, in view to be used as possible anode electrodes in solid state electrolysis (SOEC) devices. Electrochemical impedance spectroscopy investigation, coupled with an aging test at +0.15 V current load, was performed on a three-electrode system with SDC20 electrolyte.

It was found that polarization resistance decreased in both systems when an increasing anodic current load was applied. The blank BSCF electrode showed lower resistance (R_p) than the infiltrated one; then, it appeared that LSM did not give the chance to improve performance. This is possibly due to the detrimental effect of the continuous LSM layer covering most of the BSCF backbone. An LSM nano-sized discrete layer could result in increasing the surface exchange area, without subtracting appreciable BSCF surface, and consequently making the gas–electrode activity faster. This was actually found out previously by the authors.

On the other hand, the LSM layer had a positive effect on the impedance stability, since no sign of degradation appeared in the first 48 h under a voltage load of +0.150 V. Performance degradation in the first 100 h of operation is one of the major problems to be faced in SOFC/SOEC devices, and LSM infiltration was demonstrated to be a promising approach in overcoming this issue. Further attempts must be carried out in order to find the optimized parameters for a nano-sized discrete layer of LSM to be deposited. Most importantly, it was confirmed that BSCF-based electrodes are very promising candidates to be used as oxygen electrodes in rSOC.

Author Contributions: Conceptualization, M.P.C., A.B., G.C. (Giacomo Cerisola) and R.B.; methodology, M.P.C., A.B., M.V. and S.P.; software, M.P.C., M.V., D.C. (Davide Clematis); formal analysis, M.P.C., X.M.d., D.C. (Davide Clematis) and D.C. (Davide Cademartori); investigation, X.M.d.; resources, M.P.C., G.C. (Giacomo Cerisola); S.P. and R.B.; data curation, M.P.C., X.M.d.; writing—original draft preparation, M.P.C., X.M.d., M.V. and D.C. (Davide Cademartori); writing—review and editing, M.P.C., S.P., M.V., X.M.d., D.C. (Davide Cademartori), D.C. (Davide Clematis), R.B., A.B., G.C. (Giacomo Cerisola), G.C. (Gilles Caboche); visualization, M.P.C. and X.M.d.; supervision, M.P.C.; funding acquisition, A.B., G.C. (Gilles Caboche), and G.C. (Giacomo Cerisola). All authors have read and agreed to the published version of the manuscript.

Funding: This research was funded in part by the Regional Council of Bourgogne Franche-Comté and the French National Center for Scientific Research (CNRS) Graduate School EIPHI (Contract ANR-17-EURE-0002) and through a visiting researcher position granted to M.P.C.

Acknowledgments: M. Paola Carpanese acknowledges the University of Genova for funding resources coming from FRA Fondi di Ricerca di Ateneo 2018.

Conflicts of Interest: The authors declare no conflict of interest.

References

1. European Commission. A European Green Deal. Available online: https://ec.europa.eu/info/strategy/priorities-2019-2024/european-green-deal_en (accessed on 10 June 2020).
2. Gupta, R.; Soini, M.C.; Patel, M.P.; Parra, D. Levelized cost of solar photovoltaics and wind supported by storage technologies to supply firm electricity. *J. Energy Storage* **2020**, *27*, 101027. [CrossRef]
3. Zsiborács, H.; Hegedüsné Baranyai, N.; Vincze, A.; Zentkó, L.; Birkner, Z.; Kinga, M.; Pintér, G. Intermittent Renewable Energy Sources: The Role of Energy Storage in the European Power System of 2040. *Electronics* **2019**, *8*, 729. [CrossRef]
4. Mogensen, M.B.; Chen, M.; Frandsen, H.L.; Graves, C.; Hansen, J.B.; Hansen, K.V.; Hauch, A.; Jacobsen, T.; Jensen, S.H.; Skafte, T.L.; et al. Reversible solid-oxide cells for clean and sustainable energy. *Clean Energy* **2019**, *3*, 175–201. [CrossRef]
5. Venkataraman, V.; Pérez-Fortes, M.; Wang, L.; Hajimolana, Y.S.; Boigues-Muñoz, C.; Agostini, A.; McPhail, S.J.; Maréchal, F.; Van Herle, J.; Aravind, P.V. Reversible solid oxide systems for energy and chemical applications—Review & perspectives. *J. Energy Storage* **2019**, *24*, 100782.
6. Perna, A.; Minutillo, M.; Jannelli, E. Designing and analyzing an electric energy storage system based on reversible solid oxide cells. *Energy Convers. Manag.* **2018**, *159*, 381–395. [CrossRef]

7. Giarola, S.; Forte, O.; Lanzini, A.; Gandiglio, M.; Santarelli, M.; Hawkes, A. Techno-economic assessment of biogas-fed solid oxide fuel cell combined T heat and power system at industrial scale. *Appl. Energy* **2018**, *211*, 689–704. [CrossRef]
8. Carpanese, M.P.; Panizza, M.; Viviani, M.; Mercadelli, E.; Sanson, A.; Barbucci, A. Study of reversible SOFC/SOEC based on a mixed anionic- protonic conductor. *J. Appl. Electrochem.* **2015**, *45*, 657–665. [CrossRef]
9. Thorel, A.S.; Chesnaud, A.; Viviani, M.; Barbucci, A.; Presto, S.; Piccardo, P.; Ilhan, Z.; Vladikova, D.; Stynov, Z. IDEAL-cell, a high temperature innovative dual mEmbrAne fuel cell. *ECS. Trans.* **2009**, *25*, 753–762. [CrossRef]
10. Gondolini, A.; Mercadelli, E.; Sangiorgi, A.; Sanson, A. Integration of Ni-GDC layer on a NiCrAl metal foam for SOFC application. *J. Eur. Ceram. Soc.* **2017**, *37*, 1023–1030. [CrossRef]
11. Zheng, Y.; Wang, J.; Yu, B.; Zhang, W.; Chen, J.; Qiao, J.; Zhang, J. A review of high temperature co-electrolysis of H_2O and CO_2 to produce sustainable fuels using solid oxide electrolysis cells (SOECs): Advanced materials and technology. *Chem. Soc. Rev.* **2017**, *46*, 1427–1463. [CrossRef]
12. Brauns, Z.; Turek, T. Alkaline Water Electrolysis Powered by Renewable Energy: A Review. *Processes* **2020**, *8*, 248. [CrossRef]
13. Laguna-Bercero, M.A. Recent advances in high temperature electrolysis using solid oxide fuel cells: A review. *J. Power Sources* **2012**, *203*, 4–16. [CrossRef]
14. Presto, S.; Artini, C.; Pani, M.; Carnasciali, M.M.; Massardo, S.; Viviani, M. Ionic conductivity and local structural features in $Ce_{1-x}Sm_xO_{2-x/2}$. *Phys. Chem. Chem. Phys.* **2018**, *20*, 28338–28345. [CrossRef]
15. Chen, G.; Sun, W.; Luo, Y.; He, Y.; Zhang, X.; Zhu, B.; Li, W.; Liu, X.; Ding, Y.; Li, Y.; et al. Advanced Fuel Cell Based on New Nanocrystalline Structure $Gd_{0.1}Ce_{0.9}O_2$ Electrolyte. *ACS Appl. Mater. Interfaces* **2019**, *11*, 10642–10650. [CrossRef]
16. Jaiswal, N.; Tanwar, K.; Suman, R.; Kumar, D.; Upadhyay, S.; Parkash, O.; Uppadhya, S.; Parkash, O. A brief review on ceria based solid electrolytes for solid oxide fuel cells. *J. Alloy. Compd.* **2019**, *781*, 984–1005. [CrossRef]
17. Singh, S.; Singh, P.; Viviani, M.; Presto, S. Dy doped $SrTiO_3$: A promising anodic material in solid oxide fuel cells. *Int. J. Hydrogen Energy* **2018**, *43*, 19242–19249. [CrossRef]
18. Clematis, D.; Barbucci, A.; Presto, S.; Viviani, M.; Carpanese, M.P. Electrocatalytic activity of perovskite-based cathodes for solid oxide fuel cells. *Int. J. Hydrogen Energy* **2019**, *44*, 6212–6222. [CrossRef]
19. Luo, H.; Wang, H.; Chen, X.; Huang, D.; Zhou, M.; Ding, D. Cation deficiency tuning of $LaCoO_3$ perovskite as bifunctional oxygen electrocatalysts. *ChemCatChem* **2020**, *12*, 2768–2775. [CrossRef]
20. Carpanese, M.P.; Clematis, D.; Bertei, A.; Giuliano, A.; Sanson, A.; Mercadelli, E.; Nicolella, C.; Barbucci, A. Understanding the electrochemical behaviour of LSM-based SOFC cathodes. Part I—Experimental and electrochemical. *Solid State Ion.* **2017**, *301*, 106–115. [CrossRef]
21. Niu, B.; Lu, C.; Yi, W.; Luo, S.; Li, X.; Zhong, X.; Zhao, X.; Xu, B. In-situ growth of nanoparticles-decorated double perovskite electrode materials for symmetrical solid oxide cells. *Appl. Catal. B Environ.* **2020**, *270*, 118812. [CrossRef]
22. Presto, S.; Kumar, P.; Varma, S.; Viviani, M.; Singh, P. Electrical conductivity of NiMo–based double perovskites under SOFC anodic conditions. *Int. J. Hydrogen Energy* **2018**, *43*, 4528–4533. [CrossRef]
23. Rath, M.K.; Lee, K. Superior electrochemical performance of non-precious Co-Ni-Mo alloy catalyst-impregnated $Sr_2FeMoO_{6-\delta}$ as an electrode material for symmetric solid oxide fuel cells. *Electrochim. Acta* **2016**, *212*, 678–685. [CrossRef]
24. Sharma, R.K.; Burriel, M.; Dessemond, L.; Bassat, J.-M.; Djurado, E. $La_{n+1}Ni_nO_{3n+1}$ (n = 2 and 3) phases and composites for solid oxide fuel cell cathodes: Facile synthesis and electrochemical properties. *J. Power Sources* **2016**, *325*, 337–345. [CrossRef]
25. Arias-Serrano, B.I.; Kravchenko, E.; Zakharchuk, K.; Grins, J.; Svensson, G.; Pankov, V.; Yaremchenko, A. Oxygen-Deficient $Nd_{0.8}Sr_{1.2}Ni_{0.8}M_{0.2}O_{4-\delta}$ (M = Ni, Co, Fe) Nickelates as Oxygen Electrode Materials for SOFC/SOEC. *ECS Trans.* **2019**, *91*, 2387–2397. [CrossRef]
26. Wang, J.; Zhou, J.; Yang, J.; Zong, Z.; Fu, L.; Lian, Z.; Zhang, X.; Wang, X.; Chen, C.; Ma, W.; et al. Nanoscale architecture of $(La_{0.6}Sr_{1.4})_{0.95}Mn_{0.9}B_{0.1}O_4$ (B = Co, Ni, Cu) Ruddlesden—Popper oxides as efficient and durable catalysts for symmetrical solid oxide fuel cells. *Renew. Energy* **2020**, *157*, 840–850. [CrossRef]

27. Khoshkalam, M.; Tripković, Đ.; Tong, X.; Faghihi-Sani, M.A.; Chen, M.; Vang Hendriksen, P. Improving oxygen incorporation rate on $(La_{0.6}Sr_{0.4})_{0.98}FeO_{3-\delta}$ via $Pr_2Ni_{1-x}CuxO_{4+\delta}$ surface decoration. *J. Power Sources* **2020**, *457*, 228035. [CrossRef]
28. Ding, D.; Li, X.; Lai, S.Y.; Gerdes, K.; Liu, M. Enhancing SOFC cathode performance by surface modification through infiltration. *Energy Environ. Sci.* **2014**, *7*, 552–575. [CrossRef]
29. Jiang, S.P. A review of wet impregnation—An alternative method for the fabrication of high performance and nano-structured electrodes of solid oxide fuel cells. *Mat. Sci. Eng. A* **2006**, *418*, 199–210. [CrossRef]
30. Aguadero, A.; Pérez-Coll, D.; Alonso, J.A.; Skinner, S.J.; Kilner, J. A New family of Mo-Dioped $SrCoO_{3-\delta}$ Perovskites for Application in Reversible Solid State Electrochemical Cells. *Chem. Mater.* **2012**, *24*, 2655–2663. [CrossRef]
31. Giuliano, A.; Carpanese, M.P.; Clematis, D.; Boaro, M.; Pappacena, A.; Deganello, F.; Liotta, L.F.; Barbucci, A. Infiltration, Overpotential and Ageing Effects on Cathodes for Solid Oxide Fuel Cells: $La_{0.6}Sr_{0.4}Co_{0.2}Fe_{0.8}O_{3-\delta}$ versus $Ba_{0.5}Sr_{0.5}Co_{0.8}Fe_{0.2}O_{3-\delta}$. *J. Electrochem. Soc.* **2017**, *164*, F3114–F3122. [CrossRef]
32. Winkler, J.; Hendriksen, P.V.; Bonanos, N.; Mogensen, M. Geometric requirements of solid electrolyte cells with a reference electrode. *J. Electrochem. Soc.* **1998**, *145*, 1184–1192. [CrossRef]
33. Carpanese, M.P.; Giuliano, A.; Panizza, M.; Mercadelli, E.; Sanson, A.; Gondolini, A.; Bertei, A.; Sanson, A.; Gondolini, A.; Bertei, A. Experimental Approach for the study of SOFC cathodes. *Bulg. Chem. Commun.* **2016**, *48*, 23–29.
34. Meng, L.; Wang, F.; Wang, A.; Pu, J.; Chi, B.; Li, J. High performance $La_{0.8}Sr_{0.2}MnO_3$-coated $Ba_{0.5}Sr_{0.5}Co_{0.8}Fe_{0.2}O_3$ cathode prepared by a novel solid-solution method for intermediate temperature solid oxide fuel cells. *Chin. J. Catal.* **2014**, *35*, 38–42. [CrossRef]
35. Deganello, F.; Liotta, L.F.; Marcì, G.; Fabbri, E.; Traversa, E. Strontium and iron-doped barium cobaltite prepared by solution combustion synthesis: Exploring a mixed-fuel approach for tailored intermediate temperature solid oxide fuel cell cathode materials. *Mater. Renew. Sustain. Energy* **2013**, *2*, 8. [CrossRef]
36. Yang, X.; Li, R.; Yang, Y.; Wen, G.; Tian, D.; Lu, X.; Ding, Y.; Chen, Y.; Lin, B. Improving stability and electrochemical performance of $Ba_{0.5}Sr0.5Co_{0.2}Fe_{0.8}O_{3-\delta}$ electrode for symmetrical solid oxide fuel cells by Mo doping. *J. Alloy. Compd.* **2020**, *831*, 154711. [CrossRef]
37. Chen, X.J.; Khor, K.A.; Chan, S.H. Electrochemical behavior of $La(Sr)MnO_3$ electrode under cathodic and anodic polarization. *Solid State Ion.* **2004**, *167*, 379–387. [CrossRef]

Article

Glass-Ceramic Sealants for SOEC: Thermal Characterization and Electrical Resistivity in Dual Atmosphere

Hassan Javed [1,*], Antonio Gianfranco Sabato [2], Mohsen Mansourkiaei [3], Domenico Ferrero [3], Massimo Santarelli [3], Kai Herbrig [1], Christian Walter [1] and Federico Smeacetto [4]

1 Sunfire GmbH, Gasanstaltstraße 2, 01237 Dresden, Germany; kai.herbrig@sunfire.de (K.H.); christian.walter@sunfire.de (C.W.)
2 IREC-Institut de Recerca en Energia de Catalunya, 08930 Sant Adrià de Besòs, Barcelona, Spain; gsabato@irec.cat
3 Department of Energy (DENERG), Politecnico di Torino, 10129 Turin, Italy; mohsen.mansourkiaei@polito.it (M.M.); domenico.ferrero@polito.it (D.F.); massimo.santarelli@polito.it (M.S.)
4 Department of Applied Science and Technology (DISAT), Politecnico di Torino, 10129 Turin, Italy; federico.smeacetto@polito.it
* Correspondence: hassan.javed@sunfire.de

Received: 10 June 2020; Accepted: 16 July 2020; Published: 17 July 2020

Abstract: A Ba-based glass-ceramic sealant is designed and tested for solid oxide electrolysis cell (SOEC) applications. A suitable SiO_2/BaO ratio is chosen in order to obtain $BaSi_2O_5$ crystalline phase and subsequently favorable thermo-mechanical properties of the glass-ceramic sealant. The glass is analyzed in terms of thermal, thermo-mechanical, chemical, and electrical behavior. Crofer22APU-sealant-Crofer22APU joined samples are tested for 2000 h at 850 °C in a dual atmosphere test rig having reducing atmosphere of H_2:H_2O 50/50 (mol%) and under the applied voltage of 1.6 V. In order to simulate the SOEC dynamic working conditions, thermal cycles are performed during the long-term electrical resistivity test. The glass-ceramic shows promising behavior in terms of high density, suitable CTE, and stable electrical resistivity (10^6–10^7 Ω cm) under SOEC conditions. The SEM-EDS post mortem analysis confirms excellent chemical and thermo-mechanical compatibility of the glass-ceramic with Crofer22APU.

Keywords: SOEC; sealants; glass-ceramic; joining

1. Introduction

The high temperature solid oxide electrolysis cell (SOEC) technology is considered as one of the most attractive method to produce green hydrogen because of its high efficiency [1–5]. In order to maximize the efficiency, the role of sealants is crucial in planar SOEC stack configuration. Sealants are employed with the aim to avoid any gas leakage and to bond metallic interconnect with the zirconia-based electrolyte. The operating conditions of an SOEC stack such as high working temperature of 700–1000 °C, presence of oxidizing and reducing atmospheres, expected lifetime of more than 40,000 h, and applied thermal cycles during operation, make the synthesis of suitable sealants quite challenging [6–8].

Glass-ceramics are considered to be the most promising sealant candidates owing to their high electrical resistivity, thermal and chemical stability at high working temperature, and due to the possibility to tune their physical and chemical properties (i.e., viscosity, coefficient of thermal expansion) by tailoring their composition [9–12]. In the past, various glass compositions have been designed and tested mainly for solid oxide fuel cell (SOFC). Although, most of the requirements for an

SOEC sealant are similar to that of SOFC, however, in SOEC the sealant has to be electrically insulator under the applied voltage which is typically 1.3 V higher than in SOFC [8,13–15].

According to literature, silica is most commonly used as a main glass former. Besides that, different alkali and alkaline earth metals are used as modifiers to tune the properties of the resultant glass-ceramic [16–20]. The choice and relative concentration of modifiers is the most important factor in order to control the overall properties of glass-ceramic sealants. Alkali oxide-based glass-ceramic sealants are usually not very attractive because of their low electrical resistivity and detrimental reaction with Cr at high temperature and under applied voltage [21,22]. Among alkaline earth metal oxides, BaO is the most extensively used modifier as it improves the coefficient of thermal expansion (CTE) and wettability of the glass-ceramic sealant [6,23–27]. However, BaO-based glasses mostly suffer from the formation of high CTE $BaCrO_4$ phase because of chemical reaction between Ba (from glass) and Cr (from the interconnect). The high CTE phase having CTE in the range of 20–22 $\times 10^{-6}$ K^{-1}, often resulted in stress generation and consequently delamination at interconnect/sealant interface [28,29]. For instance; a SiO_2-B_2O_3-BaO-based glass system investigated by Peng et al. [29] showed strong adhesion with SS410 interface and 8YSZ (8 mol% Yttria-stabilized Zirconia) electrolyte after joining with no undesirable phase formation. However, under the thermal cycling performed at 700 °C, a gas leakage was detected. SEM-EDS post mortem analyses confirmed that the formation of undesirable phase $BaCrO_4$ took place at glass-ceramic/interconnect interface.

SiO_2-BaO-CaO-Al_2O_3-based glass systems developed by Gosh et al. [30], also showed good chemical stability and high electrical resistivity. However, the electrical resistivity was measured on glass-ceramic pellets at 800 °C up to only 100 h of steady operation. Similarly, Rost et al. [13] also investigated the electrical resistivity of different BaO-based glass systems in dual atmosphere at 850 °C up to 500 h. However, no information was reported about the long-term testing (>1000 h) and performance during thermal cycles. Schilm et al. [31] also designed different glass compositions from SiO_2-BaO-CaO-Al_2O_3 system and their electrical resistivity was measured in dual atmosphere for 300 h under the applied voltage of 0.7 V and 1.3 V. The compatibility of glasses was studied with CFY (Cr-Fe-Y) interconnect and found that a suitable BaO/SiO_2 is important in order to minimize the formation of $BaCrO_4$ phase.

In this study, a Ba-based glass-ceramic sealant was designed and studied under SOEC conditions. Besides thermal, chemical, and thermo-mechanical analysis, the electrical resistivity of joined samples was analyzed at 850 °C up to 2000 h in dual (H_2O-H_2 and air) atmosphere. Thermal cycles were also applied during resistivity analysis in order to simulate the dynamic SOEC conditions.

2. Experimental

The composition (mol%) of the glass system (further labelled as HJ28) is shown in Table 1. The glass system has SiO_2 as glass former, while BaO as a main glass modifier. According to SiO_2-BaO binary phase diagram, the SiO_2/(SiO_2 + BaO) of 0.67 is required to obtain a $BaSi_2O_5$ phase [32]. The $BaSi_2O_5$ phase has a CTE of 12–14 $\times 10^{-6}$ K^{-1} [33], therefore, is beneficial to obtain a glass-ceramic sealant with a suitable CTE for SOEC applications. Nevertheless, the presence of high concentration of Ba, can also lead to the formation of high CTE $BaCrO_4$ phase. In HJ28 glass system, a SiO_2/(SiO_2 + BaO) of 0.73 was chosen in order to form high CTE $BaSi_2O_5$ phase and to avoid the formation of $BaCrO_4$ phase. Besides that, CaO, B_2O_3, and Y_2O_3 were added in order to adjust the viscosity and the CTE of the sealant.

Table 1. Composition (mol%) of HJ28 glass system.

SiO_2	BaO	CaO	B_2O_3	Al_2O_3	Y_2O_3
60.00	22.00	6.35	7.65	3.00	1.00

For glass synthesis, different precursor powders were thoroughly mixed for 24 h and subsequently melted at 1600 °C for one hour in a Rh-Pt crucible. The molten glass was then air quenched on a brass plate followed by ball milling and sieving to obtain glass particles <25 μm. The glass transition (Tg) and crystallization temperature of as-casted glass were measured using differential thermal analysis (DTA Netzsch, Eos, Selb, Germany), while the sintering behavior was analyzed by heating stage microscope (HSM, Hesse Instruments, Osterode am Harz, Germany) at a heating rate of 5 °C/min. The coefficient of thermal expansion of as-casted glass and as-joined glass-ceramic was measured by dilatometer (Netzsch, DIL 402 PC/4). The details about the joining parameters of glass-ceramic will be discussed in the next sections of this article. For each characterization method, at least three measurements were carried out in order to ensure the reproducibility of data.

For phase analysis, the XRD (PanAlytical X'Pert Pro PW 3040/60 Philips (The Netherlands)) was performed on glass-ceramic with Cu Kα and the X'Pert software. In order to investigate the compatibility of HJ28 glass-ceramic with the Crofer22APU (from VDM® Metals) interconnect and 3YSZ electrolyte, the Crofer/glass/3YSZ joint was processed in static air in a carbolite furnace (CWF 13/5). For this purpose, the slurry composed of glass particles and ethanol (70:30 wt%) was manually deposited on the Crofer22APU and 3YSZ substrates. The SEM-EDS analysis was carried out to understand the compatibility and adhesion of glass-ceramic with Crofer22APU interconnect and 3YSZ electrolyte, and to analyze the glass-ceramic microstructure.

The electrical resistivity of the Crofer22APU/glass-ceramic/Crofer22APU joint was measured in-situ at 850 °C, under dual atmosphere and applied voltage. The glass was deposited in form of slurry on a cleaned plate of Crofer22APU (3 cm × 6 cm × 0.2 cm) to form a closed sealing frame Figure 1a) and joined to a second Crofer22APU plate of the same size. The lower plate has two holes to allow the inlet and outlet of a controlled reducing atmosphere during the experiment, while the external side of the glass sealing was exposed to static air (Figure 1b). Thermiculite™ 866 (Flexitallic, Ticengo, Italy) sealing gaskets were employed to ensure gas tightness between the lower plate of joined sample and the gas distribution/collection fixture in the furnace. A mixture of 50 mol% hydrogen and 50 mol% steam was sent to the joint sample during the experiment. Uniformly distributed weight was put on the top plate to facilitate the sintering and the adhesion of the sealant. The joining treatment described in Section 3.1 was applied before settling the temperature to 850 °C and exposing the sample to the dual atmosphere. A voltage of 1.6 V was applied between upper and lower plates, connected to a voltage generator and a measuring circuit Figure 1b by platinum wires point welded on each plate. Gas flow measurement at the reducing atmosphere outlet was periodically performed during the test to verify the sealing integrity between test fixture and lower plate and of the joint sample.

The resistivity of the joint sample—R_S—was indirectly evaluated by measuring the voltage drop—V_m—on a known resistance R_m that was put in an electrical circuit in series with the Crofer22APU/glass-ceramic/Crofer22APU (see Figure 1b). The voltage generator was regulated during the experiment in order to have a voltage of 1.6 V on the sample, i.e., the sum of V and V_m was maintained equal to 1.6 V. By solving the circuit, the resistance R_S is calculated as: $R_S = R_m(V-V_m)/V_m$, and consequently the resistivity is derived from the geometry (area and thickness) of the sealing frame.

In order to simulate the real working conditions of SOEC, three thermal cycles were performed after 500 h of resistivity test, in which the Crofer22APU/HJ28 glass-ceramic/Crofer22APU-joined sample was cooled down from 850 °C to room temperature at 2 °C/min.

After the electrical resistivity test, the SEM-EDS post mortem analysis was carried out to observe the microstructure, porosity in addition to chemical and thermo-mechanical compatibility of HJ28 glass-ceramic sealant with Crofer22APU.

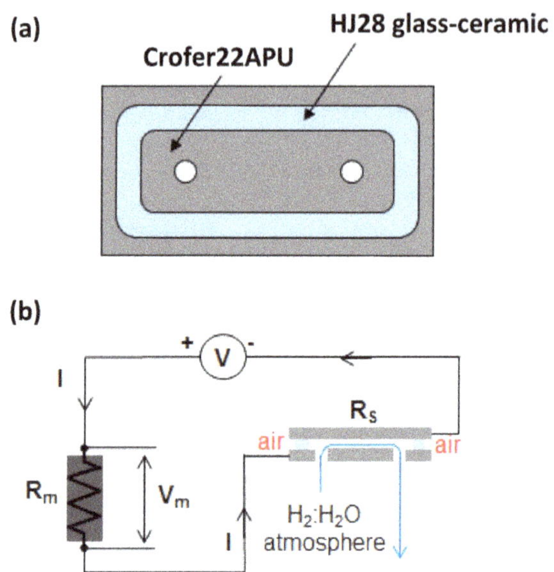

Figure 1. (**a**) Schematic of sample and (**b**) measuring setup, for the electrical resistivity analysis in dual atmosphere.

3. Results and Discussion

3.1. Thermal Analysis

The DTA and HSM curves for the HJ28 as-casted glass are shown in Figure 2. The glass transition (T_g) and peak crystallization temperature (T_p) obtained from DTA, as well as the first sintering temperature (T_{FS}) and maximum sintering temperature (T_{MS}) obtained from HSM analyses are given in Table 2.

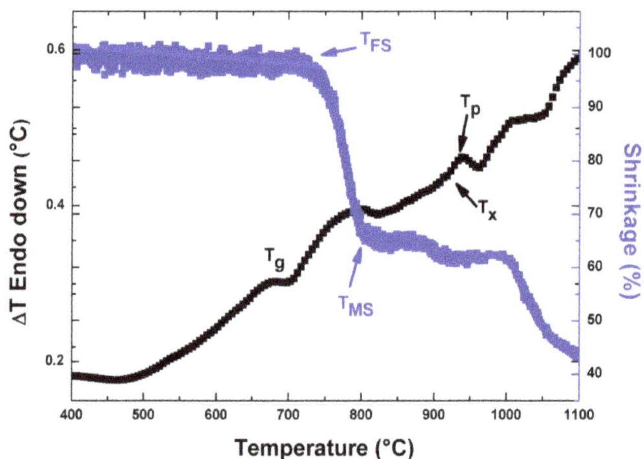

Figure 2. Differential thermal analysis (DTA) and heating stage microscope (HSM) curves of HJ28 as-casted glass. Analyses were carried out at 5 °C/min.

Table 2. Characteristic temperatures and coefficient of thermal expansion (CTE) of HJ28 glass system, obtained from DTA, HSM, and dilatometer.

Glass transition temperature T_g (°C)	685 ± 2
Onset of crystallization (°C)	907 ± 2
Peak crystallization temperature T_p (°C)	930 ± 3
First shrinkage temperature T_{FS} (°C)	717 ± 2
Maximum shrinkage temperature T_{MS} (°C)	829 ± 1
CTE of as-cast glass/1 × 10^{-6} K^{-1} (200 °C–500 °C)	8.5 ± 0.1
CTE of as-joined glass-ceramic/1 × 10^{-6} K^{-1} (200 °C–500 °C)	9.4 ± 0.1

The DTA analysis of the HJ28 glass (Figure 2) showed that the as-casted glass has T_g of 685 °C. A low intensity exothermal peak can be seen around 930 °C, labelled as T_p. The absence of sharp crystallization peak during DTA analysis of glass has been observed in our previous studies [8], and is most likely due to the fact that the degree of devitrification in the HJ28 glass is not significant to be detected. On the other hand, the HSM analysis of HJ28 (Figure 2 and Table 2) showed that the sintering process started at 717 °C, while the maximum sintering was obtained at 829 °C. After the completion of sintering process, the HJ28 glass showed a viscous flow that resulted in slightly further shrinkage.

In order to obtain a dense sealant, the sintering should be complete before the start of crystallization phenomena, otherwise, the increased viscosity due to crystallization can lead to the formation of residual pores within the glass-ceramic [18]. From the Figure 2 and characteristic temperatures mentioned in the Table 2, it is clear that the sintering process of the HJ28 glass system completed prior to the crystallization.

In order to synthesize the HJ28 glass-ceramic, a heat treatment of 950 °C having dwell time of 2 h, was chosen. The selection of heat treatment was based on the crystallization and sintering data obtained from the DTA and HSM analyses, and was chosen to obtain a maximum densification and sufficient crystallization within the resultant glass-ceramic. For glass-ceramic synthesis, the heating/cooling rate of 2 °C/min was used in order to minimize the possibilities of stress generation due to fast heating/cooling rates.

The CTEs of as-casted glass and as-joined glass-ceramic are also reported in Table 2. The as-casted HJ28 glass and as-joined glass-ceramic showed the CTEs of 8.5 × 10^{-6} K^{-1} and 9.4 × 10^{-6} K^{-1} respectively, in the temperature range of 200–500 °C. The increase in the CTE of glass-ceramic as compared with the parent glass is due to the formation of high CTE crystalline phases. Further details about the crystalline phases will be discussed in the later sections of this article. Nevertheless, the CTE of HJ28 glass-ceramic is matching with CTEs of other cell components i.e., Crofer22APU (12 × 10^{-6} K^{-1}) and 3YSZ (10.5 × 10^{-6} K^{-1}), [7,31] and is suitable for the SOEC applications.

3.2. XRD and Microstructural Analysis

The XRD pattern of the as-joined HJ28 glass-ceramic is shown in Figure 3. According to the XRD analysis, the as-joined HJ28 glass-ceramic contains only $BaSi_2O_5$ crystalline phase. The reference pattern of $BaSi_2O_5$ (PDF card # 00-026-0176) is also shown in Figure 3. The $BaSi_2O_5$ phase has CTE in the range of 12–14 × 10^{-6} K^{-1} [33] and is important to obtain a high CTE (9–12 × 10^{-6} K^{-1}) glass-ceramic sealant, for the SOEC applications [7]. These XRD results also validate the rationale behind designing the HJ28 glass composition, i.e., to have a high CTE $BaSi_2O_5$ phase and to avoid the formation of cristobalite (SiO_2) phase. Moreover, the XRD analyses are also in agreement with the dilatometer results, indicating that the formation of high CTE $BaSi_2O_5$ is responsible for the increase in CTE of HJ28 glass-ceramic as compared with as-casted glass.

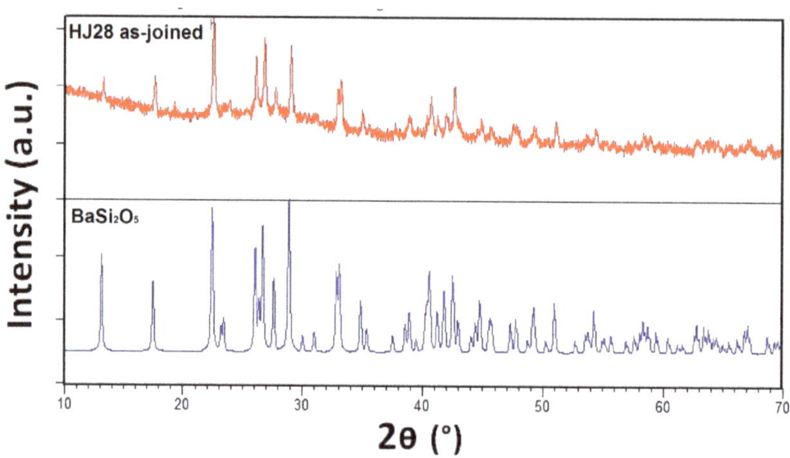

Figure 3. XRD pattern of HJ28 glass-ceramic synthesized at 950 °C for 2 h at heating rate of 2 °C/min.

The compatibility and bonding of the HJ28 glass-ceramic with the Crofer22APU interconnect and 3YSZ electrolyte were investigated by producing the Crofer22APU/glass-ceramic/3YSZ joined samples according to the heat treatments mentioned above. The SEM image of the Crofer22APU/HJ28 glass-ceramic/3YSZ joint cross section is shown in Figure 4. The HJ28 glass-ceramic showed good interfacial bonding with the Crofer22APU and 3YSZ substrates, with no crack or delamination at either interface. Some isolate pores can be seen in the as-joined glass-ceramics; however, these pores are not interconnected and could be formed as a result of manual glass deposition. Moreover, no crack within the as-joined glass-ceramics was observed.

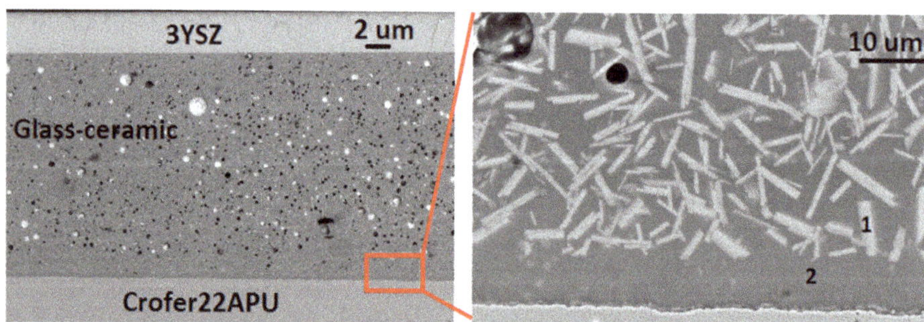

Figure 4. SEM images of as-joined Crofer22APU/HJ28/3YSZ joined samples.

The microstructure of the HJ28 glass-ceramic after joining is quite homogenous. The crystalline phases are uniformly distributed throughout the joining area. The EDS analysis reported in Table 3 confirmed that the bright phase (point 1) in the HJ28 as-joined glass-ceramic is the $BaSi_2O_5$ phase, while the dark phase (phase 2) is the residual glassy phase. The presence of 10 wt% of Ba and 3.7 wt% of Ca into the residual glassy phase is beneficial to maintain the CTE and viscosity of residual glass. The SEM-EDS analyses performed on the as-joined HJ28 glass-ceramics are in agreement with the XRD results.

Table 3. EDS point analyses (at. %) performed on the HJ28 glass-ceramics as shown in Figure 4.

Elements	Point 1	Point 2
O	48.0	49.7
Si	35.2	31.6
Ba	15.1	10.1
Ca	1.7	3.7
Al	–	4.3
Y	—	0.6

3.3. Electrical Characterization in Dual Atmosphere and Post Mortem Analysis

Figure 5 shows the electrical resistivity curve for the Crofer22APU/HJ28 glass-ceramic/Crofer22APU joined sample, as measured at 850 °C, 1.6 V and under dual atmosphere. The electrical resistivity for the Crofer22APU/HJ28 glass-ceramic/Crofer22APU joined sample was recorded in the range of 10^6–10^7 Ω cm thus higher than the minimum threshold (10^4 Ω cm) required to ensure an insulation between the two conducting Crofer22APU plates [34]. The electrical resistivity curve is uniform with small fluctuations. After 500 h, a significant increase in the electrical resistivity is due to the applied thermal cycles. Resistivity peaks occurred during cooling down phases of thermal cycles, as expected from the higher glass-ceramic resistivity at lower temperature. However, after the thermal cycles the resistivity achieved the same values as prior to the thermal cycles. The smooth and uniform resistivity curve demonstrates very good insulating characteristics of the glass-ceramic sealant. Furthermore, these values also suggest that neither corrosion phenomena nor any chemical interaction took place between the HJ28 sealant and the Crofer22APU interconnects.

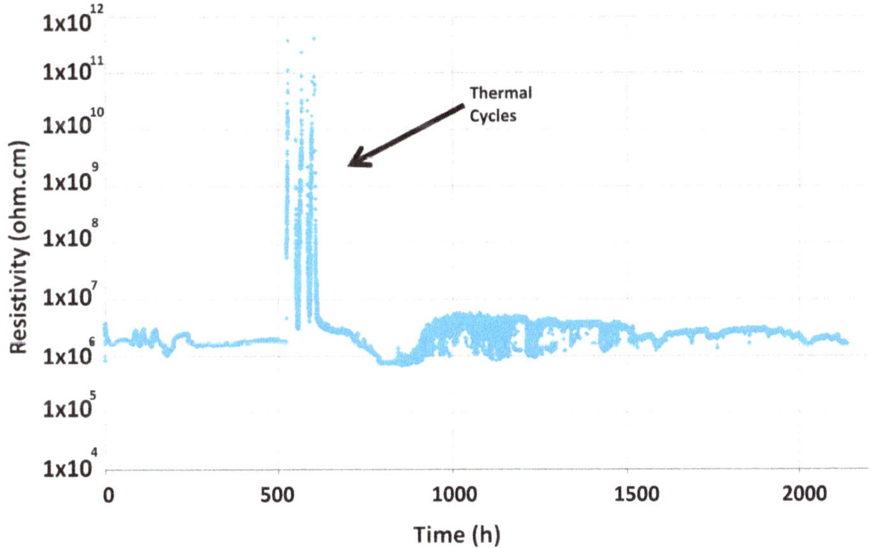

Figure 5. Electrical resistivity of Crofer22APU/HJ28 glass-ceramic/Crofer22APU joined sample, as measured for 2000 h at 850 °C under the applied voltage of 1.6 V.

Figure 6 shows the SEM-EDS post mortem analysis carried out at cathodic polarized Crofer22APU/HJ28 glass-ceramic interface at the air side after the long-term electrical resistivity test in dual atmosphere. The glass-ceramic seems to be highly dense with negligible amount of closed porosity. The formation of porosity is most likely due to the manual deposition of the glass. The glass-ceramic shows good adhesion with the cathodic polarized Crofer22APU substrate. From the

EDS-mapping shown in Figure 6, the presence of Cr rich region can be seen at glass-ceramic/air interface. These Cr containing phase corresponds to $BaCrO_4$, which was formed most likely because of the chemical reaction between the Ba from glass-ceramic and Cr from Crofer22APU. The formation of high CTE $BaCrO_4$ is commonly observed in most of the Ba-based glass, and can lead to delamination at Crofer22APU/glass-ceramic interface [25,28,29,35,36]. However, in case of HJ28 glass-ceramic, the chromates were only limited along the air side and no chromates were formed at HJ28 glass-ceramic/Crofer22APU interface, thus ensured strong bonding. This effect was due to excellent adhesion of the sealant to the Crofer22APU.

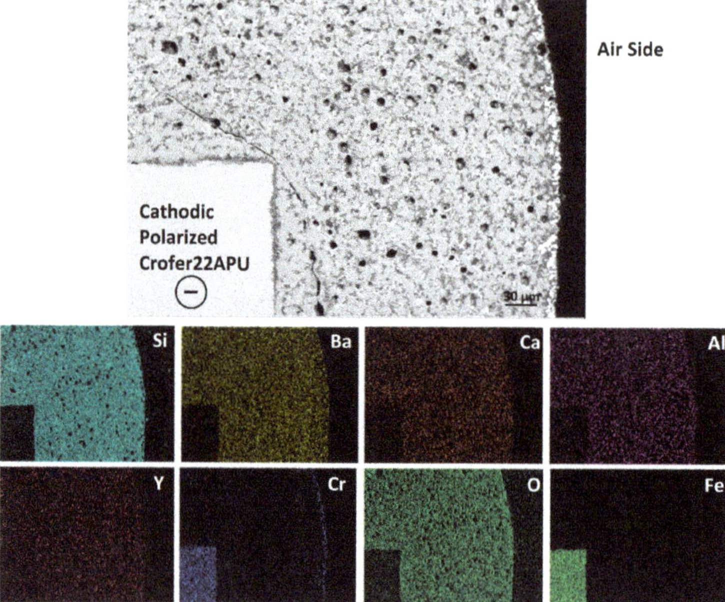

Figure 6. SEM-EDS post mortem analysis at cathodic polarized Crofer22APU/HJ28 glass-ceramic interface after the dual test, collected at air side.

The EDS analysis confirmed that there is no segregation of elements within the glass-ceramic, nor at Crofer22APU/glass-ceramic interface.

Figure 7 corresponds to the SEM-EDS analysis at the anodic polarized Crofer22APU/HJ28 glass-ceramic interface collected at the air side. The EDS analysis and uniform microstructure confirmed that under the applied voltage of 1.6 V no migration of ions has been observed toward any specific polarity. Similar to cathodic polarized Crofer22APU, a thin layer (−10 μm) of $BaCrO_4$ can be seen in Figure 7 and is localized only along the air side. A suitable $SiO_2/(SiO_2 + BaO)$ in HJ28 glass system not only resulted in the formation of the desired crystalline phase, but also minimal $BaCrO_4$ formation. Nevertheless, Figure 7 shows a negligible concentration of Si rich phases between the chromates and the Crofer22APU substrate, formed because of the fact that the formation of $BaCrO_4$ slightly imbalanced SiO_2/BaO thus, resulting in locally higher SiO_2 content.

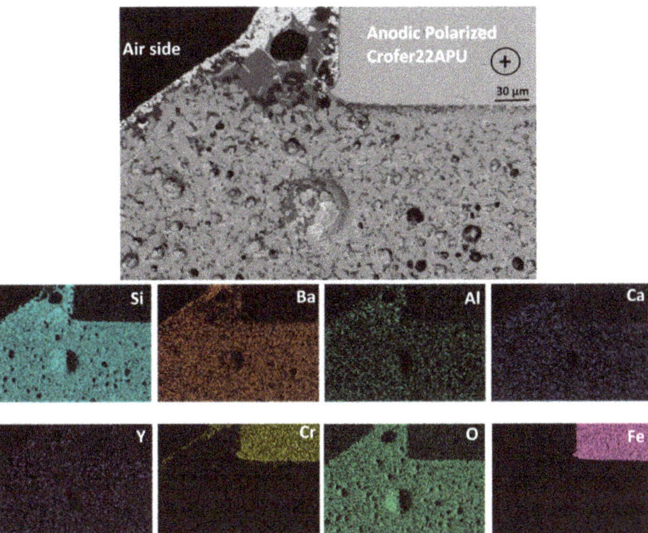

Figure 7. SEM-EDS post mortem analysis at anodic polarized Crofer22APU/HJ28 glass-ceramic interface after the dual test, collected at air side.

In a previous study, Sabato et al. [21] observed the formation of detrimental reaction between an alkali-containing glass-ceramic and Crofer22APU in similar operating conditions (dual atmosphere, 800 °C under applied voltage). This study highlighted the strong importance of the region exposed to air and the role played by the applied voltage in the presence of alkali metal oxide in the glass. These elements can react with Cr leading to degradation of the integrity of the sealant and formation of Cr_2O_3 bridges with consequent reduction in the resistivity. The present study did not detect any evidence of a similar reaction owing to the absence of alkali metal oxides. The sealant appears to be intact and no evidence of conductive "bridges" have been detected by SEM. The resistivity is also stable during all the measurement without noticeably decreasing.

Figure 8 shows the SEM post mortem analyses carried out at cathodic and anodic polarized Crofer22APU/glass-ceramic interface, on the fuel side. Likewise on air side, the HJ28 glass-ceramic seems to be strongly bonded to both the Crofer22APU plates having opposite polarity. No cracks were found within the HJ28 glass-ceramic nor at Crofer22APU/HJ28 glass-ceramic interface. The strong and crack-free interface also confirmed that the HJ28-glass-ceramic is stable under the applied thermal cycles. Moreover, the microstructure and level of closed porosity at the fuel side is also similar to that of the air side.

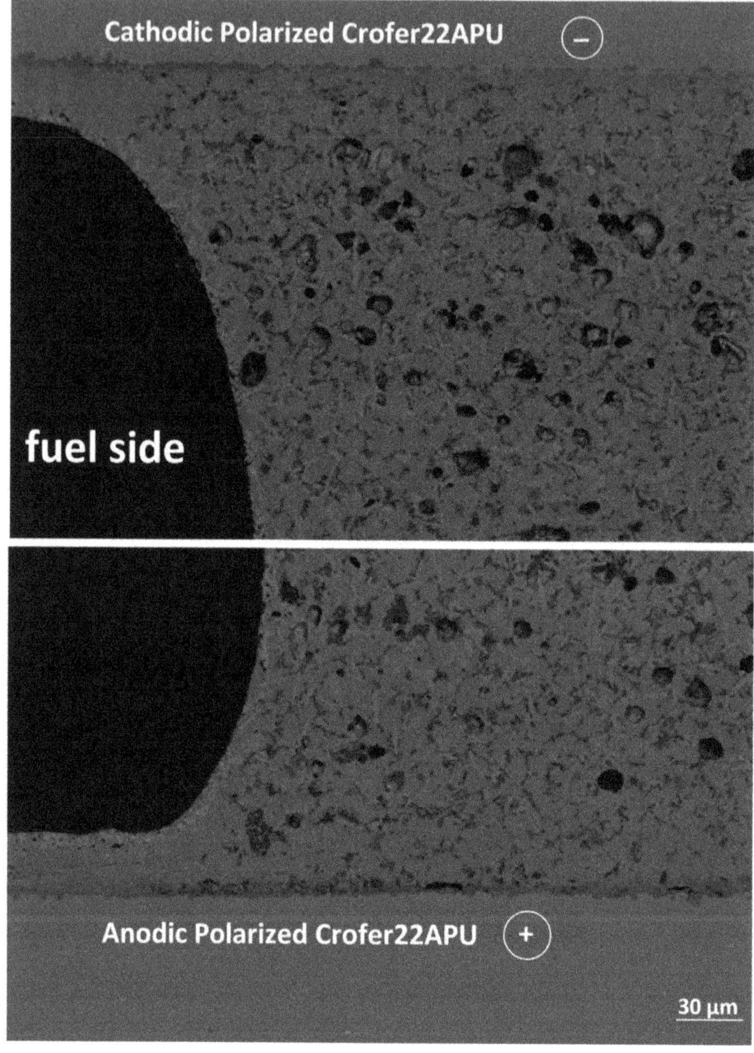

Figure 8. SEM-EDS post mortem analysis at cathodic and anodic polarized Crofer22APU/HJ28 glass-ceramic interface after the dual test, collected at fuel side.

4. Conclusions

The Ba-based glass system (HJ28) has been designed and tested for SOEC applications having a working temperature of 850 °C. The HJ28 glass-ceramic showed suitable CTE and excellent compatibility with Crofer22APU interconnect and 3YSZ electrolyte. The right sequence of crystallization and sintering processes led to the formation of dense sealant. The electrical resistivity of 10^6–10^7 Ω cm was measured for Crofer22APU/HJ28 glass-ceramic/Crofer22APU joint, as measured for 2000 h, at 850 °C and in simultaneously applied oxidizing and reducing atmospheres. Three full thermal cycles were also applied during the test at 500 h. The SEM-EDS post mortem analysis confirmed the presence of strong bonding between the Crofer22APU substrate and HJ28 glass-ceramic, subjected to long-term resistivity test and thermal cycles. A very thin layer (>10 µm) of $BaCrO_4$ was observed along

Crofer22APU/glass-ceramic/air 3-phase boundary, while no evidence of $BaCrO_4$ propagation were found along Crofer22APU/glass-ceramic interface.

Owning to excellent thermal and thermo-mechanical properties, high electrical resistivity, and chemical stability, HJ28 glass-ceramic is a promising candidate to act as a reliable sealant for long-term SOEC conditions.

Author Contributions: H.J. gave the main idea regarding glass composition and wrote the manuscript. A.G.S. conducted the thermal and SEM analysis. M.M. and D.F. performed the electrical resistivity analysis in dual atmosphere. M.S. and K.H. helped in preparing the manuscript. C.W. and F.S. supervised the whole work and helped in interpretation of data. All authors have read and agreed to the published version of the manuscript.

Funding: This research received no external funding.

Conflicts of Interest: The authors declare no conflict of interest.

References

1. Pandiyan, A.; Uthayakumar, A.; Subrayan, R.; Cha, S.W.; Krishna Moorthy, S.B. Review of solid oxide electrolysis cells: A clean energy strategy for hydrogen generation. *Nanomater. Energy* **2019**, *8*, 2–22. [CrossRef]
2. Posdziech, O.; Schwarze, K.; Brabandt, J. Efficient hydrogen production for industry and electricity storage via high-temperature electrolysis. *Int. J. Hydrog. Energy* **2019**, *44*, 19089–19101. [CrossRef]
3. Kazempoor, P.; Braun, R.J. Hydrogen and synthetic fuel production using high temperature solid oxide electrolysis cells (SOECs). *Int. J. Hydrog. Energy* **2015**, *40*, 3599–3612. [CrossRef]
4. Yan, J.; Chen, H.; Dogdibegovic, E.; Stevenson, J.W.; Cheng, M.; Zhou, X.D. High-efficiency intermediate temperature solid oxide electrolyzer cells for the conversion of carbon dioxide to fuels. *J. Power Sources* **2014**, *252*, 79–84. [CrossRef]
5. Grigoriev, S.A.; Fateev, V.N.; Bessarabov, D.G.; Millet, P. Current status, research trends, and challenges in water electrolysis science and technology. *Int. J. Hydrog. Energy* **2020**. [CrossRef]
6. Kothiyal, G.P.; Kothiyal, G.P.; Goswami, M.; Tiwari, B.; Sharma, K.; Ananthanarayanan, A.; Montagne, L. Some recent studies on glass/glass-ceramics for use as sealants with special emphasis for high temperature applications. *J. Adv. Ceram.* **2012**, *1*, 110–129. [CrossRef]
7. Fergus, J.W. Sealants for solid oxide fuel cells. *J. Power Sources* **2005**, *147*, 46–57. [CrossRef]
8. Javed, H.; Sabato, A.G.; Herbrig, K.; Ferrero, D.; Walter, C.; Salvo, M.; Smeacetto, F. Design and characterization of novel glass-ceramic sealants for solid oxide electrolysis cell (SOEC) applications. *Int. J. Appl. Ceram. Technol.* **2018**, *15*, 999–1010. [CrossRef]
9. Javed, H.; Sabato, A.G.; Dlouhy, I.; Halasova, M.; Bernardo, E.; Salvo, M.; Herbrig, K.; Walter, C.; Smeacetto, F. Shear Performance at Room and High Temperatures of Glass–Ceramic Sealants for Solid Oxide Electrolysis Cell Technology. *Materials* **2019**, *12*, 298. [CrossRef]
10. Khedim, H.; Nonnet, H.; Méar, F.O. Development and characterization of glass-ceramic sealants in the (CaO-Al_2O_3-SiO_2-B_2O_3) system for Solid Oxide Electrolyzer Cells. *J. Power Sources* **2012**, *216*, 227–236. [CrossRef]
11. Ghosh, S.; Das Sharma, A.; Kundu, P.; Basu, R.N. Glass-ceramic sealants for planar IT-SOFC: A bilayered approach for joining electrolyte and metallic interconnect. *J. Electrochem. Soc.* **2008**, *155*, B473–B478. [CrossRef]
12. Fakouri Hasanabadi, M.; Faghihi-Sani, M.A.; Kokabi, A.H.; Groß-Barsnick, S.M.; Malzbender, J. Room- and high-temperature flexural strength of a stable solid oxide fuel/electrolysis cell sealing material. *Ceram. Int.* **2019**, *45*, 733–739. [CrossRef]
13. Rost, A.; Schilm, J.; Kusnezoff, M.; Michaelis, A. Degradation of sealing glasses under electrical load. *Eur. Fuel Cell Forum* **2010**, *80*, 1–12.
14. Tulyaganov, D.U.; Reddy, A.A.; Kharton, V.V.; Ferreira, J.M.F. Aluminosilicate-based sealants for SOFCs and other electrochemical applications - A brief review. *J. Power Sources* **2013**, *242*, 486–502. [CrossRef]
15. Ferrero, D.; Lanzini, A.; Leone, P.; Santarelli, M. Dynamic reversible SOC applications: Performance and Durability with simulated load/demand profiles. In Proceedings of the 11th European SOFC and SOE Forum 2014, Lucerne, Switzerland, 1–4 July 2014; Chapter 17, Session B14. pp. 60–67.

16. Elsayed, H.; Javed, H.; Sabato, A.G.; Smeacetto, F.; Bernardo, E. Novel Glass-ceramic SOFC Sealants from Glass Powders and a Reactive Silicone Binder. *J. Eur. Ceram. Soc.* **2018**, *38*, 4245–4251. [CrossRef]
17. Smeacetto, F.; Salvo, M.; Santarelli, M.; Leone, P.; Ortigoza-Villalba, G.A.; Lanzini, A.; Ajitdoss, L.C.; Ferrarisa, M. Performance of a glass-ceramic sealant in a SOFC short stack. *Int. J. Hydrog. Energy* **2013**, *38*, 588–596. [CrossRef]
18. Sabato, A.G.; Salvo, M.; Santarelli, M.; Leone, P.; Ortigoza-Villalba, G.A.; Lanzini, A.; Ajitdoss, L.C.; Ferraris, M. Glass-ceramic sealant for solid oxide fuel cells application: Characterization and performance in dual atmosphere. *J. Power Sources* **2016**, *328*, 262–270. [CrossRef]
19. Smeacetto, F.; Salvo, M.; Leone, P.; Santarelli, M.; Ferraris, M. Performance and testing of joined Crofer22APU-glass-ceramic sealant-anode supported cell in SOFC relevant conditions. *Mater. Lett.* **2011**, *65*, 1048–1052. [CrossRef]
20. Reddy, A.A.; Tulyaganov, D.U.; Goel, A.; Pascual, M.J.; Kharton, V.V.; Tsipis, E.V.; Ferreira, J.M. Diopside–Mg orthosilicate and diopside–Ba disilicate glass-ceramics for sealing applications in SOFC: Sintering and chemical interactions studies. *Int. J. Hydrog. Energy* **2012**, *37*, 12528–12539. [CrossRef]
21. Sabato, A.G.; Rost, A.; Schilm, J.; Kusnezoff, M.; Salvo, M.; Chrysanthou, A.; Smeacetto, F. Effect of electric load and dual atmosphere on the properties of an alkali containing diopside-based glass sealant for solid oxide cells. *J. Power Sources* **2019**, *415*, 15–24. [CrossRef]
22. Chou, Y.S.; Stevenson, J.W.; Choi, J.P. Alkali Effect on the Electrical Stability of a Solid Oxide Fuel Cell Sealing Glass. *J. Electrochem. Soc.* **2010**, *157*, B348–B353. [CrossRef]
23. Arora, A.; Singh, K.; Pandey, O.P. Thermal, structural and crystallization kinetics of SiO_2–BaO–ZnO–B_2O_3–Al_2O_3 glass samples as a sealant for SOFC. *Int. J. Hydrog. Energy* **2011**, *36*, 14948–14955. [CrossRef]
24. Lin, S.E.; Cheng, Y.R.; Wei, W.C.J. BaO–B_2O_3–SiO_2–Al_2O_3 sealing glass for intermediate temperature solid oxide fuel cell. *J.Non Cryst. Solids* **2012**, *358*, 174–181. [CrossRef]
25. Laorodphan, N.; Ayawanna, J. BaO-Al_2O_3-SiO_2-B_2O_3 Glass-Ceramic SOFCs Sealant: Effect of ZnO Additive. *Key Eng. Mater.* **2017**, *751*, 455–460. [CrossRef]
26. Ferraris, M.; De la Pierre, S.; Sabato, A.G.; Smeacetto, F.; Javed, H.; Walter, C.; Malzbender, J. Torsional shear strength behavior of advanced glass-ceramic sealants for SOFC/SOEC applications. *J. Eur. Ceram. Soc.* **2020**, *40*, 4067–4075. [CrossRef]
27. Lee, H.; Kim, U.S.; Kim, S.D.; Woo, S.K.; Chung, W.J. SiO_2–B_2O_3-BaO-WO_3 glasses with varying Al_2O_3 content as a sealing material for reversible solid oxide fuel cells. *Ceram. Int.* **2020**, *46*, 18256–18261. [CrossRef]
28. Pascual, M.J.; Guillet, A.; Durán, A. Optimization of glass-ceramic sealant compositions in the system MgO-BaO-$SiO_2$for solid oxide fuel cells (SOFC). *J. Power Sources* **2007**, *169*, 40–46. [CrossRef]
29. Peng, L.; Zhu, Q. Thermal cycle stability of BaO-B_2O_3-SiO_2 sealing glass. *J. Power Sources* **2009**, *194*, 880–885. [CrossRef]
30. Ghosh, S.; Das Sharma, A.; Kundu, P.; Mahanty, S.; Basu, R.N. Development and characterizations of BaO-CaO-Al_2O_3-SiO_2 glass-ceramic sealants for intermediate temperature solid oxide fuel cell application. *J. Non Cryst. Solids* **2008**, *354*, 4081–4088. [CrossRef]
31. Schilm, J.; Rost, A.; Kusnezoff, M.; Megel, S.; Michaelis, A. Glass ceramics sealants for SOFC interconnects based on a high chromium sinter alloy. *I. J. Appl. Ceram. Technol.* **2018**, *15*, 239–254. [CrossRef]
32. Zhang, R.; Taskinen, P. Experimental investigation of liquidus and phase stability in the BaO-SiO_2 binary system. *J. Alloy. Compd.* **2016**, *657*, 770–776. [CrossRef]
33. Kerstan, M.; Rüssel, C. Barium silicates as high thermal expansion seals for solid oxide fuel cells studied by high-temperature X-ray diffraction (HT-XRD). *J. Power Sources* **2011**, *196*, 7578–7584. [CrossRef]
34. Mahapatra, M.K.; Lu, K. Seal glass for solid oxide fuel cells. *J. Power Sources* **2010**, *195*, 7129–7139. [CrossRef]
35. Kim, S.J.; Kim, K.J.; Choi, G.M. A novel solid oxide electrolysis cell (SOEC) to separate anodic from cathodic polarization under high electrolysis current. *Int. J. Hydrog. Energy* **2013**, *38*, 20–28. [CrossRef]
36. Kerstan, M.; Mueller, M.; Ruessel, C. Binary, ternary and quaternary silicates of CaO, BaO and ZnO in high thermal expansion seals for solid oxide fuel cells studied by high-temperature X-ray diffraction (HT-XRD). *Mater. Res. Bull.* **2011**, *46*, 2456–2463. [CrossRef]

Article

2D Simulation for CH_4 Internal Reforming-SOFCs: An Approach to Study Performance Degradation and Optimization

Emilio Audasso, Fiammetta Rita Bianchi * and Barbara Bosio

Department of Civil, Chemical and Environmental Engineering, University of Genova, Via Opera Pia 15, 16145 Genova, Italy; emilio.audasso@gmail.com (E.A.); barbara.bosio@unige.it (B.B.)

* Correspondence: fiammettarita.bianchi@edu.unige.it

Received: 4 July 2020; Accepted: 7 August 2020; Published: 9 August 2020

Abstract: Solid oxide fuel cells (SOFCs) are a well-developed technology, mainly used for combined heat and power production. High operating temperatures and anodic Ni-based materials allow for direct reforming reactions of CH_4 and other light hydrocarbons inside the cell. This feature favors a wider use of SOFCs that otherwise would be limited by the absence of a proper H_2 distribution network. This also permits the simplification of plant design avoiding additional units for upstream syngas production. In this context, control and knowledge of how variables such as temperature and gas composition are distributed on the cell surface are important to ensure good long-lasting performance. The aim of this work is to present a 2D modeling tool able to simulate SOFC performance working with direct internal CH_4 reforming. Initially thermodynamic and kinetic approaches are compared in order to tune the model assuming a biogas as feed. Thanks to the introduction of a matrix of coefficients to represent the local distribution of reforming active sites, the model considers degradation/poisoning phenomena. The same approach is also used to identify an optimized catalyst distribution that allows reducing critical working conditions in terms of temperature gradient, thus facilitating long-term applications.

Keywords: CH_4 internal reforming; solid oxide fuel cell; 2D local control; cell design optimization; active site degradation

1. Introduction

Solid oxide fuel cells (SOFCs) are well known electrochemical devices suitable for a sustainable energy production in both high capacity power plant and residential use [1,2]. Current research on SOFCs focuses on (i) performance and durability improvement through characterization of new innovative materials [3], (ii) experimental detection of degradation mechanisms [4], and (iii) modeling tools for system simulation and control [5]. Nevertheless, a relevant limit to their wider spread on the energy market is the lack of a proper network for distribution of hydrogen, the main cell reactant, at low price [6]. Several existing processes are repeatedly improved in order to obtain a more efficient and sustainable H_2 production route. For instance, steam reforming of hydrocarbons, the main approach used at industrial level, is enhanced through the introduction of new reactor designs to obtain a higher quality of outlet syngas, such as catalytic membranes or sorption enhanced steam reforming [7,8]. Partial oxidation reaction has been introduced in different applications thanks to the need for lower working temperature and high fuel conversion obtained through the introduction of self-sustained electrochemical promoted catalysts [9]. Gasification of biomass/waste has also gained literature attention in the last decades, considering the wide diversity and the easy availability of feed [10]. Water electrolysis is also becoming a competitive application through the integration with renewable sources in order to reduce requested external power and operating cost [11].

An alternative way to overcome the issue of H_2 production and distribution is the possibility to directly use light hydrocarbons such as CH_4, CH_3OH, gasoline, diesel, syngas, or biogas as SOFC fuel [12–14]. Differently from low temperature fuel cells, SOFCs do not require a pure H_2 feeding stream since carbon compounds such as CO and CO_2 are not poisonous for the materials used. Coupled with high operating temperatures, this allows for the in-situ production of H_2 necessary in the electrochemical mechanisms through reforming reactions. The use of alternative fuel makes SOFC safer application due to highly flammable and volatile of pure H_2. There are three possible configurations based on the integration of cells with reforming processes [15]:

- External reforming (ER): SOFCs and reforming units are two distinct blocks. In the reforming reactor, H_2 is produced from light hydrocarbons and consequently fed to the fuel cell anode inlet;
- Indirect internal reforming (IIR): SOFCs and reforming units are two neighboring blocks in order to favor thermal exchanges;
- Direct internal reforming (DIR): the reforming takes places inside SOFCs using the Ni-based anode material as catalyst.

While the first one is the easiest to operate thanks to the lack of interactions between two units, the other cases are more efficient since the requested heat for the endothermic reforming reaction is provided by the exothermic electrochemical process. At the same time, since achieved usually by flowing excess air through the cathode, the cell cooling decreases because the reforming prevents an excessive temperature rise. Moreover, capital and operating costs are reduced without separated external units [16]. Compared to IIR, DIR configuration also allows further process optimization as the continuous consumption of H_2 due to electrochemical reactions enhances the hydrocarbon conversion and results in a more uniform H_2 distribution [17]. However, this design is characterized by larger temperature gradients and relevant carbon deposition [18]. The high anodic Ni content favors a faster reforming process compared to the electrochemical one. Thus, hydrocarbon conversion is usually complete within a small distance from the inlet, leading to initial severe cooling effects. Even if proper cell materials are used in order to reduce the mismatch among thermal expansion coefficients, the subsequent temperature gradient on the fuel cell plane may induce system failure [19]. Indeed, creep formation is favored using brittle ceramics. However, it also occurs in the metallic interconnects that are characterized by a thermal behavior strongly dependent on temperature [20]. Appropriate modifications of the anodic structure, such as the partial poisoning of active sites or doping processes [21], are under investigation to decrease the reforming rate. Another common problem is the carbon deposition that is catalyzed by the presence of Ni and results in decreasing the active sites. This phenomenon is reversible and can be minimized by increasing the inlet steam to carbon ratio (S/C ratio) without a high fuel dilution to avoid the reduction of electrical efficiency [22].

Another relevant issue is the presence of different pollutant compounds in the fuel stream that can degrade cell materials and highly decrease both reforming and electrochemical performance. As for their amount, the type of such pollutants highly depends on the fuel source and usually consists of sulfur-based and chlorine-based compounds [23]. If biogas is used as fuel, siloxanes may also represent relevant issues [24]. Among these possible compounds, H_2S is the most common poison for catalyst activity. The adsorption of S and the subsequent formation of secondary Ni-S phases can cause serious but reversible degradation of the cell performance [25], making the fuel pre-treatment fundamental. To reduce the possible damage, alternative sulfur tolerant materials, such as metal sulfides [26], are under investigation.

In a view of process improvement, all three configurations of SOFC-reforming integration are analyzed in literature using different levels of details. Such experimental and theoretical studies focus on both new industrial power generation systems based on SOFC technology and on its integration into existing plants. In different works, the simulation is usually performed considering ER or IIR units or, when DIR configurations are presented, simplifying the cell through 0D [27,28] and 1D [22,29] approaches. Although effective in the first feasibility analysis, such models disregard local effects, especially in terms of temperature and current density. This may lead one to contemplate possible solutions that could bring the cell to failure. A more detailed analysis is performed through 2D simulation. Considering a planar geometry, cross-section is commonly assumed as system domain to evaluate the main changes in chemical-physical features in flow direction and along cell thickness [18,30,31]. These studies guarantee a quite good overview of cell behavior in co- and counter-flow configurations, since flow channels can be approximated as plug flow model, whereas in the case of cross-flow design, the analysis should consider occurring gradients on cell plane section that is a less common assumption [32]. The complete knowledge of the system is reached only through a 3D approach [33,34]. This permits a more detailed modeling but requires long computational times to reach a solution penalizing its use. On the other hand, when a tubular cell is simulated, 2D modeling is sufficient to describe completely system behavior [35]. A further step consists of the analysis of occurring degradation and poisoning phenomena. In literature, they have been mostly investigated with experimental tests on both single cells [36] and stacks [37] as well as in terms of regeneration [38]. However, there are few modeling efforts that evaluate coking and H_2S poisoning influence on electrochemical performance based on empirical [39] or theoretical formulation [40].

This work presents a 2D simulation tool developed for DIR-SOFC in industrial applications using a biogas type fuel (mixture of CH_4, CO, CO_2, H_2, and H_2O). To present its features, the authors consider an anode-supported planar cell design in cross-flow configuration. The chemical-physical property changes are described through local balance resolution over the cell plane, whereas material transport phenomena along cell thickness are considered only for the electrochemical kinetics following a previously validated model for H_2-feeding SOFC [41,42]. For methane DIR process, the code considers two different approaches presented in literature: one based on the thermodynamic equilibrium of the reforming reaction and one based on its reaction kinetics. Subsequently, the authors introduce a simplified way to model cell degradation phenomena affecting mainly the reforming reaction. This approach is possible thanks to the introduction of a corrective term that allows reducing the effective active catalysts surface. This coefficient is implemented in matrix form to enable a local distribution and thus identify possible different levels of degradation in each single point of the cell. Following a similar approach, this matrix is also tuned to identify an optimized catalyst distribution in order to guarantee more uniform working temperatures and reactant utilizations. Despite the modeling complexity, the results are obtained with calculations lasting only few minutes.

2. CH_4 Internal Reforming Modeling

2.1. The Cell Reactions

At the cathodic side of an anionic conductive-electrolyte SOFC, O_2 reacts electrochemically to produce $O^=$ ions (Equation (1)). These migrate through the ceramic electrolyte to the anodic side where the oxidation of H_2 (Equation (2)) or CO (Equation (3)) completes the circuit with electrical energy production, as shown by the following reaction paths.

$$\frac{1}{2}O_2 + 2e^- \rightarrow O^= \quad \text{at the cathode} \tag{1}$$

$$H_2 + O^= \rightarrow H_2O + 2e^- \quad \text{at the anode} \tag{2}$$

$$CO + O^= \rightarrow CO_2 + 2e^- \quad \text{at the anode} \tag{3}$$

Coupled with Ni catalysts usually employed, the high operating temperature allows for the direct feeding of light hydrocarbons at the SOFC anode. This is possible thanks to internal reforming reactions that convert these hydrocarbons to the H_2 necessary for cell operation.

CH_4 reforming can be carried out using H_2O (steam reforming, SR) or CO_2 (dry reforming, DR) as co-reactants. In literature, both cases are largely investigated, and different reaction paths resulting in the following formulations (Equations (4) and (5) for SR, Equations (6) and (7) for DR) have been proposed.

$$CH_4 + H_2O \leftrightarrow CO + 3H_2 \tag{4}$$

$$CH_4 + 2H_2O \leftrightarrow CO_2 + 4H_2 \tag{5}$$

$$CH_4 + CO_2 \leftrightarrow 2CO + 2H_2 \tag{6}$$

$$CH_4 + 3CO_2 \leftrightarrow 4CO + 2H_2O \tag{7}$$

Experimental remarks prove that dry reforming reactions (Equations (6) and (7)) develop when the feed has a minimum moisture level and is mainly composed of CH_4 and CO_2. Thus, in SOFC operating conditions, DR can occur only near the anode inlet due to the water generated by electrochemical reactions (Equation (2)). For this reason, in kinetic modeling related to DIR-SOFCs, only the steam reforming route is usually assumed [43].

Due to the applied electric load, parasite electrochemical reactions of CH_4 partial or total oxidation can occur (Equations (8) and (9)). However, reforming is usually faster and thus these last ones are assumed negligible [30].

$$CH_4 + O^{=} \rightarrow CO + 2H_2 + 2e^- \tag{8}$$

$$CH_4 + 4O^{=} \rightarrow CO_2 + 2H_2O + 8e^- \tag{9}$$

The presence of all requested reactants and of a suitable catalyst permits the water gas shift (WGS) reaction (Equation (10)) to occur altering the gas phase concentration.

$$CO + H_2O \leftrightarrow CO_2 + H_2 \tag{10}$$

Since Equation (10) is the preferential path for CO consumption [43], only H_2 oxidation is usually considered as anodic electrochemical reaction (Equation (2)). If both the steam reforming (Equation (4)) and the WGS (Equation (10)) are considered reversible, Equation (5) is usually neglected, resulting in the linear combination of these two reactions [44].

In addition to these, carbon deposition mechanisms such as the Boudouard reaction (Equation (11)) can occur due to the high operating temperature (Equations (12)–(16)). Carbon deposition reduces the active area of the catalyst, penalizing reforming reactions and consequently global cell performance.

$$2CO \leftrightarrow C + CO_2 \tag{11}$$

$$CH_4 \leftrightarrow C + 2H_2 \tag{12}$$

$$CO + H_2 \leftrightarrow C + H_2O \tag{13}$$

$$CO_2 + 2H_2 \leftrightarrow C + 2H_2O \tag{14}$$

$$CH_4 + 2CO \leftrightarrow 3C + 2H_2O \tag{15}$$

$$CH_4 + CO_2 \leftrightarrow 2C + 2H_2O \tag{16}$$

However, working with excess steam hinders Equations (11)–(16) that consequently are usually neglected in the modeling [30]. In addition to carbon deposition, also the presence of poisonous compounds can induce other degradation phenomena. We overlook their effects in the basic model formulation, introducing them in the catalyst deactivation simplified simulation that is discussed afterwards.

Thus, to define a kinetic model that properly describes the SOFC performance, we have to consider interactions between electrochemical (Equations (1) and (2)), SR (Equation (4)), and WGS (Equation (10)) reactions. While electrochemical and WGS processes have been already analyzed in previous studies for the simulation of high temperature fuel cells [45,46], in the following, we focus on possible SR modeling solutions.

2.2. A Model for Steam Reforming Reaction

Literature mainly follows three different modeling approaches [17]:

- equilibrium;
- power law kinetic formulation;
- surface reaction kinetic model.

The equilibrium approach assumes that the reforming reaches the thermodynamic equilibrium (Equation (17), where $K_{eq,SR}$ is the equilibrium constant, and p_i are the partial pressures of reactants [47].

$$K_{eq,SR} = \frac{p_{CO,eq} p^3_{H_2,eq}}{p_{CH_4,eq} p_{H_2O,eq}} \tag{17}$$

The equilibrium constant can be expressed according to either rigorous Van't Hoff formulation (Equation (18)) [48] or simplified semi-empirical approach (Equation (19)) [49].

$$\Delta G = -RT \ln K_{eq,SR} \tag{18}$$

$$K_{eq,SR} = \exp^{\left(30.114 - \frac{26830}{T}\right)} \tag{19}$$

where ΔG is the Gibbs free energy variation, R the ideal gas constant, and T the temperature. Experimental results show that the CH_4 conversion is lower than the equilibrium one [44]. Thus, the use of this approach may require opportune corrections to properly reproduce experimental data. Since the kinetics are independent from catalyst amount and distribution, they can be used to describe a 0D system.

A power law kinetic formulation approach is based on semi-empirical equations, as described in Equation (20). The SR reaction rate (r_{SR}) is usually expressed as the product between a kinetic constant (k_{SR}) and the partial pressure of reactants elevated to different exponents. Both kinetic constant and exponents are fitted through analysis of experimental data.

$$r_{SR} = k_{SR} S p^{\alpha}_{CH_4} p^{\beta}_{H_2O} p^{\gamma}_{H_2} p^{\delta}_{CO_2} p^{\varepsilon}_{CO} \tag{20}$$

The dependence on H_2, CO_2, and CO is usually negligible (γ, δ, and ε are close to zero), thus the reaction rate is usually assumed as a function of CH_4 and H_2O. Consequently, α and β values may vary among different studies [21,43]. Literature generally agrees that the reforming has a first order dependence on CH_4 partial pressure, while the order of water seems highly influenced by the steam to carbon ratio. It can be positive for low S/C ratio, zero for S/C close to two, and negative for higher values of S/C [50]. This is explained by negative effects that great amount of water has on the CH_4 adsorption on the catalyst surface [44]. This approach does not require the knowledge of the mechanisms involved and can be easily applied when a large number of experimental data is available for model tuning. However, results are specific to the analyzed case and cannot be applied to different systems.

The surface reaction kinetic model approach describes occurring mechanisms as a sequence of intermediate phenomena, consisting of adsorption, surface reaction, and desorption of all present gases. The rate of the total kinetics is determined by the slowest phenomenon that changes at different temperatures and reactant-product compositions. Such kinetics are usually modeled following the Langmuir–Hinshelwood (LH) or the Hougen–Watson (HW) approaches. The first one assumes a bimolecular reaction between two reactants adsorbed on neighboring sites as the rate-limiting step and the water dissociation into atomic hydrogen H and hydroxyl groups OH (Equation (21)) [51].

$$r_{SR} = \frac{k_{SR} S \prod p_i^{\varphi_i}}{\left(1 + \sum K_i p_i^{\varphi_i}\right)^{\chi}} \left(1 - \frac{Q_{SR}}{K_{eq,SR}}\right) \tag{21}$$

The second approach also takes into account the sorption and the reaction of intermediates (Equation (22)) [51].

$$r_{SR} = \frac{k_{SR} S \prod \frac{p_i^{\varphi_i}}{p_j^{\lambda_j}}}{\left(1 + \sum K_i \frac{p_i^{\varphi_i}}{p_j^{\lambda_j}}\right)^{\chi}} \left(1 - \frac{Q_{SR}}{K_{eq,SR}}\right) \tag{22}$$

In both Equations (21) and (22), the numerator shows the kinetics dependency on involved gases, while the denominator considers the availability of active sites through adsorption isotherm. The last term, expressed as ratio between the reaction quotient Q_{SR} (Equation (23)) and the equilibrium constant $K_{eq,SR}$, represents the driving force of the overall process.

$$Q_{SR} = \frac{p_{CO} p_{H_2}^3}{p_{CH_4} p_{H_2O}} \tag{23}$$

Both kinetic k_{SR} and adsorption K_i coefficients can be described by an Arrhenius type dependency on the operating temperature (Equations (24) and (25)).

$$k_{SR} = k_0 \exp^{-\frac{E_{act,\,SR}}{RT}} \tag{24}$$

$$K_i = K_{0,i} \exp^{-\frac{\Delta H_{ads,\,i}}{RT}} \tag{25}$$

where k_0 and $K_{0,i}$ are the pre-exponential coefficients, E_{act} the activation energy of SR reaction, and ΔH_{ads} the adsorption enthalpy variation. In DIR-FC modeling, Equation (22) is commonly used, considering the values detected for Ni-$MgAl_2O_3$-spinel catalysts as reference of kinetics parameters [49]. Since the process rate is strongly influenced on catalyst features, such as used support, Ni percentage, and particle size, this approach is not always effective for SR occurring inside a fuel cell due to higher Ni content compared to the traditional SR catalyst needed to guarantee a good conductivity [21].

Experimental data suggest that, for Ni/YSZ, the material usually employed as the SOFC anode, the rate-limiting step is the CH_4 dissociative adsorption. Thus, a first order expression function of only CH_4 partial pressure is formulated in accordance with power law models. Under the assumption that the surface cannot be covered by other components, the adsorption dependency is neglected, and the kinetic rate is expressed through Equation (26) as reported in different works [21,44,51].

$$r_{SR} = k_{SR} S p_{CH_4} \left(1 - \frac{Q_{SR}}{K_{eq,SR}}\right) \tag{26}$$

The kinetics constant k_{SR} usually depends on available catalyst active area. This kinetics has been validated over a wide range of temperatures and S/C ratios.

For the DIR-SOFC simulation, we implement in our code both equilibrium (Equation (17)) and surface reaction kinetics approach (Equation (26)). It is important to underline that the first is independent of the specific used catalyst, while the second one is expressed in function of its distribution on cell plane. This study takes as reference the kinetics proposed by [52] for a nickel cermet electrode (Table 1).

Table 1. Reference kinetics coefficients [52].

Kinetics Parameter	Value
k_{SR} [mol m^{-2} s^{-1} bar^{-1}]	4274
$E_{att,SR}$ [J mol^{-1}]	82,000

3. The DIR-SOFC simulation

To simulate a DIR-SOFC, the authors integrate the presented reforming kinetics into a home-made Fortran written code called SIMFC (SIMulation of Fuel Cell) that has been developed and validated in previous works. It is a 2D model that simulates the operation of high temperature molten carbonate [46,53] and solid oxide [41,42] fuel cells on the basis of local mass, energy, charge and momentum balances.

The SOFC electrochemical performance is evaluated through a semi-empirical formulation, derived by detailed experimental campaigns on H_2-feeding cell [42,45], where the main dependences of different polarization terms have been investigated. The equations proposed to describe cell voltage and used parameters are presented in Table 2. The same previous coefficients identified for H_2-feeding systems are valid also for DIR configuration, since the reactive sites for two reaction paths do not coincide [21].

Table 2. Solid oxide fuel cell (SOFC) electrochemical kinetics (for the complete description refer to [45]).

Cell voltage	$V = E - \eta_{leakage} - P_1 T \exp^{\frac{P_2}{T}} J - \frac{RT}{F}\sinh^{-1}\left(\frac{J}{2J_{0,an}}\right) - \frac{RT}{2F}\sinh^{-1}\left(\frac{J}{2J_{0,cat}}\right) - \frac{RT}{2F}\ln\left(\frac{\left(1+\frac{RTd_{an}J}{6FD^{eff}_{H_2O}p_{H_2O,an}}\right)^{2B}}{\left(1-\frac{RTd_{an}J}{6FD^{eff}_{H_2}p_{H_2,an}}\right)^{2A}}\right)$
OCV potential	$E = E^0 + \frac{RT}{2F}\ln\left(\frac{p_{H_2,an}p^{0.5}_{O_2,cat}}{p_{H_2O,an}}\right)$
Anodic exchange current density	$J_{0,an} = P_3\left(y_{H_{2,an}}\right)^A\left(y_{H_2O,an}\right)^B \exp^{-\frac{E_{act,an}}{RT}}$
Cathodic exchange current density	$J_{0,cat} = P_4\left(y_{O_2,cat}\right)^C \exp^{-\frac{E_{act,cat}}{RT}}$
Electrochemical parameters	$\eta_{leakage} = 0.03$, $P_1 = 2 \times 10^{-9}$ [Ω cm^2 K^{-1}], $P_2 = 10{,}986$ [K], $P_3 = 2.8 \times 10^5$ [A cm^{-2}], $P_4 = 4 \times 10^6$ [A cm^{-2}], $E_{act,an} = 110$ [kJ mol^{-1}], $E_{act,cat} = 120$ [kJ mol^{-1}], $A = 0.5$, $B = 0.55$, $C = 0.25$

The proposed model can be used not only to obtain overall cell information, such as cell voltage or produced power, but also to evaluate local variables, such as compositions, polarizations, and reactant utilization factors. This feature is extremely important for both theoretical and feasibility studies. Local data are fundamental for the understanding of the phenomena occurring inside SOFC as well as to avoid critical points and guarantee a more uniform utilization of the cell surface. As many properties cannot be directly measured, a detailed modeling is necessary to assess their values. Moreover, since the code needs only few minutes of computations to produce the results, it could be integrated successfully in more complex software for plant simulation (e.g., Aspen Plus).

4. Results and Discussion

The simulation tool is presented considering a single cell anode-supported SOFC in cross-flow configuration. The cell has a planar geometry with a total active area equal to 1 m^2 (anode inlet length = 71 cm, cathode inlet length = 142 cm). This design is commonly used in industrial applications due to both lower fabrication costs and a relatively more compact structure compared to tubular design [54]. In view of common used materials [55], it is assumed a Ni/YSZ porous anode provides the base for the thin dense YSZ electrolyte and the porous LSC cathode. The main cell features are presented in Table 3.

Table 3. Physical and microstructure properties of different cell layers.

Property	Anode Ni/YSZ	Electrolyte YSZ	Cathode LSC
Density [g cm^{-3}]	7.7	6.0	5.3
Heat Capacity [cal mol^{-1} K^{-1}]	50	29	34
Porosity [-]	0.40	0.01	0.35
Thickness [μm]	350	5	30
Tortuosity [-]	4	-	n.a.

The SOFC performance has been studied setting up a current density load of 0.15 A cm^{-2} coupled with an H_2 utilization of about 74%, common target value according to numerous SOFC producers [56]. The proposed feed conditions, with inlet S/C ratio of about two (safety condition to avoid carbon deposition at the cell operating temperature [57]), are represented in Table 4.

Table 4. Inlet operating conditions.

Inlet Condition	Anode	Cathode
T [K]	1023	1023
N [Nm^3 h^{-1}]	0.75	8.60
y_{CH4} [-]	0.25	-
y_{H2} [-]	0.04	-
y_{H2O} [-]	0.51	-
y_{CO2} [-]	0.13	-
y_{CO} [-]	0.03	-
y_{N2} [-]	0.04	0.79
y_{O2} [-]	-	0.21

4.1. *Analysis of DIR-SOFC Operation at Local Level*

The comparison between DIR-SOFC simulation assuming reforming reaction at the equilibrium (Figure 1A,C,E) and DIR-SOFC simulation assuming reforming reaction as kinetics driven (Figure 1B,D,F) is presented in term of local maps for the H_2 molar fraction (Figure 1A,B), the temperature of the solid structure (Figure 1C,D), and the applied current density (Figure 1E,F). In both cases, CH_4 is almost completely consumed in proximity of the anode inlet (about 10 cm), thus its mapping on cell plane is not reported here.

As verified before, the high operating temperature allows for a rapid conversion of CH_4. This is confirmed in literature, where CH_4 is observed consuming in the so called "reforming zone" [18]. The equilibrium conversion represents the maximum value that can be theoretically achieved. This is also evident in the kinetics reaction rate expression (Equation (26)), where the imbalance between actual and equilibrium composition represents the driving force of the process ($Q_{SR}/K_{eq,SR}$). The consequence is that the equilibrium case foresees a slightly faster SR to form H_2 and CO, reducing local temperature. However, in both approaches, the newly formed H_2 reacts electrochemically to produce water at the anode (Equation (2)), while CO mainly produces additional H_2 via WGS (Equation (10)). These two reactions are exothermic and balance the local temperature decrease due to the reforming. As result,

CH_4 is almost all converted in proximity of the anode inlet. H_2 is immediately formed thanks to the reforming and rapidly depletes along the cell plane to sustain electrochemical reactions (Figure 1A,B). As expected, the temperature (Figure 1C,D) decreases at the inlet due to the reforming endothermicity and increases moving towards the anode and, to a lesser degree, the cathode outlet with a peak (Figure 1E,F) where both exothermic electrochemical and WGS reactions occur [58]. Due to the cross-flow configuration, inlet cathodic gas (bottom left corner of maps) reduces the anode inlet temperature proceeding toward the cathode outlet. This in turn penalizes the reforming and the subsequent anodic reactions, forbidding the temperature increase. The assumed kinetics have a relevant influence on electrochemical processes. If the fastest equilibrium reforming kinetics induce an initial peak of H_2 production followed by a slow electrochemical conversion (Figure 1A), in the surface kinetics formulation a wider H_2 conversion zone (Figure 1B) is detected, causing lower peaks of temperature and local current density (Figure 1D,F). However, as shown in Table 5, the macroscopic results of the simulation are not particularly dissimilar, thus both approaches can be used as preliminary analysis.

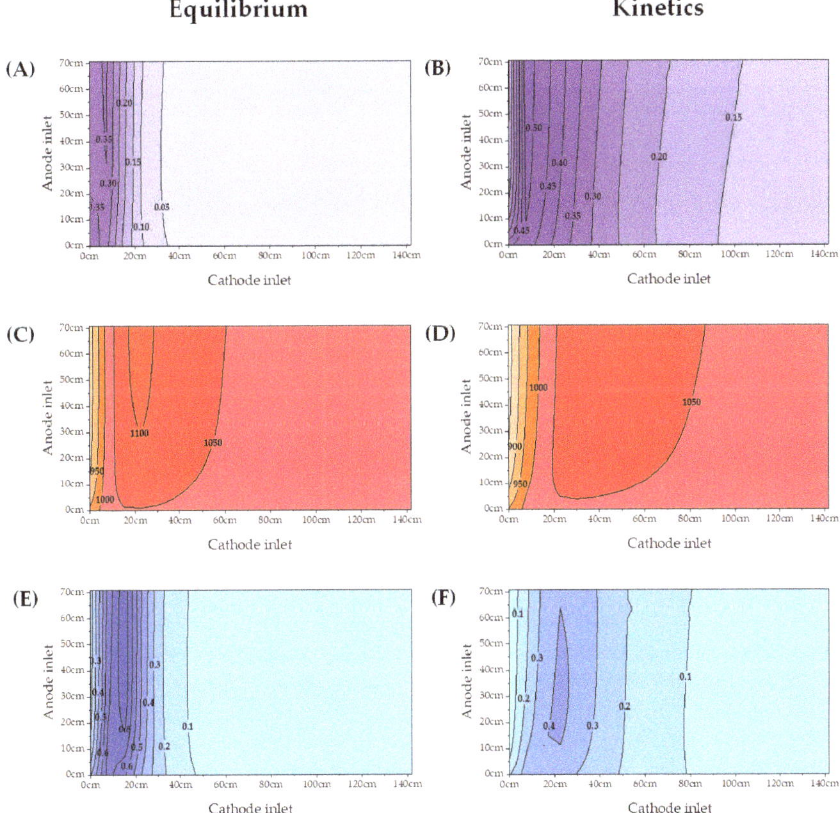

Figure 1. Results of the direct internal reforming (DIR-SOFC) simulation using steam reforming (SR) equilibrium (**A**,**C**,**E**) and kinetics formulation (**B**,**D**,**F**,); local maps of molar fraction of H_2 (**A**,**B**), temperature of the solid structure in K (**C**,**D**) and current density in A cm^{-2} (**E**,**F**).

Table 5. Main results of different simulations basing on equilibrium and surface reaction mechanism approaches.

Reforming Operating Condition	Equilibrium	Kinetics
V [V]	0.747	0.856
J_{max} [A cm^{-2}]	0.88	0.42
T_{max} [K]	1110	1084
T_{min} [K]	873	828
$T_{average}$ [K]	1037	1034
$y_{CH4,max}$ [-]	0.048	0.196
$y_{H2,max}$ [-]	0.402	0.525
Cell Power [W]	1121	1284

However, these obtained results are not completely satisfactory. As previously mentioned, the rapid conversion of CH_4 near the anode inlet induces a relevant temperature gradient with a difference between maximum and minimum temperature of more than 200 K in the equilibrium case and more than 250 K in the kinetics case. This, coupled with the high peak current density, greatly speeds up degradation processes limiting the cell operability.

For this reason, the catalyst active area is usually reduced by a partial poisoning or by changing the material microstructure [21].

4.2. Possible Applications of Local Simulation Tool

As already underlined, the local simulation has several advantages: it allows for not only a detailed knowledge of the main chemical-physical features on the cell plane but also a local description of system structure. Thus, using a local modeling approach, we can study the degradation of the reforming catalyst and how it affects the cell performance.

To simulate the catalyst deactivation that happens due to several phenomena such as sintering, poisoning by sulfur and other pollutants, or carbon deposition, the authors adjust the reforming kinetics introducing a coefficient (σ) as shown in Equation (27). This coefficient represents a corrective parameter that permits one to consider a reduced active area or an unevenly distributed catalyst. Since the equilibrium kinetics are independent of the catalysts active area; this point is added only to the surface reaction approach.

$$r_{SR} = k_{SR}\sigma S p_{CH_4}\left(1 - \frac{Q_{SR}}{K_{eq,SR}}\right) \quad (27)$$

An uneven distribution of SR active sites due to advanced degradation processes can be introduced considering local value of σ to differentiate the cell area. This is specifically relevant when there is the need to simulate long-term applications when such deactivation phenomena cannot be neglected any more. To demonstrate this possibility, the authors introduce into the code a matrix of σ, as presented in Figure 2. The cell surface is divided into a 20 × 20 mesh with each sub-cell having an area equal to 25 cm^2. The start of degradation processes is assumed at the anode inlet where the sigma value is the smallest (i.e., fewer catalyst active areas). Moving towards the anode outlet, σ increases, reflecting a milder degradation effect (i.e., more catalyst active areas). Such distribution may derive from carbon deposition (Equations (11)–(16)) that usually begins at the anode inlet, where reactions initially take place and the temperature is lower, and then it sequentially shifts to the outlet as more and more active sites are covered. Another possible cause could be the long exposition to poisoning of sulfur or other poisonous compounds present in biogas feed. Indeed, it is well known that sulfur can react with the reforming catalyst inhibiting its properties [25]. Additionally, in this case, the poisoning starts at the anode inlet and, as more sites are poisoned, it shifts towards the outlet. The local results of the simulation are reported in Figure 3.

Anode inlet

0.00001	0.00001	0.00001	0.00001	0.01	0.01	0.01	0.01	0.02	0.02	0.02	0.02	0.7	0.7	0.7	0.7	0.7	0.7	0.7	0.7
0.00001	0.00001	0.00001	0.00001	0.01	0.01	0.01	0.01	0.02	0.02	0.02	0.02	0.7	0.7	0.7	0.7	0.7	0.7	0.7	0.7
0.00001	0.00001	0.00001	0.00001	0.01	0.01	0.01	0.01	0.02	0.02	0.02	0.02	0.7	0.7	0.7	0.7	0.7	0.7	0.7	0.7
0.00001	0.00001	0.00001	0.00001	0.01	0.01	0.01	0.01	0.02	0.02	0.02	0.02	0.7	0.7	0.7	0.7	0.7	0.7	0.7	0.7
0.00001	0.00001	0.00001	0.00001	0.01	0.01	0.01	0.01	0.02	0.02	0.02	0.02	0.7	0.7	0.7	0.7	0.7	0.7	0.7	0.7
0.00001	0.00001	0.00001	0.00001	0.01	0.01	0.01	0.01	0.02	0.02	0.02	0.02	0.7	0.7	0.7	0.7	0.7	0.7	0.7	0.7
0.00001	0.00001	0.00001	0.00001	0.01	0.01	0.01	0.01	0.02	0.02	0.02	0.02	0.7	0.7	0.7	0.7	0.7	0.7	0.7	0.7
0.00001	0.00001	0.00001	0.00001	0.01	0.01	0.01	0.01	0.02	0.02	0.02	0.02	0.7	0.7	0.7	0.7	0.7	0.7	0.7	0.7
0.00001	0.00001	0.00001	0.00001	0.01	0.01	0.01	0.01	0.02	0.02	0.02	0.02	0.7	0.7	0.7	0.7	0.7	0.7	0.7	0.7
0.00001	0.00001	0.00001	0.00001	0.01	0.01	0.01	0.01	0.02	0.02	0.02	0.02	0.7	0.7	0.7	0.7	0.7	0.7	0.7	0.7
0.00001	0.00001	0.00001	0.00001	0.01	0.01	0.01	0.01	0.02	0.02	0.02	0.02	0.7	0.7	0.7	0.7	0.7	0.7	0.7	0.7
0.00001	0.00001	0.00001	0.00001	0.01	0.01	0.01	0.01	0.02	0.02	0.02	0.02	0.7	0.7	0.7	0.7	0.7	0.7	0.7	0.7
0.00001	0.00001	0.00001	0.00001	0.01	0.01	0.01	0.01	0.02	0.02	0.02	0.02	0.7	0.7	0.7	0.7	0.7	0.7	0.7	0.7
0.00001	0.00001	0.00001	0.00001	0.01	0.01	0.01	0.01	0.02	0.02	0.02	0.02	0.7	0.7	0.7	0.7	0.7	0.7	0.7	0.7
0.00001	0.00001	0.00001	0.00001	0.01	0.01	0.01	0.01	0.02	0.02	0.02	0.02	0.7	0.7	0.7	0.7	0.7	0.7	0.7	0.7
0.00001	0.00001	0.00001	0.00001	0.01	0.01	0.01	0.01	0.02	0.02	0.02	0.02	0.7	0.7	0.7	0.7	0.7	0.7	0.7	0.7
0.00001	0.00001	0.00001	0.00001	0.01	0.01	0.01	0.01	0.02	0.02	0.02	0.02	0.7	0.7	0.7	0.7	0.7	0.7	0.7	0.7
0.00001	0.00001	0.00001	0.00001	0.01	0.01	0.01	0.01	0.02	0.02	0.02	0.02	0.7	0.7	0.7	0.7	0.7	0.7	0.7	0.7
0.00001	0.00001	0.00001	0.00001	0.01	0.01	0.01	0.01	0.02	0.02	0.02	0.02	0.7	0.7	0.7	0.7	0.7	0.7	0.7	0.7
0.00001	0.00001	0.00001	0.00001	0.01	0.01	0.01	0.01	0.02	0.02	0.02	0.02	0.7	0.7	0.7	0.7	0.7	0.7	0.7	0.7

Cathode inlet

Figure 2. Matrix of σ to consider the active area reduction due to deactivation processes, such as carbon deposition and poisoning effect.

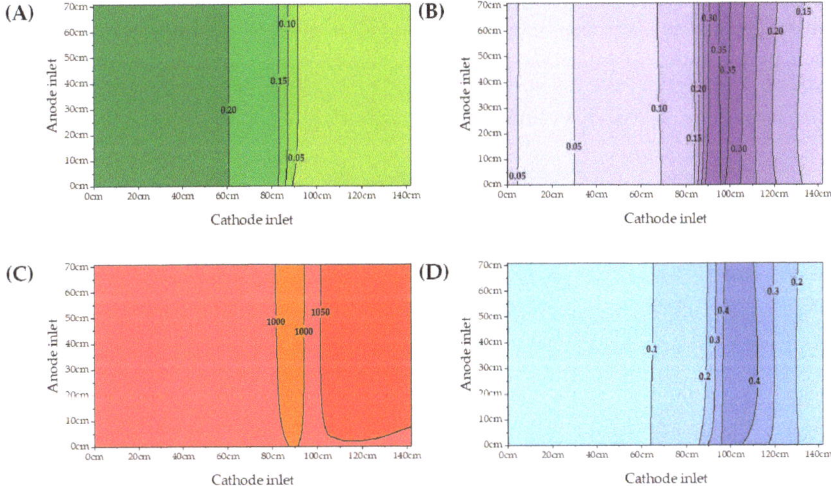

Figure 3. Results of the DIR-SOFC simulation using the SR kinetics approach correcting the catalyst actual surface area to consider the cell degradation; local maps of molar fraction of CH_4 (**A**) and H_2 (**B**), temperature of the solid structure in K (**C**), and current density in A cm^{-2} (**D**).

Close to the anode inlet, the CH_4 conversion is negligible due to the highly reduced catalytic surface area that corresponds to a value of 0.00001 in the σ matrix. This can be interpreted as an extremely deactivated or poisoned catalyst. As in Figure 3A, then moving towards the anode outlet, the conversion begins to increase when the fuel flow encounters the more active catalyst (σ from 0.01 to 0.7 progressively). The maps of H_2, current density, and temperature on the cell plane are also influenced, resulting in a more uniform distribution (Figure 3B–D). As Table 6 shows, there is not an excessive global performance variation compared to the kinetics case previously described in Table 5; still, an inefficient use of the catalyst occurs.

It is important to underline that the decreased temperature gradient (less than 150 K between maximum and minimum values) should not be interpreted as positive outcomes. It is the result of a highly deactivated catalyst that penalizes the performance in terms of power output decreasing by about 50 W in comparison with the kinetics case. In this simulation, the authors do not consider the degradation effects on electrochemical reactions, which could be added with available experimental data and would more highly affect power losses.

Table 6. Main results of the simulation using a local map of σ to describe catalyst deactivation.

Reforming Operating Condition	Degradation
V [V]	0.823
J_{max} [A cm^{-2}]	0.47
T_{max} [K]	1088
T_{min} [K]	962
$T_{average}$ [K]	1034
$y_{CH4,max}$ [-]	0.250
$y_{H2,max}$ [-]	0.373
Cell Power [W]	1234

Through a similar local approach, it is also possible to identify an optimized catalyst distribution that should yield a reduced solid temperature variation compared to the kinetics base one shown in Figure 1D. Having a lower solid temperature gradient allows for the improvement of cell stability and consequently for a longer cell durability by reducing the mechanical stress on materials. An uneven catalyst distribution can be obtained in the manufacturing process through several ways, such as different Ni particle size, addition of alkali metals in lattice, or Cu doping [21]. To simplify the optimization process, the authors decide to divide the cell surface in rectangular shaped areas of similar dimension and catalyst amount. Then, using iterative calculations, the σ matrix is tuned in order to reduce the temperature gradient, resulting in the configuration of Figure 4.

Anode inlet

0.10	0.10	0.15	0.15	0.05	0.05	0.15	0.15	0.05	0.05	0.15	0.15	0.05	0.05	0.15	0.15	0.05	0.05	0.15	0.15
0.10	0.10	0.15	0.15	0.05	0.05	0.15	0.15	0.05	0.05	0.15	0.15	0.05	0.05	0.15	0.15	0.05	0.05	0.15	0.15
0.10	0.10	0.15	0.15	0.05	0.05	0.15	0.15	0.05	0.05	0.15	0.15	0.05	0.05	0.15	0.15	0.05	0.05	0.15	0.15
0.10	0.10	0.15	0.15	0.05	0.05	0.15	0.15	0.05	0.05	0.15	0.15	0.05	0.05	0.15	0.15	0.05	0.05	0.15	0.15
0.05	0.05	0.15	0.15	0.05	0.05	0.15	0.15	0.05	0.05	0.15	0.15	0.05	0.05	0.15	0.15	0.05	0.05	0.15	0.15
0.05	0.05	0.15	0.15	0.05	0.05	0.15	0.15	0.05	0.05	0.15	0.15	0.05	0.05	0.15	0.15	0.05	0.05	0.15	0.15
0.05	0.05	0.15	0.15	0.05	0.05	0.15	0.15	0.05	0.05	0.15	0.15	0.05	0.05	0.15	0.15	0.05	0.05	0.15	0.15
0.05	0.05	0.15	0.15	0.05	0.05	0.15	0.15	0.05	0.05	0.15	0.15	0.05	0.05	0.15	0.15	0.05	0.05	0.15	0.15
0.05	0.05	0.15	0.15	0.05	0.05	0.15	0.15	0.05	0.05	0.15	0.15	0.05	0.05	0.15	0.15	0.05	0.05	0.15	0.15
0.05	0.05	0.15	0.15	0.05	0.05	0.15	0.15	0.05	0.05	0.15	0.15	0.05	0.05	0.15	0.15	0.05	0.05	0.15	0.15
0.15	0.15	0.05	0.05	0.15	0.15	0.05	0.05	0.15	0.15	0.05	0.05	0.15	0.15	0.05	0.05	0.15	0.15	0.05	0.05
0.15	0.15	0.05	0.05	0.15	0.15	0.05	0.05	0.15	0.15	0.05	0.05	0.15	0.15	0.05	0.05	0.15	0.15	0.05	0.05
0.15	0.15	0.05	0.05	0.15	0.15	0.05	0.05	0.15	0.15	0.05	0.05	0.15	0.15	0.05	0.05	0.15	0.15	0.05	0.05
0.15	0.15	0.05	0.05	0.15	0.15	0.05	0.05	0.15	0.15	0.05	0.05	0.15	0.15	0.05	0.05	0.15	0.15	0.05	0.05
0.15	0.15	0.05	0.05	0.15	0.15	0.05	0.05	0.15	0.15	0.05	0.05	0.15	0.15	0.05	0.05	0.15	0.15	0.05	0.05
0.15	0.15	0.05	0.05	0.15	0.15	0.05	0.05	0.15	0.15	0.05	0.05	0.15	0.15	0.05	0.05	0.15	0.15	0.05	0.05
0.15	0.15	0.05	0.05	0.15	0.15	0.05	0.05	0.15	0.15	0.05	0.05	0.15	0.15	0.05	0.05	0.15	0.15	0.05	0.05
0.15	0.15	0.05	0.05	0.15	0.15	0.05	0.05	0.15	0.15	0.05	0.05	0.15	0.15	0.05	0.05	0.15	0.15	0.05	0.05
0.15	0.15	0.05	0.05	0.15	0.15	0.05	0.05	0.15	0.15	0.05	0.05	0.15	0.15	0.05	0.05	0.15	0.15	0.05	0.05
0.15	0.15	0.05	0.05	0.15	0.15	0.05	0.05	0.15	0.15	0.05	0.05	0.15	0.15	0.05	0.05	0.15	0.15	0.05	0.05

Cathode inlet

Figure 4. Matrix of σ to optimize cell performance in terms of uneven distribution of temperature and current density.

Using this σ matrix, the cell performance is reevaluated, as presented in Figures 5 and 6. A more uniform temperature distribution is obtained on the cell plane, which maintains good global performance (Table 7). For an easy comparison, the temperature profiles of the solid structure from anode inlet to anode outlet are presented for kinetics base case (Figure 6A,C,E) and optimized approach (Figure 6B,D,F) in different cell positions in correspondence with the cathode inlet (Figure 6A,B), in the middle between cathode inlet and outlet (Figure 6C,D), and in correspondence with the cathode outlet (Figure 6E,F). As can be seen, the temperature change in optimized catalyst distribution is much less sharpened in all analyzed points. Moreover, the difference between the lower and the higher temperature is less than 100 K, whereas it is more than 250 K in the kinetics base case. Since it is estimated that about 30% of occurring stresses are due to thermal gradients, this configuration offers a relevant improvement to guarantee long-term operations [59]. This result is obtained by expanding the CH_4 reforming zone in order to have a more homogeneous conversion and consequently a minor temperature drop caused by the reforming endothermicity. Compared with previous solution, the current density and the H_2 molar fraction also have lower maximum values (J_{max} = 0.29 A cm^{-2} vs. 0.42 A cm^2 in kinetics base case, $y_{H2,max}$ = 0.322 vs. 0.525 in kinetics base case), signifying less stressed working conditions. However, the measured voltage and the total output power are slightly decreased in this configuration; reductions of 9 mV and 14 W are obtained.

Table 7. Main results of the simulation using local maps of σ to optimize cell performance.

Reforming Operating Condition	Optimized Configuration
V [V]	0.847
J_{max} [A cm^{-2}]	0.29
T_{max} [K]	1059
T_{min} [K]	963
$T_{average}$ [K]	1034
$y_{CH4,max}$ [-]	0.233
$y_{H2,max}$ [-]	0.322
Cell Power [W]	1270

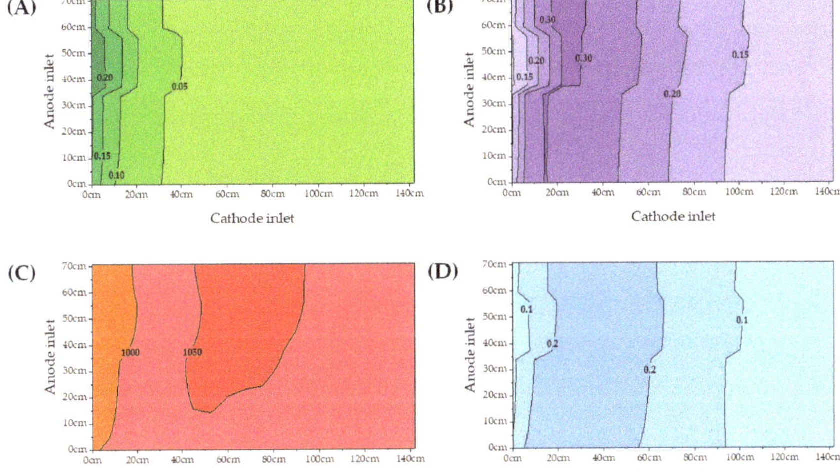

Figure 5. Results of the DIR-SOFC simulation using the SR kinetics approach correcting the catalyst actual surface area to obtain cell performance optimization; local maps of molar fraction of CH_4 (**A**) and H_2 (**B**), temperature of the solid structure in K (**C**), and current density in A cm^{-2} (**D**).

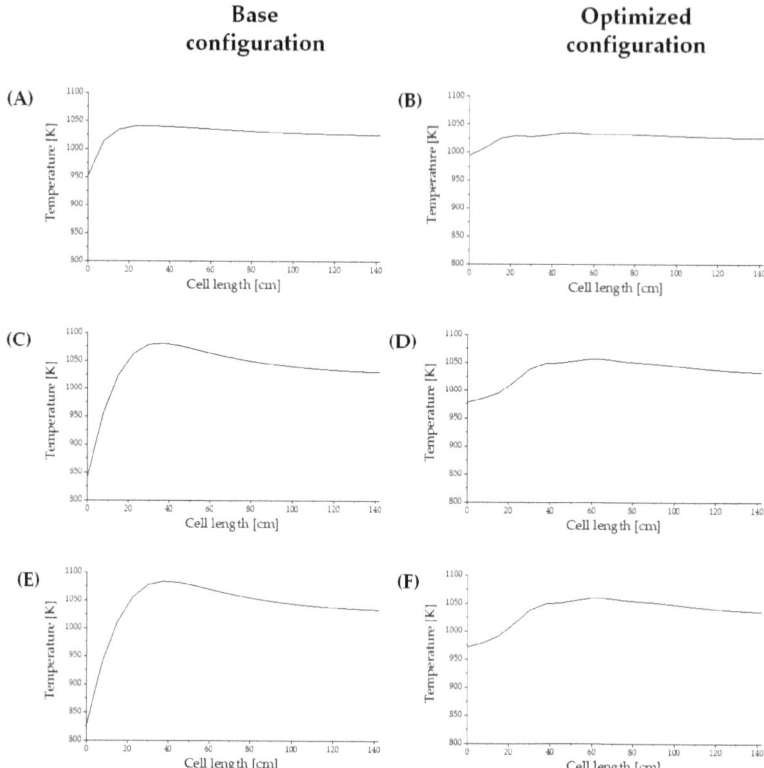

Figure 6. Temperature profile along the cell from anode inlet to anode outlet at different distances from the cathode inlet for the kinetics base case (**A**,**C**,**E**) and the optimized approach (**B**,**D**,**F**) in correspondence with the cathode inlet (**A**,**B**), in the middle between cathode inlet and outlet (**C**,**D**), and in correspondence with the cathode outlet (**E**,**F**).

5. Conclusions

This work aims at presenting a suitable tool for the simulation of DIR-SOFCs for industrial applications, with regard to local effects on the main chemical-physical variables (e.g., composition of reactants, current density, and solid temperature) that are fundamental for both feasibility studies and to favor long-term operations.

The SIMFC code, a home-made tool previously validated for H_2-feeding high temperature fuel cells, is improved, introducing the possibility to consider internal CH_4 reforming. Based on experimental observations and literature modeling, both thermodynamic equilibrium and surface reaction kinetic approaches are implemented to simulate the direct internal reforming using a biogas as feeding. In both cases, similar results are obtained; the CH_4 conversion occurs in the initial 10 cm, and the obtained voltage is around 0.8 V.

Being independent of the catalysts used, the equilibrium approach does not permit to differentiate catalyst operation on the cell plane. However, this is possible using the kinetic approach that is consequently more suitable for a detailed local modeling. As examples of possible applications, the authors analyze (i) how to simulate SOFC performance affected by the deactivation of reforming catalysts and (ii) how to use also local control in order to detect an optimized catalyst distribution that would provide minimized thermal stresses on the cell. This is achieved with the introduction of a specific coefficient (σ) in matrix form to describe the catalyst amount on the cell surface.

When catalyst deactivation is considered, the term σ indicates the degree of degradation or poisoning of reforming active sites, with lower values corresponding to reduced catalytic activity. At such conditions, the reforming reaction is deeply penalized, reaching the complete conversion over half along cathode inlet direction in the presented example.

When used to determine an optimized catalysts distribution, the value of σ indicates the quantity of catalyst or, in other terms, the active surface available for the reactions. In order to reduce the solid temperature gradient on the cell plane, in the example, the authors tune the matrix of σ coefficients to obtain an optimized distribution that can be easily realized in manufacturing process. A steep temperature profile induces higher mechanical stresses that can result in creep formation and ultimately global anode degradation. Thanks to the identified simple distribution, it is possible to sensitively reduce the temperature gradient by about 60% compared to the kinetics base example. The proposed configuration also permits one to halve the peak of current density. At the same time, the complete CH_4 conversion is guaranteed without a relevant power penalization that is reduced by only 14 W.

Author Contributions: Conceptualization, E.A., F.R.B., and B.B.; Methodology, E.A., F.R.B., and B.B.; Software, E.A., F.R.B., and B.B.; Writing—original draft, E.A., F.R.B., and B.B.; Writing—review & editing, E.A., F.R.B., and B.B. All authors have read and agreed to the published version of the manuscript.

Funding: This research received no external funding.

Conflicts of Interest: The authors declare no conflict of interest.

Nomenclature

A, B, C	Kinetics orders for exchange current density [–]
D	Diffusion coefficient [$m^2\ s^{-1}$]
d	Thickness [m]
E	OCV potential [V]
E^0	Reversible voltage [V]
E_{act}	Activation energy [$J\ mol^{-1}$]
F	Faraday constant [$C\ mol^{-1}$]
G	Gibbs free energy [$J\ mol^{-1}$]
H_{ads}	Adsorption enthalpy [$J\ mol^{-1}$]
J	Current density [$A\ m^{-2}$]
J_0	Exchange current density [$A\ m^{-2}$]
K	Adsorption coefficient
K_0	Adsorption pre-exponential coefficient
K_{eq}	Equilibrium constant [atm^2]
k	Kinetics constant
k_0	Kinetics pre-exponential constant
N	Volumetric flow rate [$Nm^3\ h^{-1}$]
P	Electrochemical parameter
p	Pressure [atm]
Q	Reaction quotient [atm^2]
R	Gas constant [$J\ mol^{-1}\ K^{-1}$]
r	Reaction rate [$mol\ s^{-1}$]
S	Active area [m^2]
T	Temperature [K]
V	Cell voltage [V]
y	Molar fraction [-]
Greek letters	
$\alpha, \beta, \gamma, \delta, \varepsilon$	Power law kinetic coefficients [-]
λ, φ, χ	Steam reforming LH-HW kinetic orders [-]
η	Overpotential [V]
σ	Active area corrective coefficient [-]

Subscript	
an	Anode
cat	Cathode
eff	Effective
eq	Equilibrium
SR	Steam Reforming
Abbreviations	
DIR	Direct Internal Reforming
DR	Dry Reforming
ER	External Reforming
IIR	Indirect Internal Reforming
LSC	Lanthanum Strontium Cobaltite
OCV	Open Circuit Voltage
SIMFC	SIMulation of Fuel Cell
SOFC	Solid Oxide Fuel Cell
SR	Steam Reforming
WGS	Water Gas Shift
YSZ	Yttria-Stabilized Zirconia

References

1. Al-Khori, K.; Bicer, Y.; Boulfrad, S.; Koç, M. Techno-Economic and Environmental Assessment of Integrating SOFC with a Conventional Steam and Power System in a Natural Gas Processing Plant. *Int. J. Hydrog. Energy* **2019**, *44*, 29604–29617. [CrossRef]
2. Mehrpooya, M.; Sadeghzadeh, M.; Rahimi, A.; Pouriman, M. Technical Performance Analysis of a Combined Cooling Heating and Power (CCHP) System Based on Solid Oxide Fuel Cell (SOFC) Technology—A Building Application. *Energy Convers. Manag.* **2019**, *198*, 111767. [CrossRef]
3. Singh, S.; Jha, P.A.; Presto, S.; Viviani, M.; Sinha, A.S.K.; Varma, S.; Singh, P. Structural and Electrical Conduction Behaviour of Yttrium Doped Strontium Titanate: Anode Material for SOFC Application. *J. Alloys Compd.* **2018**, *748*, 637–644. [CrossRef]
4. Barelli, L.; Barluzzi, E.; Bidini, G. Diagnosis Methodology and Technique for Solid Oxide Fuel Cells: A Review. *Int. J. Hydrog. Energy* **2013**, *38*, 5060–5074. [CrossRef]
5. van Biert, L.; Visser, K.; Aravind, P.V. A Comparison of Steam Reforming Concepts in Solid Oxide Fuel Cell Systems. *Appl. Energy* **2020**, *264*, 114748. [CrossRef]
6. Dispenza, G.; Sergi, F.; Napoli, G.; Antonucci, V.; Andaloro, L. Evaluation of Hydrogen Production Cost in Different Real Case Studies. *J. Energy Storage* **2019**, *24*, 100757. [CrossRef]
7. Saidi, M. Performance Assessment and Evaluation of Catalytic Membrane Reactor for Pure Hydrogen Production via Steam Reforming of Methanol. *Int. J. Hydrog. Energy* **2017**, *42*, 16170–16185. [CrossRef]
8. Capa, A.; García, R.; Chen, D.; Rubiera, F.; Pevida, C.; Gil, M.V. On the Effect of Biogas Composition on the H_2 Production by Sorption Enhanced Steam Reforming (SESR). *Renew. Energy* **2020**, *160*, 575–583. [CrossRef]
9. Zhou, X.; Huang, H.; Liu, H. Study of Partial Oxidation Reforming of Methane to Syngas over Self-Sustained Electrochemical Promotion Catalyst. *Int. J. Hydrog. Energy* **2013**, *38*, 6391–6396. [CrossRef]
10. Cao, L.; Yu, I.K.M.; Xiong, X.; Tsang, D.C.W.; Zhang, S.; Clark, J.H.; Hu, C.; Ng, Y.H.; Shang, J.; Ok, Y.S. Biorenewable Hydrogen Production through Biomass Gasification: A Review and Future Prospects. *Environ. Res.* **2020**, *186*, 109547. [CrossRef] [PubMed]
11. Mastropasqua, L.; Pecenati, I.; Giostri, A.; Campanari, S. Solar Hydrogen Production: Techno-Economic Analysis of a Parabolic Dish-Supported High-Temperature Electrolysis System. *Appl. Energy* **2020**, *261*, 114392. [CrossRef]
12. Ru, Y.; Sang, J.; Xia, C.; Wei, W.-C.J.; Guan, W. Durability of Direct Internal Reforming of Methanol as Fuel for Solid Oxide Fuel Cell with Double-Sided Cathodes. *Int. J. Hydrog. Energy* **2020**, *45*, 7069–7076. [CrossRef]
13. Schluckner, C.; Subotić, V.; Lawlor, V.; Hochenauer, C. Three-Dimensional Numerical and Experimental Investigation of an Industrial-Sized SOFC Fueled by Diesel Reformat—Part II: Detailed Reforming Chemistry and Carbon Deposition Analysis. *Int. J. Hydrog. Energy* **2015**, *40*, 10943–10959. [CrossRef]

14. Lanzini, A.; Leone, P. Experimental Investigation of Direct Internal Reforming of Biogas in Solid Oxide Fuel Cells. *Int. J. Hydrog. Energy* **2010**, *35*, 2463–2476. [CrossRef]
15. Bao, C.; Wang, Y.; Feng, D.; Jiang, Z.; Zhang, X. Macroscopic Modeling of Solid Oxide Fuel Cell (SOFC) and Model-Based Control of SOFC and Gas Turbine Hybrid System. *Prog. Energy Combust. Sci.* **2018**, *66*, 83–140. [CrossRef]
16. Ahmed, K.; Foger, K. Kinetics of Internal Steam Reforming of Methane on Ni/YSZ-Based Anodes for Solid Oxide Fuel Cells. *Catal. Today* **2000**, *63*, 479–487. [CrossRef]
17. Luo, X.J.; Fong, K.F. Development of 2D Dynamic Model for Hydrogen-Fed and Methane-Fed Solid Oxide Fuel Cells. *J. Power Sources* **2016**, *328*, 91–104. [CrossRef]
18. Sohn, S.; Baek, S.M.; Nam, J.H.; Kim, C.-J. Two-Dimensional Micro/Macroscale Model for Intermediate-Temperature Solid Oxide Fuel Cells Considering the Direct Internal Reforming of Methane. *Int. J. Hydrog. Energy* **2016**, *41*, 5582–5597. [CrossRef]
19. Chiang, L.-K.; Liu, H.-C.; Shiu, Y.-H.; Lee, C.-H.; Lee, R.-Y. Thermal Stress and Thermo-Electrochemical Analysis of a Planar Anode-Supported Solid Oxide Fuel Cell: Effects of Anode Porosity. *J. Power Sources* **2010**, *195*, 1895–1904. [CrossRef]
20. Greco, F.; Frandsen, H.L.; Nakajo, A.; Madsen, M.F.; Van herle, J. Modelling the Impact of Creep on the Probability of Failure of a Solid Oxide Fuel Cell Stack. *J. Eur. Ceram. Soc.* **2014**, *34*, 2695–2704. [CrossRef]
21. Mogensen, D.; Grunwaldt, J.-D.; Hendriksen, P.V.; Dam-Johansen, K.; Nielsen, J.U. Internal Steam Reforming in Solid Oxide Fuel Cells: Status and Opportunities of Kinetic Studies and Their Impact on Modelling. *J. Power Sources* **2011**, *196*, 25–38. [CrossRef]
22. Aguiar, P.; Adjiman, C.S.; Brandon, N.P. Anode-Supported Intermediate Temperature Direct Internal Reforming Solid Oxide Fuel Cell. I: Model-Based Steady-State Performance. *J. Power Sources* **2004**, *138*, 120–136. [CrossRef]
23. Papurello, D.; Lanzini, A. SOFC Single Cells Fed by Biogas: Experimental Tests with Trace Contaminants. *Waste Manag.* **2018**, *72*, 306–312. [CrossRef] [PubMed]
24. Madi, H.; Lanzini, A.; Diethelm, S.; Papurello, D.; Van herle, J.; Lualdi, M.; Gutzon Larsen, J.; Santarelli, M. Solid Oxide Fuel Cell Anode Degradation by the Effect of Siloxanes. *J. Power Sources* **2015**, *279*, 460–471. [CrossRef]
25. Lee, H.S.; Lee, H.M.; Park, J.-Y.; Lim, H.-T. Degradation Behavior of Ni-YSZ Anode-Supported Solid Oxide Fuel Cell (SOFC) as a Function of H_2S Concentration. *Int. J. Hydrog. Energy* **2018**, *43*, 22511–22518. [CrossRef]
26. Vorontsov, V.; Luo, J.L.; Sanger, A.R.; Chuang, K.T. Synthesis and Characterization of New Ternary Transition Metal Sulfide Anodes for H2S-Powered Solid Oxide Fuel Cell. *J. Power Sources* **2008**, *183*, 76–83. [CrossRef]
27. Badur, J.; Lemański, M.; Kowalczyk, T.; Ziółkowski, P.; Kornet, S. Zero-Dimensional Robust Model of an SOFC with Internal Reforming for Hybrid Energy Cycles. *Energy* **2018**, *158*, 128–138. [CrossRef]
28. Hou, Q.; Zhao, H.; Yang, X. Thermodynamic Performance Study of the Integrated MR-SOFC-CCHP System. *Energy* **2018**, *150*, 434–450. [CrossRef]
29. Menon, V.; Banerjee, A.; Dailly, J.; Deutschmann, O. Numerical Analysis of Mass and Heat Transport in Proton-Conducting SOFCs with Direct Internal Reforming. *Appl. Energy* **2015**, *149*, 161–175. [CrossRef]
30. Andersson, M.; Yuan, J.; Sundén, B. SOFC Modeling Considering Hydrogen and Carbon Monoxide as Electrochemical Reactants. *J. Power Sources* **2013**, *232*, 42–54. [CrossRef]
31. Chalusiak, M.; Wrobel, M.; Mozdzierz, M.; Berent, K.; Szmyd, J.S.; Kimijima, S.; Brus, G. A Numerical Analysis of Unsteady Transport Phenomena in a Direct Internal Reforming Solid Oxide Fuel Cell. *Int. J. Heat Mass Transf.* **2019**, *131*, 1032–1051. [CrossRef]
32. Li, J.; Cao, G.-Y.; Zhu, X.-J.; Tu, H.-Y. Two-Dimensional Dynamic Simulation of a Direct Internal Reforming Solid Oxide Fuel Cell. *J. Power Sources* **2007**, *171*, 585–600. [CrossRef]
33. Ho, T.X.; Kosinski, P.; Hoffmann, A.C.; Vik, A. Transport, Chemical and Electrochemical Processes in a Planar Solid Oxide Fuel Cell: Detailed Three-Dimensional Modeling. *J. Power Sources* **2010**, *195*, 6764–6773. [CrossRef]
34. Nikooyeh, K.; Jeje, A.A.; Hill, J.M. 3D Modeling of Anode-Supported Planar SOFC with Internal Reforming of Methane. *J. Power Sources* **2007**, *171*, 601–609. [CrossRef]
35. Lanzini, A.; Leone, P.; Pieroni, M.; Santarelli, M.; Beretta, D.; Ginocchio, S. Experimental Investigations and Modeling of Direct Internal Reforming of Biogases in Tubular Solid Oxide Fuel Cells. *Fuel Cells* **2011**, *11*, 697–710. [CrossRef]

36. Stoeckl, B.; Subotić, V.; Reichholf, D.; Schroettner, H.; Hochenauer, C. Extensive Analysis of Large Planar SOFC: Operation with Humidified Methane and Carbon Monoxide to Examine Carbon Deposition Based Degradation. *Electrochim. Acta* **2017**, *256*, 325–336. [CrossRef]
37. Fang, Q.; Blum, L.; Batfalsky, P.; Menzler, N.H.; Packbier, U.; Stolten, D. Durability Test and Degradation Behavior of a 2.5 KW SOFC Stack with Internal Reforming of LNG. *Int. J. Hydrog. Energy* **2013**, *38*, 16344–16353. [CrossRef]
38. Subotić, V.; Schluckner, C.; Stoeckl, B.; Preininger, M.; Lawlor, V.; Pofahl, S.; Schroettner, H.; Hochenauer, C. Towards Practicable Methods for Carbon Removal from Ni-YSZ Anodes and Restoring the Performance of Commercial-Sized ASC-SOFCs after Carbon Deposition Induced Degradation. *Energy Convers. Manag.* **2018**, *178*, 343–354. [CrossRef]
39. Klein, J.-M.; Bultel, Y.; Pons, M.; Ozil, P. Modeling of a Solid Oxide Fuel Cell Fueled by Methane: Analysis of Carbon Deposition. *J. Fuel Cell Sci. Technol.* **2007**, *4*, 425–434. [CrossRef]
40. Yan, M.; Zeng, M.; Chen, Q.; Wang, Q. Numerical Study on Carbon Deposition of SOFC with Unsteady State Variation of Porosity. *Appl. Energy* **2012**, *97*, 754–762. [CrossRef]
41. Conti, B.; Bosio, B.; McPhail, S.J.; Santoni, F.; Pumiglia, D.; Arato, E. A 2-D Model for Intermediate Temperature Solid Oxide Fuel Cells Preliminarily Validated on Local Values. *Catalysts* **2019**, *9*, 36. [CrossRef]
42. Bianchi, F.R.; Spotorno, R.; Piccardo, P.; Bosio, B. Solid Oxide Fuel Cell Performance Analysis Through Local Modelling. *Catalysts* **2020**, *10*, 519. [CrossRef]
43. Kazempoor, P.; Braun, R.J. Model Validation and Performance Analysis of Regenerative Solid Oxide Cells for Energy Storage Applications: Reversible Operation. *Int. J. Hydrog. Energy* **2014**, *39*, 5955–5971. [CrossRef]
44. Timmermann, H.; Sawady, W.; Reimert, R.; Ivers-Tiffée, E. Kinetics of (Reversible) Internal Reforming of Methane in Solid Oxide Fuel Cells under Stationary and APU Conditions. *J. Power Sources* **2010**, *195*, 214–222. [CrossRef]
45. Bianchi, F.R.; Bosio, B.; Baldinelli, A.; Barelli, L. Optimization of a Reference Kinetic Model for Solid Oxide Fuel Cells. *Catalysts* **2020**, *10*, 104. [CrossRef]
46. Audasso, E.; Nam, S.; Arato, E.; Bosio, B. Preliminary Model and Validation of Molten Carbonate Fuel Cell Kinetics under Sulphur Poisoning. *J. Power Sources* **2017**, *352*, 216–225. [CrossRef]
47. Sanchez, D.; Chacartegui, R.; Munoz, A.; Sanchez, T. On the Effect of Methane Internal Reforming Modelling in Solid Oxide Fuel Cells. *Int. J. Hydrog. Energy* **2008**, *33*, 1834–1844. [CrossRef]
48. Green, D.W.; Perry, R.H. *Perry's Chemical Engineers' Handbook*, 8th ed.; McGraw-Hill Education: New York, NY, USA, 2007.
49. Xu, J.; Froment, G.F. Methane Steam Reforming, Methanation and Water-Gas Shift: I. Intrinsic Kinetics. *AIChE J.* **1989**, *35*, 88–96. [CrossRef]
50. Nagel, F.P.; Schildhauer, T.J.; Biollaz, S.M.A.; Stucki, S. Charge, Mass and Heat Transfer Interactions in Solid Oxide Fuel Cells Operated with Different Fuel Gases—A Sensitivity Analysis. *J. Power Sources* **2008**, *184*, 129–142. [CrossRef]
51. van Biert, L.; Visser, K.; Aravind, P.V. Intrinsic Methane Steam Reforming Kinetics on Nickel-Ceria Solid Oxide Fuel Cell Anodes. *J. Power Sources* **2019**, *443*, 227261. [CrossRef]
52. Achenbach, E.; Riensche, E. Methane/Steam Reforming Kinetics for Solid Oxide Fuel Cells. *J. Power Sources* **1994**, *52*, 283–288. [CrossRef]
53. Audasso, E.; Bosio, B.; Bove, D.; Arato, E.; Barckholtz, T.; Kiss, G.; Rosen, J.; Elsen, H.; Gutierrez, R.B.; Han, L.; et al. New, Dual-Anion Mechanism for Molten Carbonate Fuel Cells Working as Carbon Capture Devices. *J. Electrochem. Soc.* **2020**, *167*, 084504. [CrossRef]
54. Irshad, M.; Siraj, K.; Raza, R.; Ali, A.; Tiwari, P.; Zhu, B.; Rafique, A.; Ali, A.; Kaleem Ullah, M.; Usman, A. A Brief Description of High Temperature Solid Oxide Fuel Cell's Operation, Materials, Design, Fabrication Technologies and Performance. *Appl. Sci.* **2016**, *6*, 75. [CrossRef]
55. da Silva, F.S.; de Souza, T.M. Novel Materials for Solid Oxide Fuel Cell Technologies: A Literature Review. *Int. J. Hydrog. Energy* **2017**, *42*, 26020–26036. [CrossRef]
56. McPhail, S.J.; Kiviaho, J.; Conti, B. The Yellow Pages of SOFC Technology International Status of SOFC Deployment 2017. VTT Technical Research Centre of Finland Ltd.: Espoo, Finland, 2017.
57. Chen, T.; Wang, W.G.; Miao, H.; Li, T.; Xu, C. Evaluation of Carbon Deposition Behavior on the Nickel/Yttrium-Stabilized Zirconia Anode-Supported Fuel Cell Fueled with Simulated Syngas. *J. Power Sources* **2011**, *196*, 2461–2468. [CrossRef]

58. Ahmed, K.; Föger, K. Analysis of Equilibrium and Kinetic Models of Internal Reforming on Solid Oxide Fuel Cell Anodes: Effect on Voltage, Current and Temperature Distribution. *J. Power Sources* **2017**, *343*, 83–93. [CrossRef]
59. Jiang, C.; Gu, Y.; Guan, W.; Zheng, J.; Ni, M.; Zhong, Z. 3D Thermo-Electro-Chemo-Mechanical Coupled Modeling of Solid Oxide Fuel Cell with Double-Sided Cathodes. *Int. J. Hydrog. Energy* **2020**, *45*, 904–915. [CrossRef]

Article

Single-Cell Tests to Explore the Reliability of Sofc Installations Operating Offshore

Nelson Thambiraj [1,2,†], Ivar Waernhus [2,†], Crina Suciu [2,*,†], Arild Vik [2,†] and Alex C. Hoffmann [1,†]

1 Department of Physics and Technology, University of Bergen, 5007 Bergen, Norway; nelson.thambiraj@student.uib.no (N.T.); alex.hoffmann@uib.no (A.C.H.)
2 CMR Prototech AS, Fantoft, 5072 Bergen, Norway; ivar.warnhus@prototech.no (I.W.); arild.vik@prototech.no (A.V.)
* Correspondence: crina.silvia.ilea@prototech.no Tel.: +47-942-57-802
† These authors contributed equally to this work.

Received: 3 March 2020; Accepted: 25 March 2020; Published: 2 April 2020

Abstract: This paper studies the robustness of off-shore solid oxide fuel cell (SOFC) installations and the nature and causes of possible cell degradation in marine environments. Two important, cathode-related, impediments to ensuring SOFC reliability in off-shore installations are: cathode degradation due to salt contamination and oxygen depletion in the air supply. Short-term and long-term tests show the effect of salt contamination in the cathode feed on cell performance, and reveal the underlying cause of the degradation seen. SEM/X-ray Diffraction /(XRD) analyses made it possible to identify salt taken up in the cathode microstructure after the short-term testing while the macroscopic cell structure remained intact after the short-term tests. The long-term degradation was found to be more severe, and SEM images showed delamination at the cathode/electrolyte interface with salt present, something that was not seen after long-term testing without salt. The effect of oxygen depletion on the performance was also determined at three different temperatures using I-V curves.

Keywords: SOFC; reliability; contamination; salt; oxygen starvation; concentration polarization

1. Introduction

Fuel cells represent an important innovation within the energy conversion industry, converting the chemical energy in fuels, e.g., hydrogen or, in the case of high-temperature fuel cells, hydrocarbon fuels, directly to electricity without, as in combustion, the detour around heat energy with the need to convert this heat energy to mechanical or electrical energy, a process which is subject to the Carnot efficiency limitation. Fuel cell installations are easily scalable, they are noise-free and they emit no particulates and virtually no NOx, the energy conversion taking place at much lower temperatures than for combustion.

Fuel cells have longer lifetime and offer significantly higher power densities than batteries and are also smaller and lighter. Like batteries, they are suitable also for applications which require only low power output, such as laptop computers and other portable electronic devices and mobile phones [1].

For investors to invest in large-scale power conversion installations based on fuel cells, however, fuel cell installations need to be shown to be robust and reliable in the field. Herein lies an important challenge in fuel cell research and development today [2,3].

Since the issue of robustness is so important for implementation of fuel cell technology in the main-stream energy conversion industry, this has received some attention in the research literature recently. Much of the literature focuses on the effects of contaminants in the fuel on the anode or the effect of oxygen starvation on the cells as a whole. Below we review some of the literature dedicated to the effect of impurities on fuel cell performance, concentrating on solid oxide fuel cells.

Cayan et al. [4] investigated the effects of the presence of coal-derived impurities in the anode feed to solid oxide fuel cells. They found that Hg, Si, Zn and NH_3 do not lead to significant degradation. Cl, Sb, As, and P, on the other hand, led to significant degradation of the fuel cell performance, attacking the Ni-YSZ anode.

Bao et al. [5] similarly investigated the effects of eight kinds of coal-derived contaminants on the working of Ni-YSZ/YSZ/LSM solid oxide fuel cells (SOFC) and found that As and P in the feed resulted in significant loss of performance at the lower operating temperatures due to the Ni phase in the anode being compromised. Microstructural analyses helped the authors to characterize the mechanisms behind the degradation. Zn, Hg and Sb led to little degradation.

Cheng et al. [6] studied the effect of sulphur on the working of YSZ anodes in solid oxide fuel cells. Also Haga et al. [7] studied the effect of impurities, among other things sulphur compounds, on the working of solid oxide fuel cell cermet anodes. They found a direct initial drop in voltage when introducing the impurities, and also found a gradual additional drop in performance when the impurity was CH_3SH. They also found that Cl_2 in the feed gave rise to nickel nanoparticles via the formation of gaseous $NiCl_2$.

Horita et al. [8] studied the effects of impurities on the stability of solid oxide fuel cells. They studied the durability of a fuel cell stack for more than 5000 h. They found a gradual loss of performance (linear in time) of about 1.5% per 1000 h during the operation. They found by secondary ion mass spectrometry that the contaminants were present in the cathode, their concentration increasing with increasing operation time.

Yan et al. [9] modelled numerically degradation of the performance of solid oxide fuel cells due to carbon deposition. They found that the bilayer interconnect that they were testing was more robust under carbon deposition than a conventional interconnect, and that the performance of the cell only degraded slightly due to carbon deposition.

Many studies understandably concentrate on the effect of impurities in the fuel supply on the anode side. There are, however, some studies that focus on the effect of impurities on the cathode side.

Xiong et al. [10] studied the degradation due to the presence of trace SOx in the air on the cathode of solid oxide fuel cells. They found rapid degradation of the performance of an SSC cathode both in the form of ohmic and concentration polarization due to the introduction of SO_2 in the feed, while the performance of an LSM cathode decreased only slightly. They found that the formation of secondary phases, e.g., as strontium sulphate, was seen in the SSC cathode upon introduction of SO_2, while this was not detected in the LSM cathode.

Two studies of the effect of salt in the feed to an fuel cell cathode, one in solid oxide fuel cells, were found in the literature. Mikkola et al. [11] studied the effect of salt in the air stream to a cathode of a PEM fuel cell. They injected NaCl in the air supply to the cell. They found that the degradation due to salt in the air feed was less than they had anticipated, and studied the microstructure of the cathode after the testing to pinpoint the reason for the degradation that they did see.

Liu et al. [3] studied the effect of salt, at two different concentrations, 30 and 600 ppm, in the air feed to solid oxide fuel cells operated at a fixed current density of 200 mA/cm^2. With salt at a concentration of 30 ppm they found only moderate reduction in the cell voltage. When the salt concentration was increased to 600 ppm and the cell operated for 24 h, the cell voltage closely resembled that obtained with 30 ppm, i.e., only moderate reduction of the performance. In the conclusions of the paper the authors suggest that partial substitution of the Sr with Na could be responsible for the moderate reduction in the cell performance that they saw. They emphasize the need for longer-term testing.

We conclude from this study of the literature that most of the literature naturally enough focuses on the effects of impurities introduced with the fuel on the anode side of cells. For the operation of maritime fuel cell installations impurities on the cathode side may well pose a problem for the long-term performance of the installation. To our knowledge there are no studies in addition to that of Liu et al. [3] in the literature focusing on the likely disturbances leading to degradation in fuel cell installations offshore, and therefore no studies wherein complete I-V curves are presented or studies that involve long-term testing.

In this paper the focus is on SOFC installations working in a marine environment, namely:

1. To which extent individual cells will be degraded by the presence of salt (NaCl) in the feed to an SOFC cathode, and the underlying reasons for any such degradation.
2. The effect of oxygen depletion in the air supply to a cell.

I-V curves were generated to determine the cell performance both in short-term and long-term degradation tests with different salt concentrations in the cathode feed at three different temperatures, and the cell microstructure before and after the tests were investigated using scanning electron microscopy (SEM) in order to determine the underlying reasons for any changes in the performance.

I-V curves were also generated under conditions of varying degrees of oxygen depletion in the feed to the cathode at three different realistic operating temperatures.

2. Experimental

2.1. Test Set-Up

The experiments use the set-up shown in Figure 1, details of the cell are also shown in Figure 1. The components used in the set-up are described below.

2.1.1. Oven

One of the main components of the experimental set-up is the oven. It is used to heat the cell to the required temperature. The oven is programmed with ramping (rate at which cell is heated: 1.6 °C/min), set temperature (required temperature) and dwell time (time period during which the temperature is maintained).

2.1.2. The Cell Itself

The overall performance of SOFCs are determined by activation, ohmic and concentration polarisations. The electrolyte is the main contributor to the ohmic losses. The desired characteristics of an electrolyte to minimize ohmic losses are high ionic conductivity and small thickness. In this work, the electrolyte material used is yttria stabilised zirconia, or YSZ, which is widely used due to its excellent stability in both reducing and oxidising environments, even though its ionic resistivity, especially for the doping level used here, is not the lowest of the alternative electrolyte materials.

The single cell used for the experimental work is purchased from KERAFOL Keramische Folien GmbH, Germany, the commercial name of this cell type being "KeraCell I". It is an Electrolyte-Supported Cell with NiO/YSZ type anode and ScSZ/LSM type cathode. The electrolyte is made of 3 % Yttrium stabilised Zirconia (YSZ) and has a thickness of 85 μm. The diameters of the electrolyte and electrodes are 40 mm and 30 mm respectively (shown in Figure 2).

The key features of the cell are as follows:

- High power density
- Robust in nature for long term usage and in redox environments

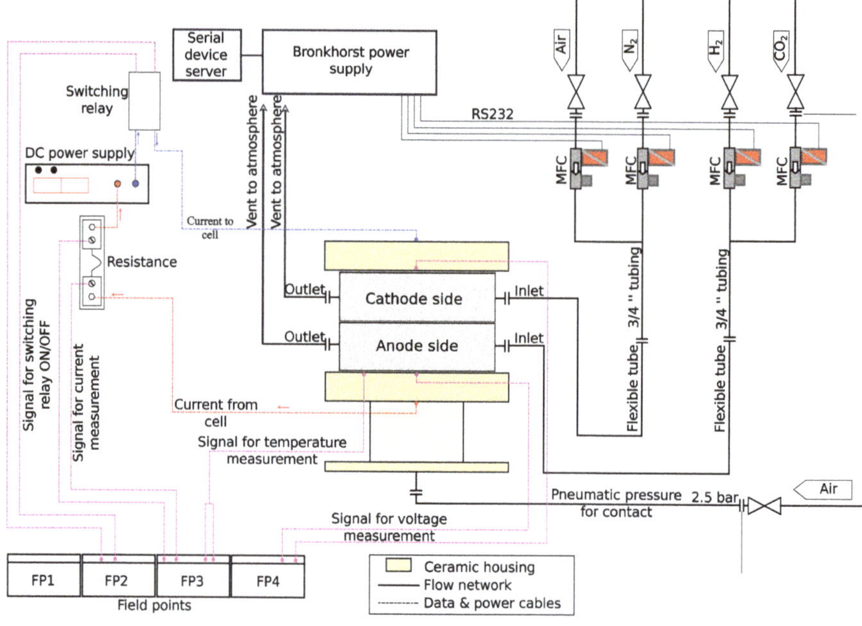

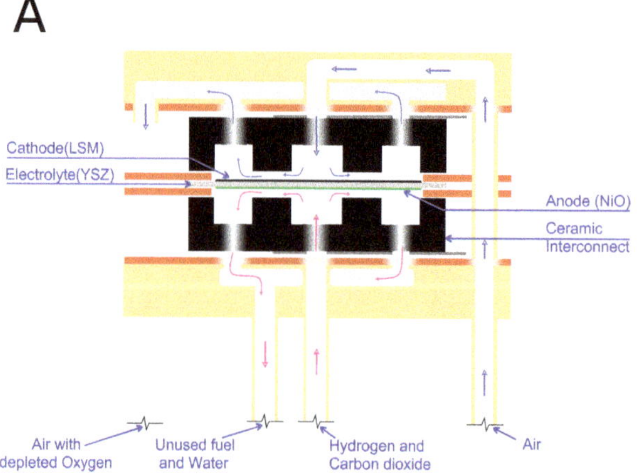

Figure 1. Piping and instrumentation diagram for the experimental set-up. (A) shows the whole setup, (B) a detail of the configuration around the cell itself

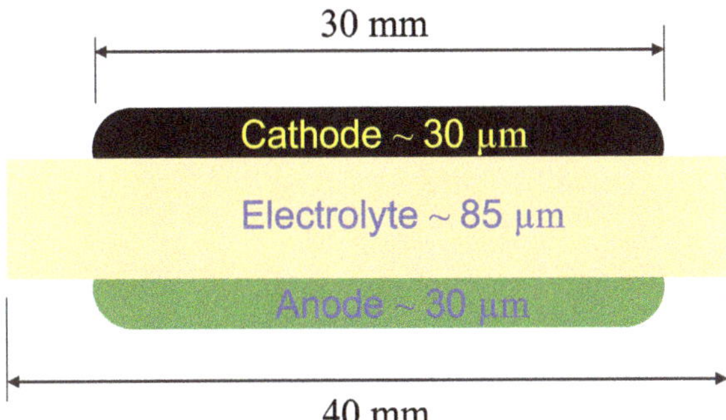

Figure 2. The circular, electrolyte-supported cell used in the experiments, the thicknesses of the electrodes and the electrolyte are indicated.

2.1.3. Gas Handling System

H_2 and CO_2 gases are supplied to the anode, while air and nitrogen are supplied to the cathode. Bronkhorst mass flow meters/controllers are installed on each of the gas lines and a common tube is connected to the fuel supply point of the set-up. A pneumatic compression force of 123×10^6 N was applied on the cell via the interconnects in order to achieve a better electrical contact. The air was supplied by 3/4 inch tubing from the main valve to a pressure-regulating valve. From this valve a flexible hose was connected to the pneumatic piston cylinder arrangement on the test set-up.

2.1.4. Field Points

The voltage and the current across the cell and the temperature of the cell were measured using field-points. Field-points gave capabilities that improved the reliability of the test system. It provided online diagnostics and innovative architecture that modularized I/O functions and communications. Easy connectivity was also provided by screw terminals.

An external DC device was used for current control. This device was controlled by LabVIEW programs and made it possible to move along the I-V curve in a controlled manner. The voltage across the cell was measured using an NI 16-channel analogue voltage input module. The OCV was measured without allowing any current to pass through the cell and this was done by using the external device. This relay was controlled by a 8-channel fused sourcing digital output module. The current from the anode was measured using a known resistor and an NI 8-channel module.

The temperature was measured using a thermocouple which was connected to a field point serial device. Shared variables were created in LabVIEW for reading the voltage, current and cell temperature and to control the current supply using the relay. The Bronkhorst gas flow controller communicates digitally with the gas flow meter/controller through an RS232 and the gas flow controller device is controlled remotely from LabVIEW.

2.1.5. Data Acquisition

The experimental set-up used makes a variety of measurements that require signal conditioning before the raw signal is digitized by the data acquisition system. Each individual cell may generate up to 1 V and hence calibration is important to ensure accurate measurements. Monitoring output current requires signal conditioning and scaling to convert the data back to a current reading. Temperature is another variable that is monitored by thermocouples. Hydrogen, oxygen, nitrogen and airflow rates are measured by mass flow meters (see Figure 3). These meters generate pulses at a rate proportional to the gas flow rate. A counter monitors the pulse rate and this is converted into flow rate.

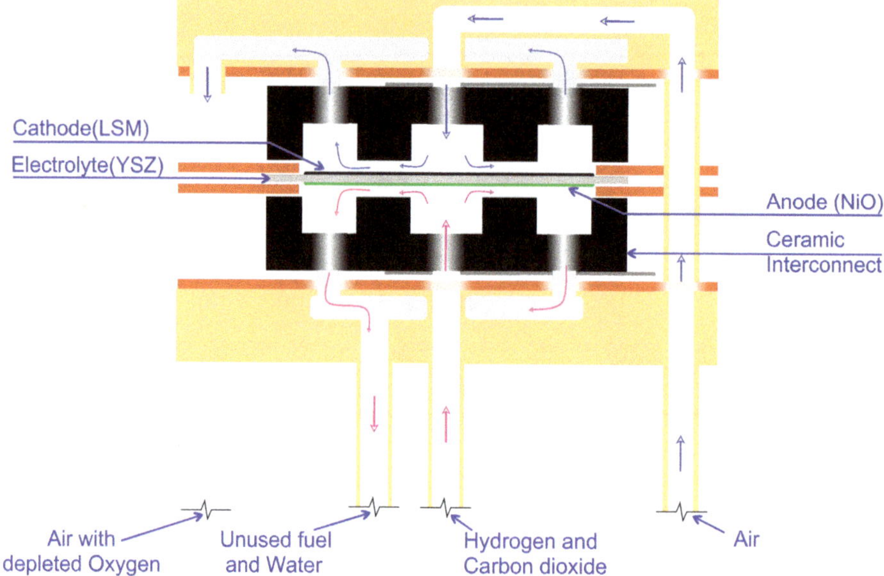

Figure 3. Simplified arrangement of the single cell experimental set-up used in the experiments, the flow directions are indicated.

The other tools used to prepare and characterize the cell are explained in the subsequent sections.

2.1.6. Current-Collection Layers

Additional layers can be applied to planar cells using screen printing (SP). The cells used for the experiments are provided with primary anode and cathode layers by the supplier, while additional current-collecting layers were formed on both electrodes in-house by SP [12]. Using efficient anode/cathode materials and structures will optimize the SOFC performance. The preparation process and the microstructure of the cathode strongly affect the concentration polarization resistance of the cathode [13].

2.1.7. Characterization of the Layers

In order to determine the local composition and morphology of the materials, two microscopic analysis methods, namely SEM and TEM (transmission electron microscopy) were used. SEM was used to analyse the surface of a sample, while TEM was used to analyse the internal structure by permeating a thin sliver or powder [14].

2.2. Method

2.2.1. Short-Term Salt Degradation Tests

In these tests the effect of salt on the cathode was evaluated for a relatively short period with the cell operating at a temperature of 850 °C. The tests were conducted in three stages. For the first 70 h, the cell was operated without any salt contamination. The salt was then introduced in the feed line and heated to a temperature of 600 °C (giving a molar concentration of 1.6 ppm in the gaseous feed to the cathode, see [15]) and the cell operated for 14 h. During the next stage the temperature of the solid salt in the feed line was increased to 770 °C (giving 250 ppm) for 50 h plus 18 h after the solid salt was replenished.

The flow parameters during these tests were: Cathode feed: Air: 600 mL/min, N_2: 0 mL/min. Anode feed: H_2: 250 mL/min, CO_2: 75 mL/min.

2.2.2. Long-Term Salt Degradation Tests

In these tests the performance of cathodes subjected to salt contamination for a long period was studied. Three single cells were studied, two of which were operated for 850 h, one of these two cells was contaminated with salt. The SEM images from these cells were compared to that of the third cell operated for a short duration without contamination or any load as a reference.

We should mention that the preparation of the cells for these tests was slightly different from those used for the short-term tests described above. Screen printing of NiO/LSM on the respective electrodes with embedded platinum mesh was avoided and this resulted in a reduction of losses thanks to thinner electrodes. In order to increase the current collecting layers on the cell, layers were screen printed onto the interconnects instead of on the electrodes themselves; two layers of LSM30 were screen printed on the cathode side, and one layer of NiO on the anode side using a precursor nanopowder prepared by a sol-gel process developed in-house [16]. The voltage across the cell was measured by means of a platinum wire making point contact directly on both the electrodes.

The flow parameters during these tests were: Cathode feed: Air: 2000 mL/min, N_2: 0 mL/min. Anode feed: H_2: 250 mL/min, CO_2: 75 mL/min.

The procedure was as follows. After the cell was heated up to the operating temperature in an electric furnace, the anode was reduced by supplying CO_2 at the rate of 75 mL/min and gradually increasing the flow of H_2 to the anode in steps as follows: 10 mL/min, 20 mL/min, 50 mL/min, 100 mL/min, 150 mL/min and 250 mL/min. Between each step sufficient time intervals were allowed for the cell voltage to stabilize. At the cathode the air flow was maintained at 2000 mL/min. After reduction of the anode the I-V characteristics of the cell were drawn and then the cell was operated at a constant current density of 285 mA/cm^2 while the cell voltage was monitored.

During the first test the operating temperature of the cell was maintained at 900 °C and the long-term performance without salt contamination was measured at a constant current load of 285 mA/cm^2 for 850 h. The next test was conducted on a new cell with the same operating conditions with salt introduced to the cathode at 1 ppm. The SEM images and I-V characteristics of these two cells were then, as mentioned, compared with a third, reference, cell operated for a short time, namely 12 h, without any current load at 900 °C.

2.2.3. Oxygen Depletion Tests

The performance of an SOFC cathode under decreasing partial pressure of oxygen (P_{O_2}) in the oxidant air supplied to the cathode was studied and analysed. This experiment was performed to evaluate the effect on the cell performance if it is coupled in series with other cells, the oxygen concentration reducing in successive stages, or the supply of oxygen in air is reduced, e.g., due to leakage of a seal or is stopped for short periods, e.g., due to air supply interruption.

The experiments were conducted at three different cell temperatures of 800 °C, 850 °C and 900 °C and the effects were characterized by I-V curves. For planar SOFCs with LSM electrodes, the normal operating temperature range is between 800 °C and 1000 °C, hence we chose to conduct the experiments in this range.

Heating of the cell for the testing was carried out under a constant supply of air at 200 mL/min at the cathode side. The cell was heated to 850 °C at a rate of 1.6 °C/min. Heating and cooling rates were determined in such a way as to prevent cracking in the cell.

Reduction of the cell was carried out at 850 °C at a constant supply of CO_2 of 75 mL/min and a stepwise addition of H_2 of 10, 20, 50, 100, 150, 250 mL/min at the fuel side. Oxidant air was supplied at the rate of 600 mL/min to the cathode side.

A sequence of measurements at a given cathode oxygen concentration included a measurement of the OCV with no external current applied and a series of measurements of the voltage over the cell at step-wise increasing external current (each time allowing a suitable time interval to reach steady-state), until the current exceeded a previously chosen maximum value of 3.5 A or the voltage over the cell decreased below a chosen minimum value of 0.5 V.

The oxygen concentration at the cathode was decreased step-wise by decreasing the oxygen flow—and increasing the nitrogen flow to make the total flow constant—in steps of 80 mL/min. The initial flow parameters are shown in Table 1. Also here time was allowed for establishment of steady-state after each change in flowrates.

Table 1. Initial flow parameters for the oxygen depletion tests.

Cathode:	Air	600 mL/min	N_2	0 mL/min
Anode:	H_2	250 mL/min	CO_2	75 mL/min

All measurements were performed with a fuel composition of CO_2 = 75 mL/min and H_2 = 250 mL/min. This rate avoids the depletion of fuel influencing the I-V measurements at higher fuel utilisation rates. A comparison of the electrochemical performance at different P_{O_2} supplied to the cathode of the single cell and at different operating temperatures (800, 850, 900 °C) are analysed by the I-V characteristics. The I-V curve was measured by increasing the current load starting from 0 A, incrementing step-wise with 0.25 A/cm^2, until a current load of 3.5 A/cm^2 was reached or the cell voltage dropped below 0.5 V.

The initial oxidant air composition had 21% oxygen and 79% nitrogen and the total flow rate was 600 mL/min. The oxygen in air was reduced in steps by reducing the supply of air and adding nitrogen in such a way as to keep the total flow constant at 600 mL/min. After completing one step, the flow parameters for air and nitrogen were changed automatically.

3. Results and Discussion

3.1. Salt Degradation Tests

3.1.1. Short-Term Salt Degradation Tests

The effect of salt contamination on the cell operating at 850 °C is shown in Figure 4. This cell was initially operated without any salt contamination for 70 h (marked initial in Figure 4) and the cell voltage remained almost unchanged during this period. Also after salt had been introduced to the cathode at a concentration of 1.6 ppm for 14 h (stage 2 in the figure), the performance of the cell remained almost the same as it was without contamination.

After the salt contamination had been increased to 250 ppm for 50 h, significant degradation in the performance of the cell was seen (stage 3 in the figure). The voltage dropped by 25 mV at 325 mA/cm^2. New salt crystals of 0.512 g were then added to the feed pipe and charged to the cathode at the rate of 250 ppm for 18 h.

The microstructure of the cell was examined using SEM, see Figure 5 for an SEM image of the whole cell cross-section, and traces of Na present at the surface of the cathode and diffusion of Na into the cathode was analysed by performing X-ray Diffraction (XRD) associated with the SEM on the cathode cross-section.

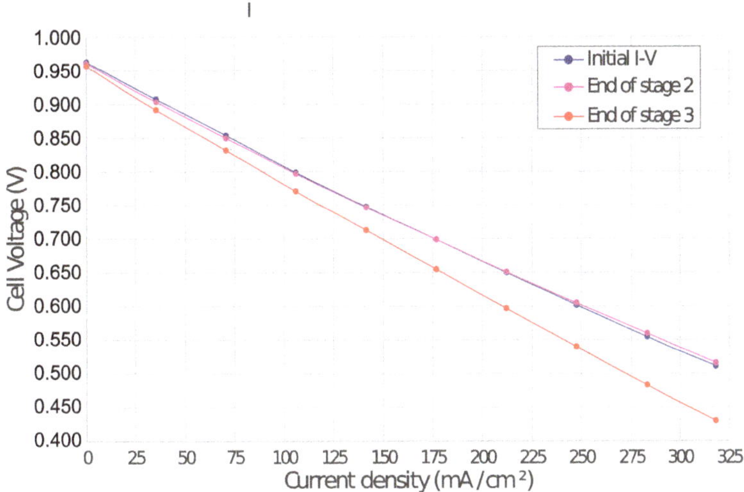

Figure 4. The performance of the cell at different stages of the short-term degradation test: **Stage 1** is the I-V of the cell having been operated for a period of 50 h without any salt contamination. **Stage 2** is the I-V for the cell after having been operated with salt contamination for 14 h at the rate of 1.6 ppm. **Stage 3** is the I-V for the cell after having been operated with increased salt contamination of 250 ppm for 50 h.

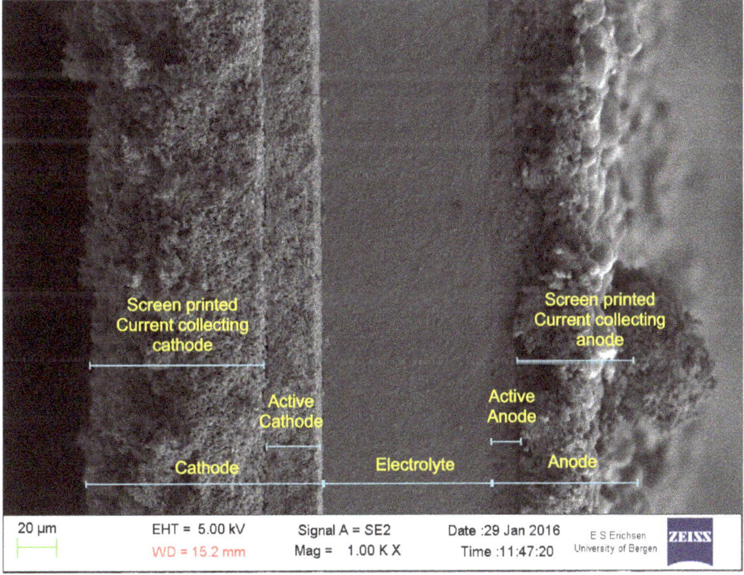

Figure 5. Cross-sectional SEM image of the entire cell operated at 850 °C, after a short-term degradation test.

The points at which the XRD was done are shown in Figure 6 and the result of the XRD analyses is shown in Figure 7, the concentrations measured are shown in Table 2. It is clear that the concentrations of Na in weight % increase gradually from point 2 to point 5 (1.06%) and some traces of Na can be found in the cathode functional layer.

Table 2. Mass % of elements.

pt.	B-K	C-K	O-K	Na-K	Al-K	Ca-K	Sc-K	Mn-K	Ni-K	Sr-L	Zr-L	La-L
1		7.77	7.47	0.00	15.76			3.75	45.36		8.50	11.38
2		1.85	11.67	0.25	1.90	0.32		28.72	5.30	7.66		42.33
3		1.36	10.66	0.26	0.45	0.27		30.36	4.04	7.77		44.82
4		1.56	13.64	0.44				30.16	1.23	8.89		30.26
6		2.03	17.86	0.91				28.66		8.99		42.04
7	0.65	2.47	27.27	0.30			3.55	12.36		2.64	34.78	15.99
8	0.85	2.88	29.39	0.28			2.83	10.94	0.64	2.05	34.99	15.15

The SEM images obtained from the cell operated with salt contamination for a short term duration do not show any sign of delamination. This indicates that the cell mechanically tolerates salt contamination for a short duration. However, as shown by Figure 4 the cell voltage dropped considerably, without us being able to say categorically why this should be so.

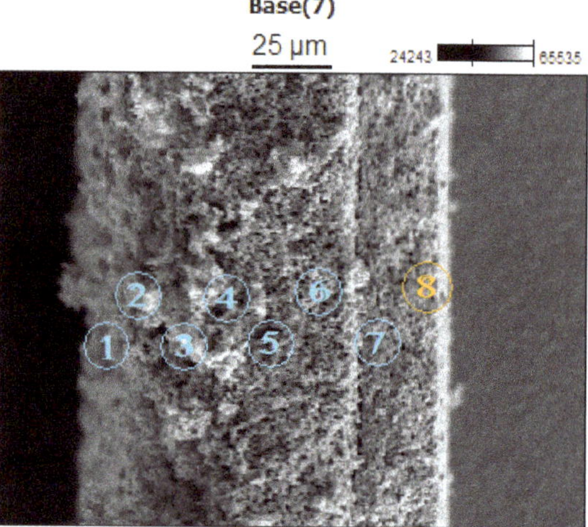

Figure 6. Cross-sectional SEM image of the cathode showing the points at which X-ray diffraction was performed to analyse the components present.

We thus do not see any sudden degradation of the cell performance upon introduction of the impurity, like that seen by Cheng et al. [6] for some of the impurities that they tested, but rather a gradual decrease in the performance as also seen by them in some of their tests. This is confirmed by the long-term tests discussed below.

The results of the short-term tests are qualitatively similar to those obtained by Liu et al. [3], but in contrast to them we did find more reduction in cell performance, although still moderate, with the higher salt concentration in the cathode feed than with the lower concentration.

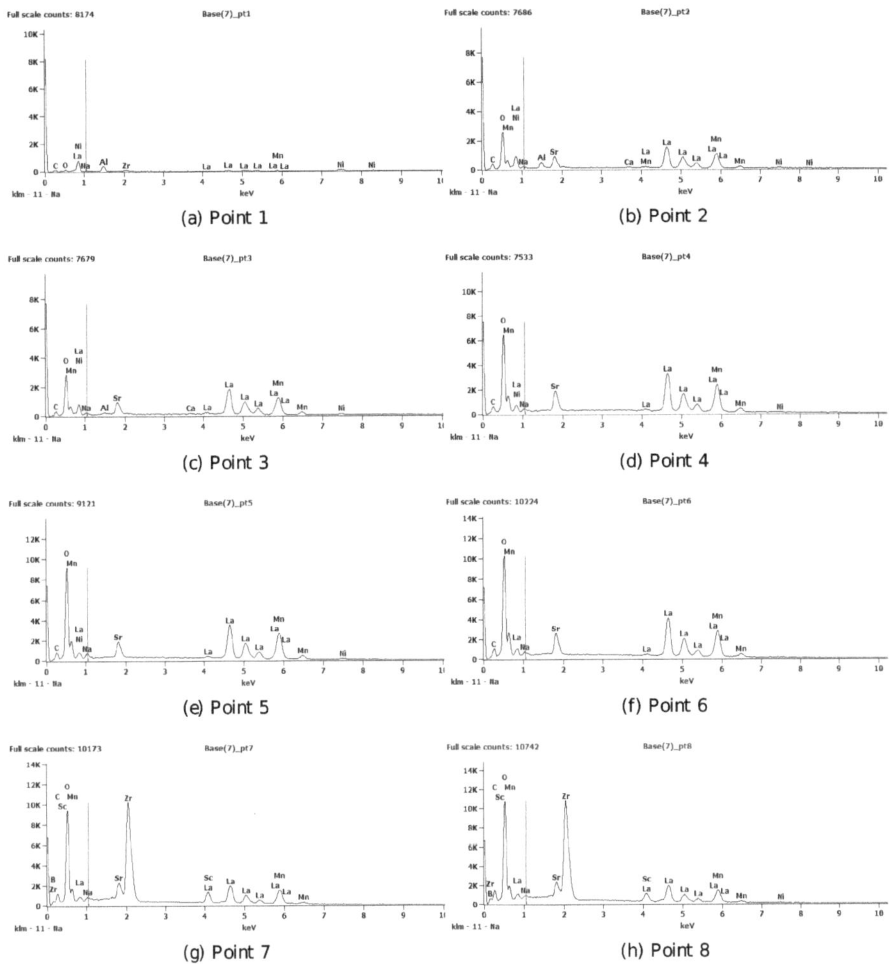

Figure 7. X-ray Diffraction (XRD) patterns of LSM current collecting layer in the cathode and the ScSZ/LSM cathode electro-catalytic layer.

3.1.2. Long-Term Salt Degradation Tests

The long-term degradation effects on a single cell are shown in Figure 8. The degradation curve for a cell without any salt contamination at 900 °C shows that there was an initial degradation from 0.8375 V to 0.825 V (12.5 mV) taking place over a period of 40 h. After this, the cell voltage stabilised and there was a drop of only 10 mV during the next 460 h of operation. After an initial 550 h of operation, a bad contact was observed in the current-collecting wire from the anode. For this reason, the cell had to be cooled down and, after fixing the connections, the cell was operated again. During this period the anode had oxidised and the anode redox reactions started to take place when the cell was operated again. Hence, there was a drop in cell voltage at around 560 h (23rd day of cell operation). The total drop in cell voltage over 850 h was 25 mV, as observed from the degradation curve.

The performance of the cell with salt contamination at 900 °C was studied in the second test. Here a continually decreasing performance of the cell was observed and the cell voltage dropped by 200 mV over a period of about 660 h (27.5 days). During the operation of this test with salt contamination for a long duration, a bad contact was observed at the voltage measurement probe of the anode side. The cell operation was limited to 660 h due to time constraints.

Figure 9 shows the initial and final performance of the two cells with and without salt contamination. The curves marked initial indicates the baseline I-V characteristics observed immediately after reducing the anode. The curves marked final show the characteristics of the cell after operating it under a steady-state current load for a long time (respectively 850 h and 660 h without and with salt contamination). The initial baseline I-V curves for both the cells show that their performance before the testing were identical.

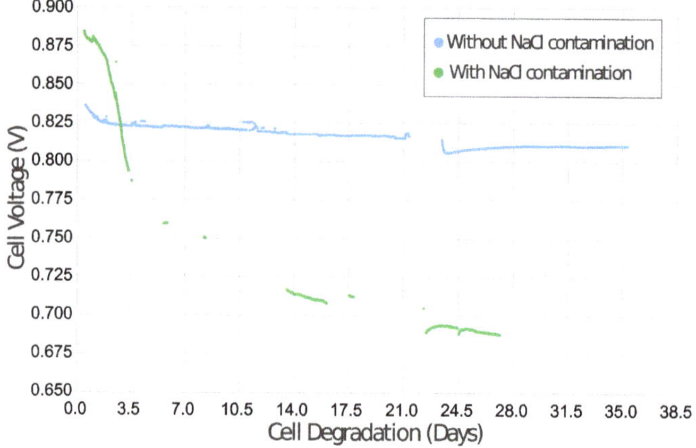

Figure 8. Long-term degradation of the single cell at 900 °C and a current density of 285 mA/cm^2 without salt contamination and with 1 ppm of continuous salt contamination.

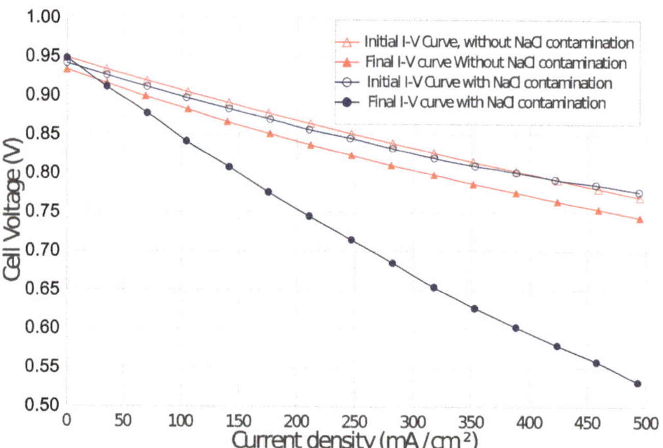

Figure 9. The baseline I-V curves and final I-V curves of the cell at 900 °C, without salt contamination for 850 h and with salt contamination of 1 ppm for 660 h.

The I-V curves obtained after 850 h of cell operation without salt contamination exhibited a drop of 25 mV cell voltage when compared to the initial baseline curve.

The final I-V curve obtained after long-term testing with salt contamination to the cell shows a large increase in ohmic losses when compared to the loss seen in the final I-V curve without salt contamination. At a current load of 500 mA/cm^2, the drop in cell voltage is about 200 mV.

When searching for the reason for the severe drop in long-term cell performance, the microstructure of the cell was studied using SEM. The results of this are shown in Figure 10.

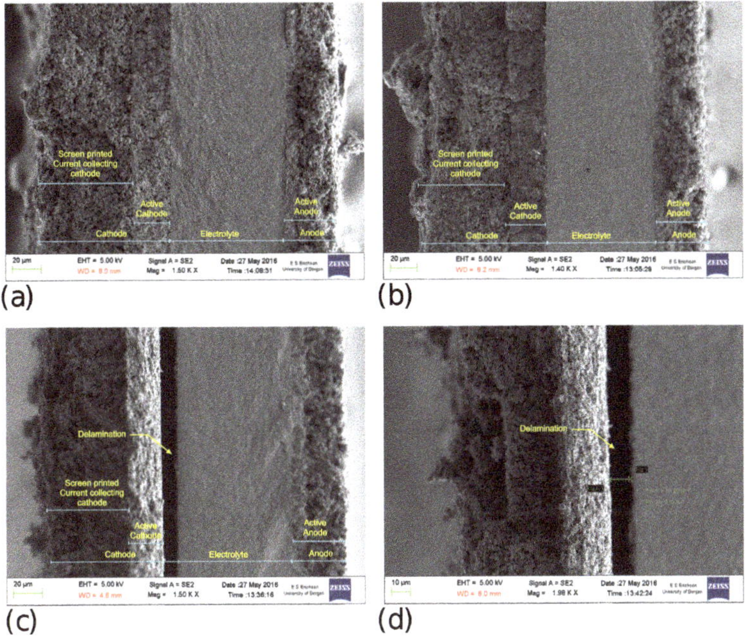

Figure 10. Cross sectional SEM images of the cells after long-term testing at 900 °C. (**a**) Shows the cell in its initial state. (**b**) shows the cell that has been operated for 850 h without salt contamination. (**c**) shows the cell after operation with salt contamination (at 1 ppm) for 660 h, the cell can be seen to have delaminated leading to severe and irreversible degradation. (**d**) shows a detail of the delamination.

The SEM images show that there is no delamination of the cell operated without salt contamination, while the cell operated with salt contamination develops delamination. This is clearly the reason that the cell operated for a long term without any salt contamination shows better performance than the cell operated with salt contamination.

Liu et al. [3] did not perform long-term tests and could therefore not have found the same severe reduction in cell performance due to delamination as we found in this study.

3.2. Oxygen Depletion Tests

The I-V tests to study the effect of oxygen depletion in the air supply to the cathode have three main elements:

- Characterising the open circuit voltage (OCV) of the cell to ensure that there was very little cross-leak or electronic conductivity in the electrolyte and that the components were sealed properly [17].
- Examining the shape of the I-V curve at a specific temperature in order to determine the amount of the polarizations: activation, ohmic and (importantly) concentration polarization at the cathode due to varying oxygen partial pressure at three temperatures (800 °C, 850 °C and 900 °C).
- Studying the electrochemical performance of the cell at high current density and at reducing partial pressure of oxygen.

We would expect that the I-V characteristics of the SOFC will show that the ohmic loss increases almost linearly as the current density increases. At a relatively large current density, the potential of the cell would be expected to decrease more rapidly and in a non-linear fashion, indicating the dominance of concentration polarization. A high current density requires fast consumption of the reactant gas (oxygen) and also fast exhaustion of the products, possibly exceeding the gas transport rate in the electrode, both due to the oxygen-depleted diffusion film adjacent to the active sites and due to transport limitation in the fine pores [18,19].

The equation for the EMF of the cell from fundamental thermodynamics is:

$$E = E^{\circ} + \frac{RT}{2F} \ln \left(\frac{\alpha \cdot \beta^{\frac{1}{2}}}{\delta} \right) + \frac{RT}{4F} \ln \left(\frac{P}{P^{\circ}} \right) \tag{1}$$

where, if the gaseous reactants and products behave ideally, which is probably a good approximation at the low pressures used here, α, β and δ are the mole fractions of H_2, O_2 and H_2O, respectively.

The effect of the oxygen concentration on the EMF of the cell can thus be expressed:

$$E = E_0 + \frac{RT}{4F} \ln \beta \tag{2}$$

or the effect on the cell voltage:

$$\Delta V = \frac{RT}{4F} \ln \beta \tag{3}$$

where β is the mole fraction of O_2

As mentioned, we can roughly distinguish between ohmic polarization (OP) and concentration polarization (CP) from the shape of the I-V curves.

Figures 11 and 12 show the I-V curves for cells working at 800 °C and 850 °C, respectively. The I-V curve is first determined at 21% oxygen to establish a reference curve. Keeping all the operating parameters constant, it is clear that the OP, which is the slope of the I-V curve, is higher in the cell operating at 800 °C and decreases with increasing temperature (Figure 13) . This agrees very well with the literature that the oxygen ion conductivity through the 3YSZ electrolyte increases with temperature in the range 600 °C to 1000 °C [20].

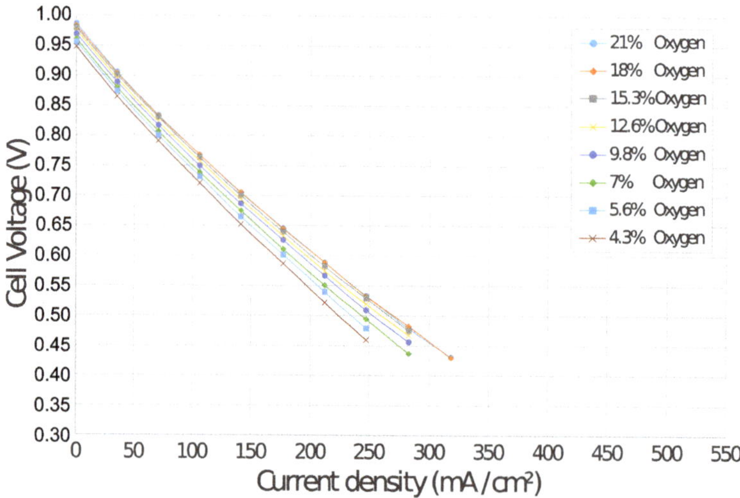

Figure 11. The performance of the cell at 800 °C with decreasing partial pressure of oxygen. The ohmic losses are high and the drop in voltage is significant as the partial pressure of oxygen at the cathode decreases.

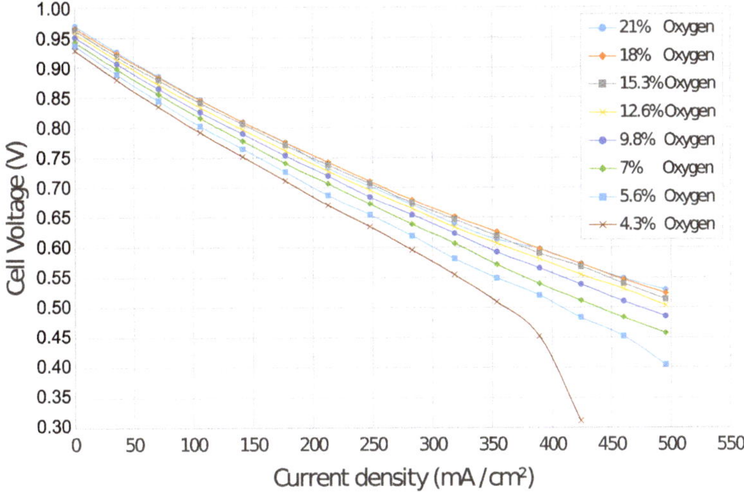

Figure 12. The performance of the cell at 850 °C with decreasing oxygen partial pressure. The performance decreases for lower partial pressures, under 12.6% oxygen. However, the loss is less than that at 800 °C.

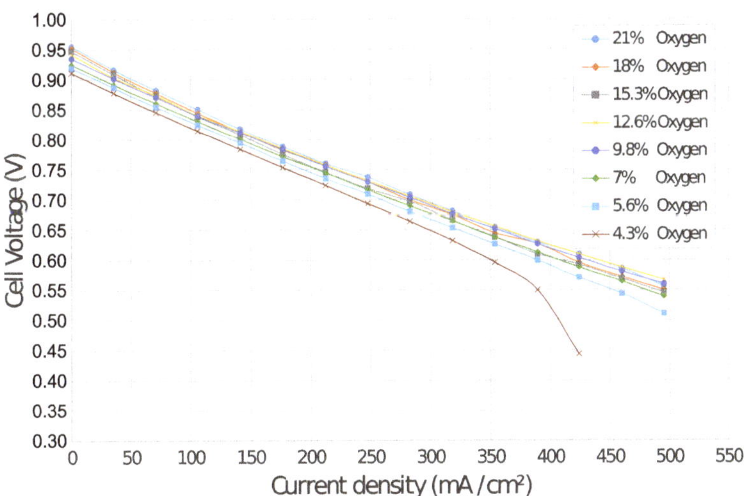

Figure 13. The performance of the cell operating at 900 °C with decreasing partial pressure of oxygen—At 900 °C, the performance of the cell is better than that of the cells operated at 800 °C and 850 °C. But its performance decreases for very low partial pressure of oxygen, smaller than 5.6% oxygen.

A comparison of CP at 850 °C and 900 °C at lower partial pressures of oxygen in the feed supplied to the cathode is shown in Figure 14. This plot indicates that as the operating temperature of the cell is increased, the effect of decreasing partial pressure is reduced. However, the current densities at these higher temperatures could be made so high that the effect of CP could be seen to set in at the lowest oxygen concentrations.

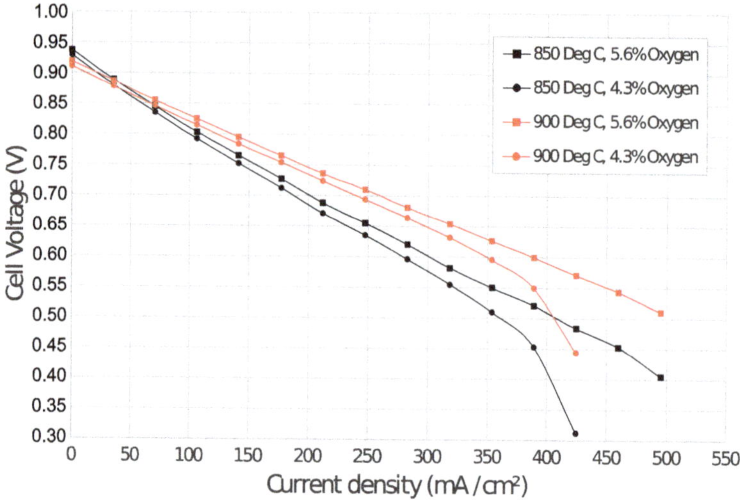

Figure 14. I-V curves for the two highest operating temperatures at the two lowest oxygen partial pressures. This shows that as the operating temperature of the cell is increased, the effect of decreasing partial pressure of oxygen is mitigated.

Figure 15 gives information about the cell performance at the three different temperatures and shows voltage-current density and power-current density characteristics, from which it can be seen that at 900 °C the cell has the higher power density of 280 mW/cm^2. This is consistent with the literature [21,22].

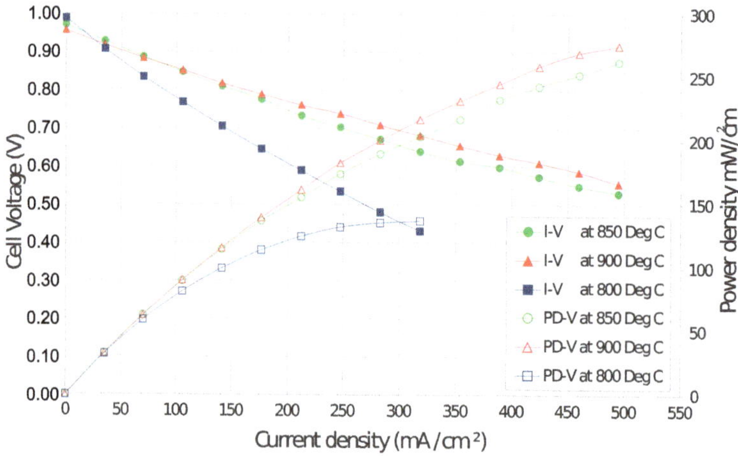

Figure 15. Voltage-current density and power-current density plots of the planar SOFC using air as oxidant—While using air as an oxidant (21% oxygen), it is observed that at low current densities very little difference in the performances of the cells operating at 850 °C and at 900 °C is observed. However, the cell operated at 800 °C exhibits very high ohmic losses.

Figure 16 shows the effect on the cells of carrying out the testing program. The figure shows I-V curves with 21% oxygen to the cathode both before and after the testing program. It is seen that the performance of the cell before carrying out the testing is not much different from the performance after the testing. Only at 850 °C the cell performance after the testing is significantly lower than its initial performance, which might indicate that the oxygen concentration to the cathode at this temperature, which was reduced to lower values (4%) than at the other two temperatures, reached a critically low level during the experimental program whereby the cathode changed.

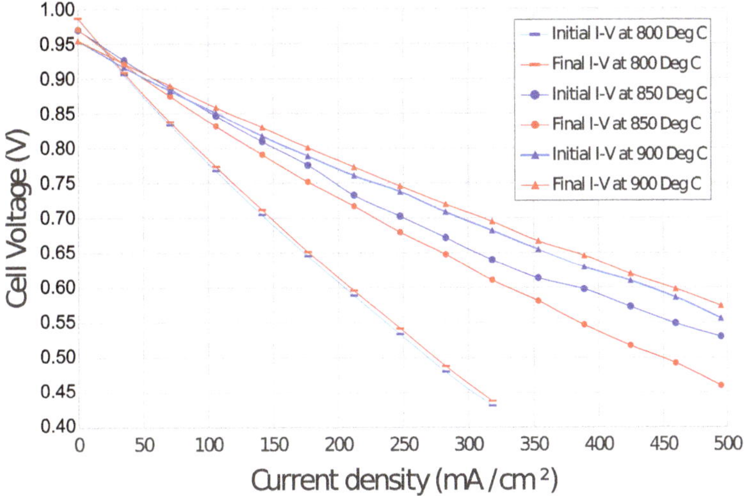

Figure 16. Voltage-current density comparison of the cell at different temperatures before and after the testing program—At 800 °C, 850 °C and at 900 °C

4. Further Discussion

Figure 8 clearly shows that there were some equipment-related problems involved in the long-term tests, but although the curve is interrupted it does show a plausible continuous development in the cell performance as a function of time. We were able to show a clear difference in cells operated under the same conditions, apart from salt contamination in the flow to the anode, obtaining delamination only in the contaminated cell. It is important to carry out work to reproduce this result and to determine the reason for the delamination, perhaps testing whether expansion or contraction of one the cell components due to the presence of the salt in the feed to the cathode could cause the delamination.

The effect of the oxygen depletion in the cathode feed to cause concentration polarization is quite clear and the evidence shown in this paper is in accord with the classical I-V response to concentration polarization. The results for the different operating temperatures also demonstrate that these effects are reproducible.

5. Conclusions

- When the experiments were conducted without any salt contamination (both short-term and long-term), the final I-V curve of the cell was very close to that of the reference I-V characteristics observed immediately after reducing the anode showing that the testing method itself did not lead to significant degradation of the cell.
- During the short-duration tests, when salt crystals were introduced at the rate of 250 ppm, there was a drop in cell voltage by 25 mV. The SEM images after operating the cell with salt contamination for about 50 h, did not show any sign of delamination, but salt is detectable in the cathode microstructure.
- During the long-duration testing, when salt was introduced at the rate of 1 ppm, a drop in cell voltage of 200 mV was seen. The SEM images show that after operating the cell with salt contamination for about 660 h, delamination had taken place. This indicates that there is a severe and irreversible degradation of the cell when salt pollutes the cathode for a long period.

- As the operating temperature of the cell is increased from 800 °C to 900 °C, the contribution to losses from ohmic polarization decreases. An electrolyte supported cell (YSZ) should therefore be operated at higher temperatures.
- The effect of decreasing oxygen partial pressure is more significant when the oxygen concentration is decreased below 12.6%. The fall in cell voltage at higher current densities and very low oxygen concentration indicates the development of concentration polarisation.
- While using air as an oxidant (21% O_2), after operating the cell at reducing partial pressure, it is observed that the cell recovers immediately and there is very little difference in cell losses after testing when compared to before. When testing at 850 °C the cell was starved of oxygen to a greater extent (4% O_2) and in this case a significant drop in voltage due to the experimental program having been carried out was seen. The existence of a threshold low oxygen concentration, beyond which the cell appears to be damaged irreversibly, and the reason for the existence of such a threshold is worth further investigation.

Author Contributions: Conceptualization I.W., A.V., C.S. and A.C.H.; methodology, C.S.I.W.; software, N.T. and I.W.; validation, C.S., N.T. and A.C.H.; formal analysis, C.S. and N.T. and A.C.H.; investigation, N.T. and C.S.; resources, A.V. and I.W.; data curation, N.T. and C.S.; writing–original draft preparation, N.T. and A.C.H.; writing–review and editing, C.S. and A.C.H.; visualization, X.X.; supervision, C.S. and A.C.H.; project administration, C.S. and A.C.H.; funding acquisition, A.V. and I.W. All authors have read and agreed to the published version of the manuscript.

Funding: This research was funded by Gassnova grant number CHEOP-CC616060, and The Norwegian Research Council grant number CHEOP 245489/E30.

Acknowledgments: The authors would like to thank Gassnova and The Norwegian Research Council for funding this project and Shell and Statoil for helping as partners.

Conflicts of Interest: The authors declare no conflict of interest.

References

1. Ormerod, R.M. Solid oxide fuel cells. *Chem. Soc. Rev.* **2003**, *32*, 17–28. [CrossRef]
2. Nakajo, A.; Mueller, F.; Brouwer, J.; van Herle, J.; Favrat, D. Mechanical reliability and durability of SOFC stacks. Part I : Modelling of the effect of operating conditions and design alternatives on the reliability. *Int. J. Hydrog. Energy* **2012**, *37*, 9249–9268. [CrossRef]
3. Liu, R.; Wang, D.; Jing, L.; Wang, L.; Bai, X. NaCl Effect on SOFC Cathode. In Proceedings of the International Conference on Civil, Materials and Environmental Sciences (CMES 2015), London, UK, 14–15 March 2015.
4. Cayan, F.N.; Zhi, M.J.; Pakalapati, S.R.; Celik, I.; Wu, N.Q.; Gemmen, R. Effects of coal syngas impurities on anodes of solid oxide fuel cells. *J. Power Sources* **2008**, *185*, 595–602. [CrossRef]
5. Bao, J.; Krishnan, G.N.; Jayaweera, P.; Perez-Mariano, J.; Sanjurjo, A. Effect of various coal contaminants on the performance of solid oxide fuel cells: Part I. Accelerated testing. *J. Power Sources* **2009**, *193*, 607–616. [CrossRef]
6. Cheng, Z.; Zha, S.W.; Liu, M.L. Influence of cell voltage and current on sulfur poisoning behavior of solid oxide fuel cells. *J. Power Sources* **2007**, *172*, 688–693. [CrossRef]
7. K..; Adachi, S.; Shiratori, Y.; Itoh, K.; Sasaki, K. Poisoning of SOFC anodes by various fuel impurities. *Solid State Ion.* **2008**, *179*, 1427–1431. [CrossRef]
8. Horita, T.; Kishimoto, H.; Yamaji, K.; M. E. Brito, ME (Brito, M.E.; Xiong, Y.P.; Yokokawa, H.; Hori, Y.; Miyachi, I. Effects of impurities on the degradation and long-term stability for solid oxide fuel cells. *J. Power Sources* **2009**, *193*, 194–198. [CrossRef]
9. Yan, M.; Fu, P.; Chen, Q.Y.; Wang, Q.W.; Zeng, M.; Pandit, J. Electrical Performance and Carbon Deposition Differences between the Bi-Layer Interconnector and Conventional Straight Interconnector Solid Oxide Fuel Cell. *Energies* **2014**, *7*, 4601–4613. [CrossRef]
10. Xiong, Y.P.; Yamaji, K.; Horita, T.; Yokokawa, H.; Akikusa, J.; Eto, H.; Inagaki, T. Sulfur poisoning of SOFC cathodes. *J. Electrochem. Soc.* **2009**, *156*, B588–B592. [CrossRef]
11. Mikkola, M.S.; Rockward, T.; Uribe, F.A.; B. S. Pivovar, B. The effect of NaCl in the cathode air stream on PEMFC performance. *Fuel Cells* **2007**, *7*, 153–158. [CrossRef]

12. Rotureau, D.; Viricelle, J.P.; Pijolat, C.; Caillol, N.; Pijolat, M. Development of a planar SOFC device using screen-printing technology. *J. Eur. Ceram. Soc.* **2005**, *25*, 2633–2636. [CrossRef]
13. Hildenbrand, N.; Boukamp, B.A.; Nammensma, P.; Blank, D.H. Improved cathode/electrolyte interface of SOFC. *Solid State Ion.* **2011**, *192*, 12–15. [CrossRef]
14. Goodhew, P.J.; Humphreys, J.; Beanland, R. *Electron Microscopy and Analysis*; CRC Press: Boca Raton, FL, USA, 2000.
15. Thambiraj, N.; Suciu, C.; Waernhus, I.; Vik, A.; Hoffmann, A.C. SOFC cathode degradation due to salt contamination. *ECS Transactions* **2017**, *78*, 915–925. [CrossRef]
16. Suciu, C.; Hoffmann, A.C.; Dorolti, E.; Tetean, R. NiO/YSZ nanoparticles obtained by new sol-gel route. *Chem. Eng. J.* **2008**, *140*, 586–592. [CrossRef]
17. Waldbillig, D.; Wood, A.; Ivey, D.G. Electrochemical and microstructural characterization of the redox tolerance of solid oxide fuel cell anodes. *J. Power Sources* **2005**, *145*, 206–215. [CrossRef]
18. Yonekura, T.; Tachikawa, Y.; Yoshizumi, T.; Shiratori, Y.; Ito, K.; Sasaki, K. Exchange Current Density of Solid Oxide Fuel Cell Electrodes. *ECS Transactions* **2011**, *35*, 1007–1014.
19. Bevc, F. Advances in solid oxide fuel cells and integrated power plants. *Proc. Inst. Mech. Eng. Part A: J. Power Energy* **1997**, *211*, 359–366. [CrossRef]
20. Shekhawat II, D.; Spivey, J.J.; Berry, D.A. *Fuel Cells: Technologies for Fuel Processing: Technologies for Fuel Processing*; Elsevier: Amsterdam, The Netherlands, 2011.
21. Sasaki, K.; Tamura, J.; Hosoda, H.; Lan, T.N.; Yasumoto, K.; Dokiya, M. Pt–perovskite cermet cathode for reduced-temperature SOFCs. *Solid State Ion.* **2002**, *148*, 259–271. [CrossRef]
22. Tietz, F. Peculiarities in the thermal expansion behavior of ceramic fuel cell materials. *Adv. Sci. Technol.* **1999**, 61–70.

Article

A Transient Behavior Study of Polymer Electrolyte Fuel Cells with Cyclic Current Profiles

Yan Shi [1,*], Holger Janßen [1] and Werner Lehnert [1,2]

1 Forschungszentrum Jülich GmbH, Institute of Energy and Climate Research (IEK-3), 52425 Jülich, Germany; h.janssen@fz-juelich.de (H.J.); w.lehnert@fz-juelich.de (W.L.)
2 Faculty of Mechanical Engineering, RWTH Aachen University, 52072 Aachen, Germany
* Correspondence: y.shi@fz-juelich.de; Tel.: +49-2461-61-1902

Received: 22 May 2019; Accepted: 19 June 2019; Published: 20 June 2019

Abstract: This paper reports on the effects of different load profiles on the transient behavior of a polymer electrolyte fuel cell (PEFC). A protocol of six tests, each with different current density ramps, was conducted. The corresponding cell voltage, pressure drop response, and ohmic resistance were then experimentally investigated. The time-dependent voltage profiles were applied to represent the cell performance. The cathodic pressure drop and ohmic resistance were utilized to analyze the water dynamic behavior inside the cell. The voltage overshoot and undershoot behavior were observed throughout the experiment. It was found that with an increase of the current change rates, the magnitude of voltage over/undershoots also increased. When the holding time at the constant current density was zero, the overshoot or undershoot behavior disappeared. The results of the pressure drop analysis showed that the load ramp did not have a significant effect on the average pressure drop in the tests. During the load cyclic operation in each test, the two-phase flow tended to reach equilibrium in the cell. Impedance analysis showed that the ohmic resistance changed with the change in the current density; however, the difference between the tests was not obvious.

Keywords: polymer electrolyte fuel cell; cyclic current profile; transient behavior; pressure drop; Ohmic resistance

---

1. Introduction

The internal combustion engine has been used for decades as a means of locomotion, but it releases many pollutants, such as NO_x, CO, and soot, etc. These contribute to environmental problems, including climate change and local pollution in cities [1]. Thus, car makers supported by research groups have focused their research on new technologies, such as fuel cell electric vehicles, to displace traditional internal combustion engine locomotion [2]. By using fuel cell technology, the energy-carrying fuel can be electrochemically transformed into electric energy [3]. Amongst many different kinds of fuel cells, the polymer electrolyte fuel cell (PEFC), with a relatively low operating temperature of up to 80 °C, has attracted significant attention due to its high efficiency, high power density, and 100 percent environmentally friendly output when operated with hydrogen. PEFC systems were regarded as promising alternatives and clean energy converters for automotive applications [4,5]. A variety of studies on PEFCs have focused on steady-state behavior at different operating conditions, including the effect of temperature, the humidification level, and the stoichiometric ratio of reactants, as well as the flow-field structure of single cells or stack performance [6–8]. However, in the real application of PEFC systems for vehicles, the fuel cell should satisfy the requirements of fast response during the start-up, stop, acceleration, and braking processes in a short time. Thus, thorough research on the PEFC transient response to continuous load change is necessary.

In 1996, Amphlett et al. [9] created a dynamic model to investigate the transient behavior of PEFCs. This model was coupled to an electrochemical and a thermal model, which takes the mass and heat transfer features into consideration. By means of this, the researchers calculated the cell voltages and heat losses for predicting fuel-cell performance. Um et al. [10] numerically studied the transient behavior of PEFCs. The model they proposed takes the electrochemical reaction kinetics, current density distributions, as well as multi-substance transport, into consideration. The results showed that the current density increased correspondingly as the voltage decreased from 0.6 V to 0.55 V, and the current overshoot behavior was observed in the first few seconds. They determined that compared to the electrochemical reaction, the reactant gas concentration required more time to reach its equilibrium, causing this transient behavior. Hamelin et al. [11] experimentally studied the transient behavior of PEFCs. They performed experiments on Ballard Mk5-E fuel cells and observed rapid voltage transient behavior, which was determined to be an overshoot and undershoot behavior. The system response time was less than 0.15 s.

In 2004, Kim et al. [12] experimented on a PEFC with a 25 cm^2 triple serpentine flow field. They studied the effects of voltage change and stoichiometric ratio on the current density overshoot and undershoot behavior. The results showed that an increase in the stoichiometric ratio caused a decrease in the current overshoot amplitude. It proved the experimental results from Hamelin et al. [11], as well as the numerical simulation results from Um et al. [10]. Shen et al. [13] investigated the effect of an air stoichiometric ratio on the voltage undershoot behavior and found that increasing the stoichiometric ratio reduced the voltage undershoot tendency. The reason for this is that an increased stoichiometric ratio leads to reduced water accumulation in the gas diffusion layer and the flow channel, which consequently increases the reactant concentration on the reaction site and improves the reactant distribution.

In 2005, Yan et al. [14] numerically analyzed the effects of flow channel width and catalyst layer overpotential on the mass transport transient behavior. The results showed that a higher value of the channel width and catalyst layer overpotential caused a faster dynamic response behavior during the startup phase. Yan et al. [15] experimentally studied the transient behavior of single cell during cyclic load changes and different operating conditions, including inlet gas temperature, operating temperature, stoichiometric ratio, and inlet gas pressure. They found that a higher cathodic reactant humidity caused a lower voltage undershoot amplitude. Also, a higher stoichiometric ratio and higher pressure resulted in a reduction in the transient behavior of the cell.

In 2006, Liu et al. [16] investigated the long-term durability of PEFCs under different current cycles to simulate the real driving conditions for vehicular applications. Cell polarization curves, electrochemical impedance spectra, as well as hydrogen crossover rates, were used to characterize the MEAs (Membrane Electrode Assembly). They found that the hydrogen crossover rate was kept nearly stable during constant current operation. However, under cyclic current conditions, a pinhole was formed on the MEA after 500 h of operation, which led to dramatically increased hydrogen crossover. In 2009, Lin et al. [17] investigated the degradation of PEFC performance under cyclic load operating conditions. It was noticed that after 280 h of operation, the cell performance starts to decrease. Furthermore, the gaps and cracks were formed on the catalyst layer after 370 h of operation. In 2014, Lin et al. [18] studied the influence of dynamic load changes on the durability of a segmented cell. Their results showed that the performance of the cell decreased dramatically after 200 cycles of operation. Moreover, the current density decreased much faster at the inlet and outlet region than in other positions. By analyzing the impedance spectroscopy, they found that the ohmic resistance increased after 200 cycles, especially on the cathode side.

These mentioned above can be classified into electrochemical transients. Banerjee and Kandlikar [19] confirmed that, apart from the electrochemical transients, the PEFCs' transient behavior also included the thermal and two-phase flow transients. The two-phase behavior of PEFCs has been investigated by several researchers [20–22]. However, most of their work has focused on steady-state instead of transient behavior in PEFCs. Therefore, an investigation of the two-phase behavior of PEFCs for automotive drivetrains is necessary.

The pressure drop inside fuel cells has been used as an effective tool to analyze the dynamic water behavior in channels [23–25]. In 2014, Banerjee and Kandlikar [26] investigated the impact of two-phase flow pressure drops on cell performance. They studied the effects of temperatures and load changes on cathodic pressure drops and observed the pressure drop overshoot and undershoot behavior. During their later work [27,28], the pressure drop overshoot behavior could only be observed at lower temperatures due to the generated liquid water presented in the flow channels. In comparison, the overshoot behavior became insignificant at higher temperatures because of the higher water vapor fraction. They also found that the current change amplitude had more effects on the overshoot behavior of a pressure drop compared to the current change ramps.

Despite the changing load and operating conditions having been studied by many research groups, limited attention has focused on the relationship between voltage responses, two-phase flow pressure drops, and cell resistance under transient conditions. Furthermore, the effect of different load cycles has also not been investigated. In this paper, different current density profiles were imposed on the cell, and its effects on the voltage response, pressure drop, as well as cell ohmic resistance, were observed and analyzed.

2. Experimental Preparation

2.1. Test Cell Components

The test cell, as shown in Figure 1, was manufactured and assembled at Forschungszentrum Jülich. The active area of the Gore catalyst-coated membrane (CCM) is 17.64 cm^2. The non-woven Gas Diffusion Layer (GDL), Freudenberg H23Cx165, which was coated with a microporous layer (MPL) with 40 wt% polytetrafluoroethylene (PTFE) loading, was added on both sides of the CCM. The detailed drawing of the in-house-designed monopolar plate applied in the present work is shown in Figure 2, where the flow field is also presented. The monopolar plate was fabricated from graphite composite. The flow field consists of three parallel serpentine channels, each of which has a quadratic cross-section with 1 mm width and 1 mm depth, and a land width that is also 1 mm. The flow fields at both monopolar plates have symmetrical designs. The reactant gas flow direction was arranged as a co-flow configuration. The end plate was also used as the current collector plate and was fabricated from stainless steel type 1.4571 (316 Ti).

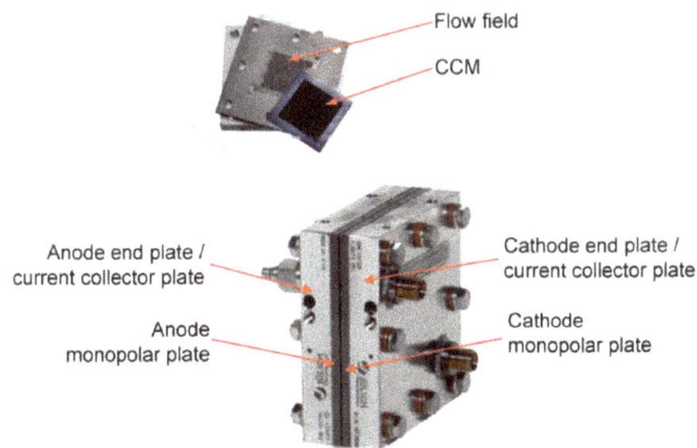

Figure 1. The single cell used in the experiment.

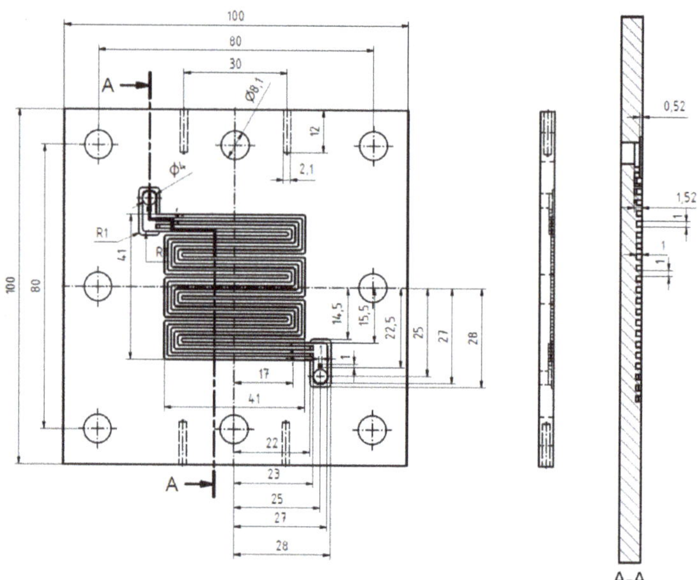

Figure 2. Detailed drawing of the monopolar plate and the flow field.

2.2. *Test Equipment*

The experimental operating conditions were controlled by Greenlight Technology's G40 test station, which could supply a stable reactant gas flow rate, temperature, and relative humidity based on the experimental requirements. As is shown in Figure 3, the voltage test ports are inserted into the monopolar plate to record the cell voltage. Two heater tubes are inserted into the end plates to control the cell temperature. Two thermocouples are inserted into each end plate to measure the cell temperature. A manometer is used to measure the two-phase flow pressure drops on the cathode side. The ohmic resistance of the cell was measured by means of an Electrochemical Impedance Spectroscopy (EIS) device (Zahner Zennium, ZAHNER-elektrik GmbH & Co. KG, Kronach, Germany).

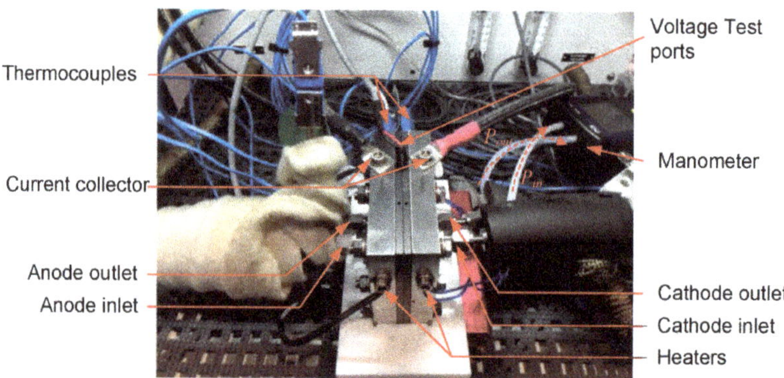

Figure 3. Picture of test cell connected to the test stand.

2.3. *Test Procedures*

In order to maintain stable performance of the test cell for subsequent testing, a repeatable break-in procedure needed to be performed for the new CCM [29]. During the break-in procedure [30], the test cell was conditioned at 70 °C, with fully saturated (relative humidity (RH) = 100%) hydrogen at 1.2 stoichiometric ratio and fully saturated air at 2.5 stoichiometric ratio. The current was increased until the cell voltage reached 0.6 V. Then, the voltage was cycled between 0.6 V (30 min), 0.4 V (30 min), and OCV (1 min) approximately 6–8 times until there was no further increase in cell performance.

The conditioning procedure was performed before each test in order to set the equal start condition. During the conditioning process, the test cell temperature was set at 70 °C, with fully saturated hydrogen at a stoichiometric ratio of 1.2, and fully saturated air at $\lambda = 2.5$. The cell voltage was set to 0.6 V for the 30 min of operation. This step was used to make sure that the membrane was fully hydrated and that the test cell had the same historic situation before each test [31]. During the testing process, the voltage response, cathodic pressure drop response, and ohmic resistance of the cell under each operating condition were recorded. The cell temperature and reactant relative humidity were considered to be constant in the test, while the flow rate of the gas was correspondingly adjusted to the change of the current density.

2.4. Test Conditions

During the test process, the cell operating temperature was set to 60 °C and the reactant's relative humidity was set to 90% on both sides. The stoichiometric ratio of hydrogen (λ_{H2}) and air (λ_{Air}) were set to 2. The test operating conditions are shown in Table 1.

Table 1. Test conditions.

Anode Reactant Gas	Hydrogen
Rel. Humidity	90%
λ_{H2}	2
Cathode reactant gas	Air
Rel. Humidity air	90%
λ_{Air}	2
Cell Temperature	60 °C
Flow Direction	Co-flow, top to bottom

Six sets of experiments were conducted in this study. The maximum duration of each test was no more than two hours. The maximum number of cycles of each test was no more than 10. Figure 4 shows the first load-cycle curve of each load-cycling profile. The starting current density is always 0.2 A/cm^2 and remains so for five minutes, which is the so-called startup phase. The load ramps in test No. 1 can be described by four repeated steps, each with a duration of five minutes (indicated by 5-5-5-5):

1. Linear current increase from 0.2 A/cm^2 to 0.6 A/cm^2
2. Constant current at 0.6 A/cm^2
3. Linear current decrease from 0.6 A/cm^2 to 0.2 A/cm^2
4. Constant current at 0.2 A/cm^2

This is the first load cycle for test No. 1. The other tests have similar load-cycling profiles but with different times for the load ramps and constant current phases. The load ramps in test No. 2 are represented by 2.5-5-2.5-5, in test No. 3 it is 1-5-1-5, in test No. 4 it is 0-5-0-5, in test No. 5 it is 5-0-5-0, and in test No. 6 it is 1-0-1-0.

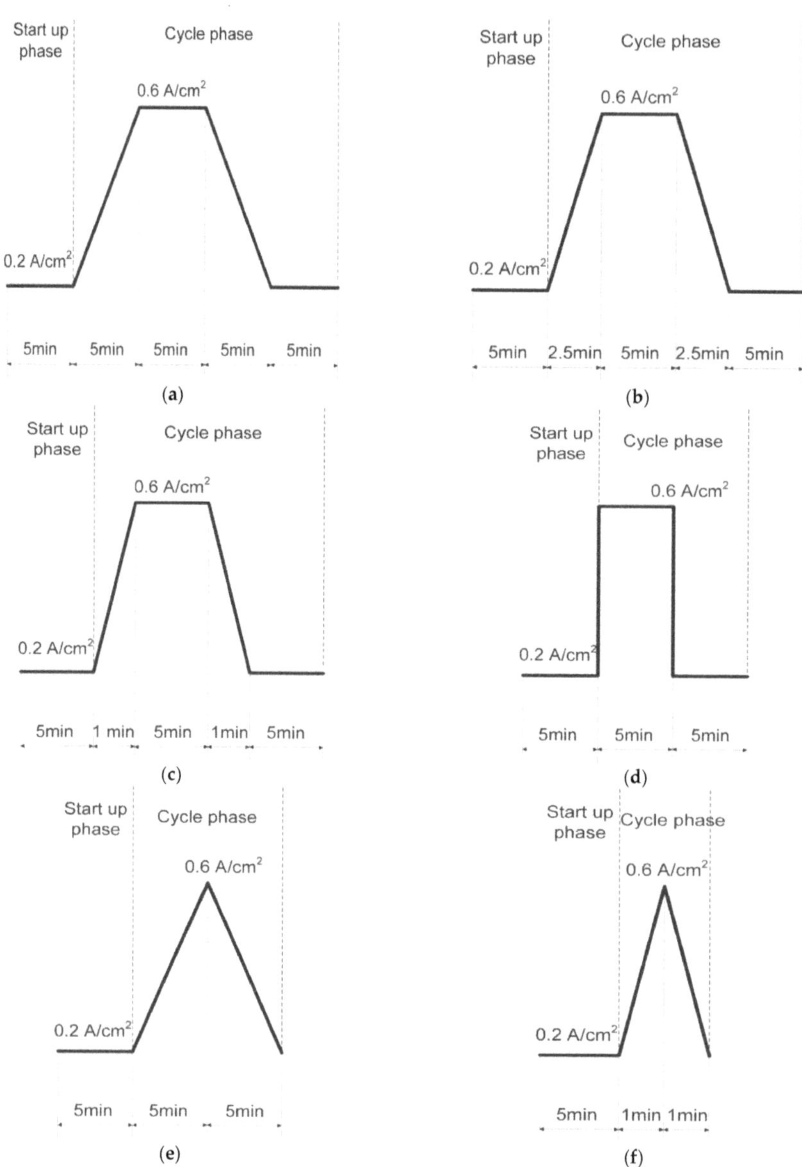

Figure 4. The first cycle of the load-cycling profile in (**a**) test No. 1; (**b**) test No. 2; (**c**) test No. 3; (**d**) test No. 4; (**e**) test No. 5; and (**f**) test No. 6.

3. Results and Discussion

The cathodic pressure drop response and the cell voltage profile of test No. 1 to test No. 6 are shown in Figure 5. In the first cycle of each test, the cell voltages changed in correspondence with the changes of the current density, as well as the cathodic pressure drop. When the load decreased from 0.6 A/cm^2 to 0.2 A/cm^2, the pressure drop on the cathode side began to decrease. It reached the minimum value when the load reached 0.2 A/cm^2. The cell voltage also reached the maximum corresponding value. A similar pattern was also observed in the subsequent cycles in each test.

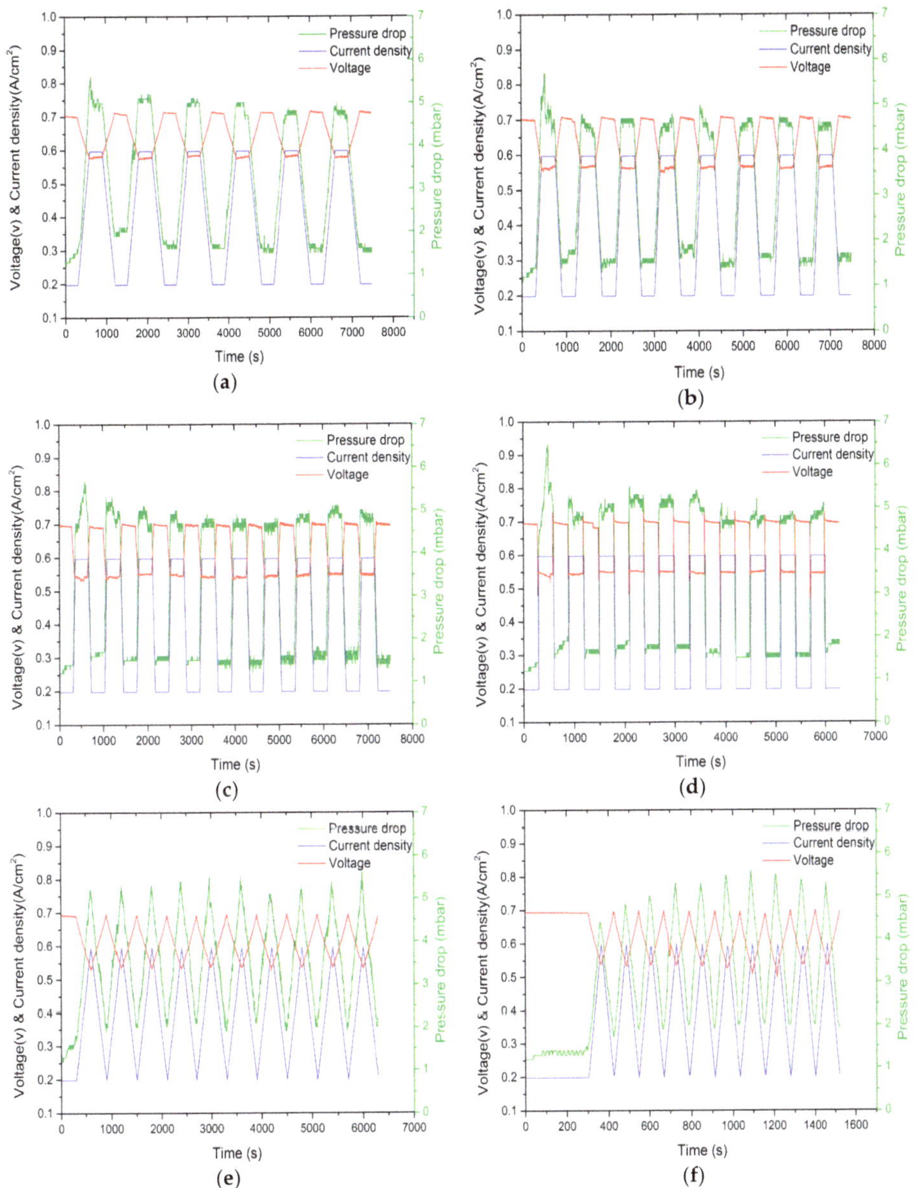

Figure 5. The results of the cathodic pressure drop response and the cell voltage response according to the load-cycling profile in (**a**) test No. 1; (**b**) test No. 2; (**c**) test No. 3; (**d**) test No. 4; (**e**) test No. 5; and (**f**) test No. 6.

3.1. Effect of Load Changes on Voltage Overshoot and Undershoot Behavior

The voltage overshoot and undershoot behavior were observed from test No. 1 to test No. 4. The reason for this is that the mass transfer is slower than the electrochemical reaction [19]. When increasing the current density, the gas consumption rate increases alongside the water generation rate. The increase in gas consumption leads to a decrease in the gas concentration on the catalyst layer.

At the same time, the generated water accumulates in the GDL so that the reactant transport pathways are blocked and unable to reach the reaction site on time. This will cause sudden starvation as the voltage starts to decrease. The voltage achieves a constant value again as soon as the mass transfer process reaches a new equilibrium. When the current suddenly decreases, the gas consumption rate decreases, as well as the water generation rate, with excess water generated by the larger current making the membrane sufficiently hydrated. At the same time, the oxygen concentration on the cathode's electrode is relatively large, which leads to the voltage overshoot behavior. When the mass transfer process reaches the new balance, the voltage achieves a constant value. When the holding time at the constant current density was zero, the overshoot or undershoot behavior was no longer observed—see test No. 5 and test No. 6.

The average overshoot/undershoot magnitude (V_{AOM}/V_{AUM}) was studied to characterize the impact of load ramps on the voltage overshoot and undershoot behavior. The overshoot magnitude (V_{OM}) is defined as the difference between the peak voltage and the steady state voltage. Similarly, the magnitude of undershoot (V_{UM}) is defined as the difference between the minimum voltage value and the steady state voltage value. The V_{AOM} can be calculated by the following equation:

$$V_{AOM} = \sum_{i=1}^{n} V_{OMi}/n \quad (1)$$

where n means the number of cycles in each test. Equation (1) can also be used for the V_{AUM} calculation. Figure 6 shows the impact of load ramps on the V_{AOM} and V_{AUM}. In Figure 6a, it can be seen that the V_{AOM} of tests No. 1, No. 2, and No. 3 are in the same range, but for test No. 4, the V_{AOM} is increased. Similarly, in Figure 6b, the V_{AUM} of tests No. 1, No. 2, and No. 3 are nearly unchanged, while for test No. 4, the V_{AUM} is increased with a large deviation. It can be noticed that when the load ramp rates ($|\frac{\Delta i}{\Delta t}|$) are not infinite (test No. 1 to test No. 3), this has a strong impact on the voltage overshoot and undershoot behavior. The load profiles with non-infinite ramps can help the mass transfer process to easily and quickly achieve equilibrium.

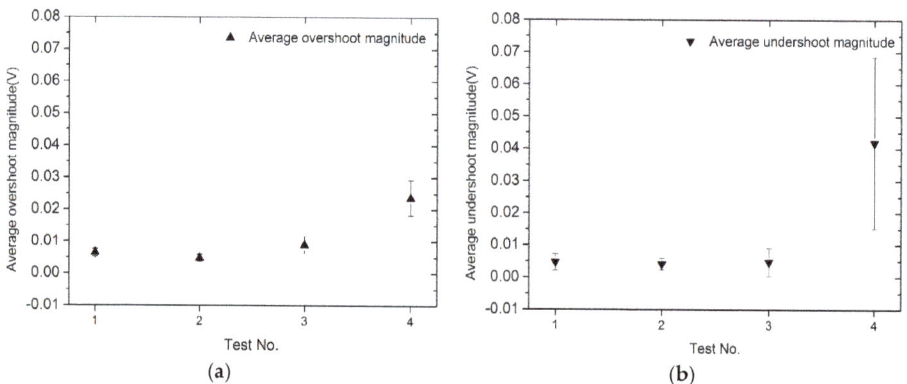

Figure 6. Effect of load ramps on the average magnitude (**a**) average overshoot magnitude, (**b**) average undershoot magnitude.

3.2. Effect of Load Ramps on Cathodic Pressure Drop Change

In order to characterize the cathodic pressure drop of the cell at different load ramps, we studied the top average pressure drop value (ΔP_{TopA}) at 0.6 A/cm^2 (constant current phase) and the bottom average pressure drop value (ΔP_{BotA}) at 0.2 A/cm^2 (constant current phase) in each test. As is shown in Figure 7, in tests No. 2 to No. 4, the ΔP_{TopA} increases slightly with the increase of the load ramp rates ($|\frac{\Delta i}{\Delta t}|$). However, the corresponding changes in ΔP_{BotA} were not obvious, nor were the changes in the difference between ΔP_{TopA} and ΔP_{BotA}. Figure 8 shows the standard deviation of ΔP_{TopA} and ΔP_{BotA} with different load ramp rates. It can be observed that with an increase in the load ramp, the difference between the top and bottom pressure drop standard deviation becomes more significant. In tests No. 2 to No. 4 and tests No. 5 to No. 6, an increase of load ramp rates will lead to a increase in the standard deviation of ΔP_{TopA}. The reason for this might be that the smaller load ramp rates cause slower changing rates in the pressure drop, and the pressure drop reaches its equilibrium in the constant current phase. However, with the higher load ramp rates, the change of water flow produced is also higher, so that the equilibrium of the two-phase flow is harder to reach. This causes a higher pressure drop fluctuation.

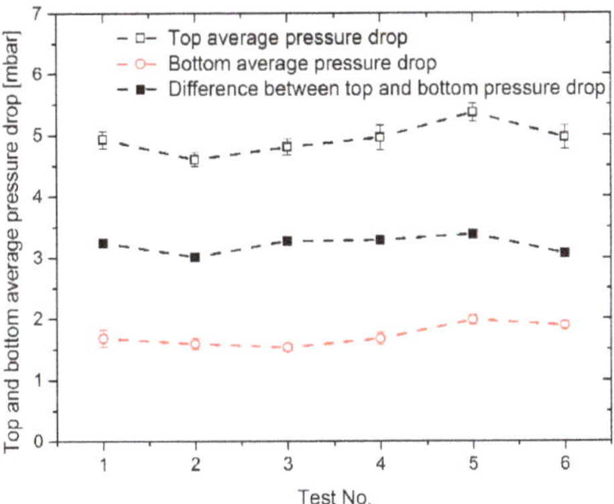

Figure 7. Top and bottom average pressure drop and the difference between them.

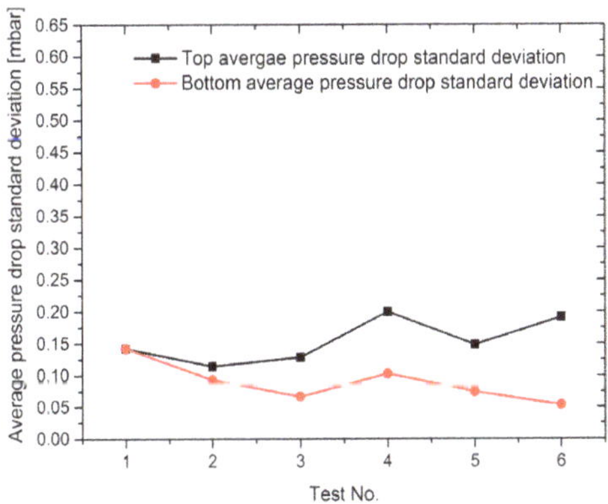

Figure 8. Top and bottom average pressure drop standard deviation.

Figure 9 shows the top and bottom pressure drop standard deviation in tests No. 1 to No. 4. In test No. 1, it can be observed that in the first cycle, the difference between the top and bottom pressure drop standard deviation is the largest, with cycles going on, then starting to decrease. A similar pattern can also be found in tests No. 2 to No. 4, although there was some fluctuation in this trend. It can be noted that during the cyclic operation, the two-phase flow tends to reach equilibrium in the cell. It was impossible to calculate the top and bottom pressure drop standard deviation in test No. 5 and test No. 6 because of the lack of a constant current phase in these cycles.

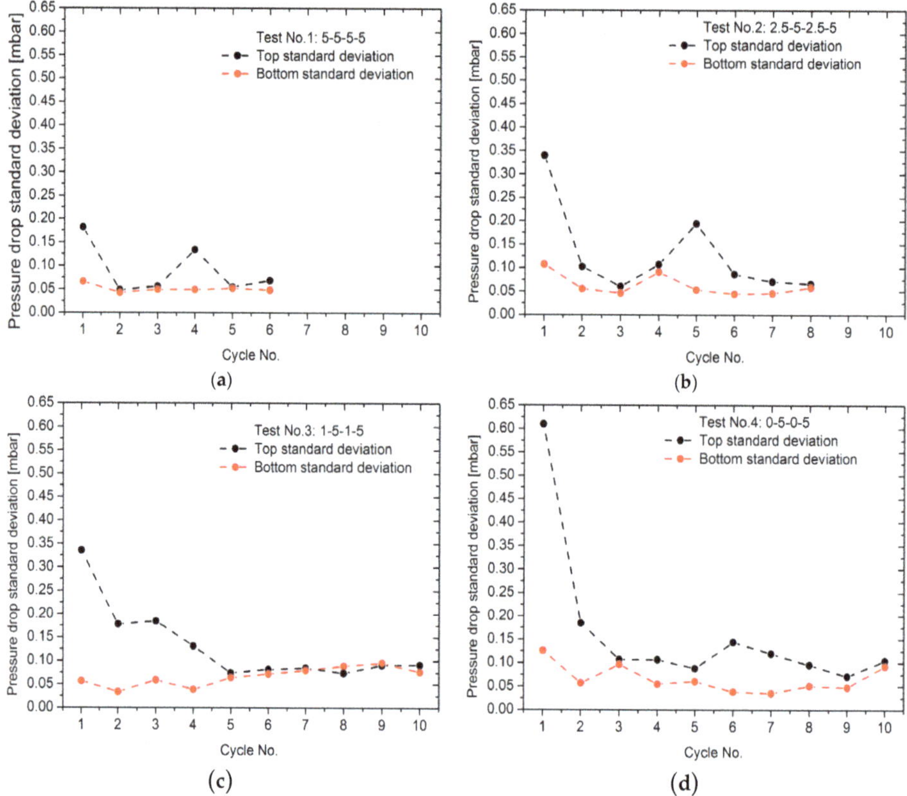

Figure 9. Top and bottom pressure drop standard deviation in (**a**) test No. 1, (**b**) test No. 2, (**c**) test No. 3, and (**d**) test No. 4.

3.3. *Effect of Load Ramps on Ohmic Resistance*

The galvanostatic mode was applied during resistance measurements. The amplitude of the AC voltage was set to 10 mV. The cell impedance was measured at a frequency of 3 kHz and can be considered as the cell ohmic resistance, which is the intercept on the high-frequency part of the real axis in the impedance spectra. Figure 10 shows the ohmic resistance in each cycle (indicated by the white and grey stripes) for tests No. 1 to No. 4. The ohmic resistance was measured in each cycle, first at 0.6 A/cm^2 and then at 0.2 A/cm^2. Before the first cycle was started, the ohmic resistance was measured during the start phase of each cycle at 0.2 A/cm^2. It was clearly observed that the ohmic resistance is significantly higher at the lower current density. This is expected because at a higher current density value, more water is generated and the membrane has higher water content. This results in higher protonic conductivity. It was also observed that the ohmic resistance difference between each test is less than obvious. The reason for this might be that there is enough water inside the cell that the membrane is sufficiently humidified. The ohmic resistance was only measured from test No. 1 to No. 4, as it was impossible to measure the ohmic resistance in test No. 5 and No. 6 because of the lack of constant current phase in these cycles.

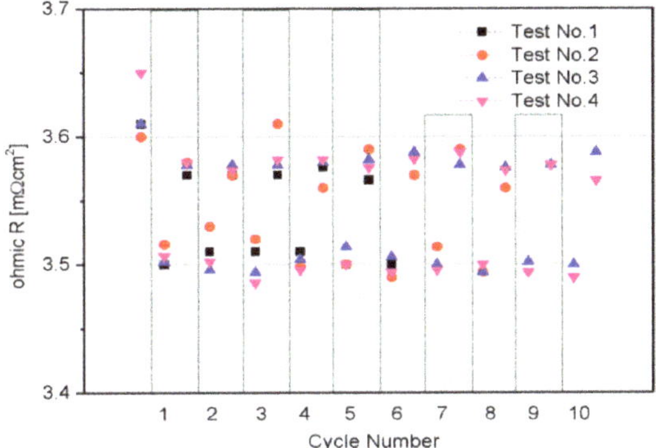

Figure 10. Ohmic resistance of different tests and cycles (upper values at 0.2 A/cm^2, lower values at 0.6 A/cm^2).

4. Conclusions

In this paper, experiments on the transient behavior of a PEFC under different load profiles were conducted. The cell voltage, pressure drop response, and ohmic resistance were measured and studied.

Voltage overshoot and undershoot behavior were observed during the tests. When the load ramp rate was infinite (load change time was zero), it had a strong effect on the voltage overshoot and undershoot behavior compared to other load ramp rates, for which the load change time was not zero. When the constant load holding time was zero, the voltage overshoot or undershoot behavior was no longer observed.

The load ramp did not have an obvious impact on the average pressure drop differences between each test. During the cyclic operation in each test, the difference between the top and bottom pressure drop standard deviation was gradually decreasing, which meant that the two-phase flow tended to reach equilibrium in the cell with the cyclic operation.

Finally, the impedance analysis showed that the ohmic resistance changes with the change of the current density and the lower current caused higher ohmic resistance. The larger current will cause lower ohmic resistance of the membrane. The reason for this was that under the higher load, the cell generated more water. The membrane was sufficiently humidified and increased the ionic conductivity of the membrane. Furthermore, the difference between each test was not obvious. The reason for this may have been that there was always enough water inside the cell.

Overall, this study provides information about how the cyclic load profile influences the cell performance and water dynamics behavior in the cell. It can be concluded that the cyclic operation seemed to have no obvious influence on the cell's overall performance. The pressure drop had a larger fluctuation in the first cycle and became smaller as the cycles went on. This represented the water-related dynamic behavior of the cell without affecting the cell's performance. Other operational parameters or cell and flow field designs than chosen in this study led to different results and conclusions. Furthermore, when changing the load, it was necessary to adjust operating conditions such as relative humidity in order to keep the membrane at a lower ohmic resistance. All of this information could help improve the control strategy of the fuel cell system for the real vehicular application.

Author Contributions: Conceptualization, W.L.; Methodology: Y.S., H.J. and W.L.; Investigation, Y.S. and H.J.; Writing—Original Draft Preparation, Y.S.; Writing—Review & Editing, Y.S., H.J. and W.L.; Funding Acquisition: Y.S. and W.L.; Supervision, H.J. and W.L.

Funding: This research was funded by the China Scholarship Council (CSC), grant number 201608110140.

Acknowledgments: We are thankful to Christopher Wood for proofreading the manuscript.

Conflicts of Interest: The authors declare no conflict of interest. The funders had no role in the design of the study; in the collection, analyses, or interpretation of data; in the writing of the manuscript, or in the decision to publish the results.

References

1. Liu, Y.; Jia, X.; Pei, P.; Lu, Y.; Yi, L.; Shi, Y. Simulation and experiment for oxygen-enriched combustion engine using liquid oxygen to solidify CO_2. *Chin. J. Mech. Eng.* **2016**, *29*, 188–194. [CrossRef]
2. Niakolas, D.K.; Daletou, M.; Neophytides, S.G.; Vayenas, C.G. Fuel cells are a commercially viable alternative for the production of "clean" energy. *J. Hum. Environ.* **2016**, *45*, 32–37. [CrossRef] [PubMed]
3. Liu, Y.; Lehnert, W.; Janßen, H.; Samsun, R.C.; Stolten, D. A review of high-temperature polymer electrolyte membrane fuel-cell (HT-PEMFC)-based auxiliary power units for diesel-powered road vehicles. *J. Power Sources* **2016**, *311*, 91–102. [CrossRef]
4. Wang, F.-C.; Gao, C.-Y.; Li, S.-C. Impacts of power management on a PEMFC electric vehicle. *Int. J. Hydrog. Energy* **2014**, *39*, 17336–17346. [CrossRef]
5. Zhang, C.; Zhou, W.; Zhang, L.; Chan, S.H.; Wang, Y. An experimental study on anode water management in high temperature PEM fuel cell. *Int. J. Hydrog. Energy* **2015**, *40*, 4666–4672. [CrossRef]
6. Urbani, F.; Barbera, O.; Giacoppo, G.; Squadrito, G.; Passalacqua, E. Effect of operative conditions on a PEFC stack performance. *Int. J. Hydrog. Energy* **2008**, *33*, 3137–3141. [CrossRef]
7. Ahn, S.-Y.; Shin, S.-J.; Ha, H.Y.; Hong, S.-A.; Lee, Y.-C.; Lim, T.W.; Oh, I.H. Performance and lifetime analysis of the kW-class PEMFC stack. *J. Power Sources* **2002**, *106*, 295–303. [CrossRef]
8. Scholta, J.; Häussler, F.; Zhang, W.; Küppers, L.; Jörissen, L.; Lehnert, W. Development of a stack having an optimized flow field structure with low cross transport effects. *J. Power Sources* **2006**, *155*, 60–65. [CrossRef]
9. Amphlett, J.C.; Mann, R.F.; Peppley, B.A.; Roberge, P.R.; Rodrigues, A. A model predicting transient responses of proton exchange membrane fuel cells. *J. Power Sources* **1996**, *61*, 183–188. [CrossRef]
10. Um, S.; Wang, C.-Y.; Chen, K.S. Computational Fluid Dynamics Modeling of Proton Exchange Membrane Fuel Cells. *J. Electrochem. Soc.* **2000**, *147*, 4485. [CrossRef]
11. Hamelin, J.; Agbossou, K.; Laperrière, A.; Laurencelle, F.; Bose, T.K. Dynamic behavior of a PEM fuel cell stack for stationary applications. *Int. J. Hydrog. Energy* **2001**, *26*, 625–629. [CrossRef]
12. Kim, S.; Shimpalee, S.; van Zee, J.W. The effect of stoichiometry on dynamic behavior of a proton exchange membrane fuel cell (PEMFC) during load change. *J. Power Sources* **2004**, *135*, 110–121. [CrossRef]
13. Shen, Q.; Hou, M.; Yan, X.; Liang, D.; Zang, Z.; Hao, L.; Shao, Z.; Hou, Z.; Ming, P.; Yi, B. The voltage characteristics of proton exchange membrane fuel cell (PEMFC) under steady and transient states. *J. Power Sources* **2008**, *179*, 292–296. [CrossRef]
14. Yan, W.-M.; Soong, C.-Y.; Chen, F.; Chu, H.-S. Transient analysis of reactant gas transport and performance of PEM fuel cells. *J. Power Sources* **2005**, *143*, 48–56. [CrossRef]
15. Yan, Q.; Toghiani, H.; Causey, H. Steady state and dynamic performance of proton exchange membrane fuel cells (PEMFCs) under various operating conditions and load changes. *J. Power Sources* **2006**, *161*, 492–502. [CrossRef]
16. Liu, D.; Case, S. Durability study of proton exchange membrane fuel cells under dynamic testing conditions with cyclic current profile. *J. Power Sources* **2006**, *162*, 521–531. [CrossRef]
17. Lin, R.; Li, B.; Hou, Y.P.; Ma, J.M. Investigation of dynamic driving cycle effect on performance degradation and micro-structure change of PEM fuel cell. *Int. J. Hydrog. Energy* **2009**, *34*, 2369–2376. [CrossRef]
18. Lin, R.; Xiong, F.; Tang, W.C.; Técher, L.; Zhang, J.M.; Ma, J.X. Investigation of dynamic driving cycle effect on the degradation of proton exchange membrane fuel cell by segmented cell technology. *J. Power Sources* **2014**, *260*, 150–158. [CrossRef]
19. Banerjee, R.; Kandlikar, S.G. Two-phase flow and thermal transients in proton exchange membrane fuel cells–A critical review. *Int. J. Hydrog. Energy* **2015**, *40*, 3990–4010. [CrossRef]
20. Sergi, J.M.; Lu, Z.; Kandlikar, S.G. In Situ Characterization of Two-Phase Flow in Cathode Channels of an Operating PEM Fuel Cell With Visual Access. In Proceedings of the ASME 2009 7th International Conference on Nanochannels, Microchannels and Minichannels (ASME), Pohang, Korea, 22–24 June 2009; pp. 303–311.
21. Zhang, L.; Du, W.; Bi, H.T.; Wilkinson, D.P.; Stumper, J.; Wang, H. Gas–liquid two-phase flow distributions in parallel channels for fuel cells. *J. Power Sources* **2009**, *189*, 1023–1031. [CrossRef]

22. Adroher, X.C.; Wang, Y. Ex situ and modeling study of two-phase flow in a single channel of polymer electrolyte membrane fuel cells. *J. Power Sources* **2011**, *196*, 9544–9551. [CrossRef]
23. Pei, P.; Li, Y.; Xu, H.; Wu, Z. A review on water fault diagnosis of PEMFC associated with the pressure drop. *Appl. Energy* **2016**, *173*, 366–385. [CrossRef]
24. Stumper, J.; Löhr, M.; Hamada, S. Diagnostic tools for liquid water in PEM fuel cells. *J. Power Sources* **2005**, *143*, 150–157. [CrossRef]
25. Chen, J. Dominant frequency of pressure drop signal as a novel diagnostic tool for the water removal in proton exchange membrane fuel cell flow channel. *J. Power Sources* **2010**, *195*, 1177–1181. [CrossRef]
26. Banerjee, R.; Kandlikar, S.G. Experimental investigation of two-phase flow pressure drop transients in polymer electrolyte membrane fuel cell reactant channels and their impact on the cell performance. *J. Power Sources* **2014**, *268*, 194–203. [CrossRef]
27. Banerjee, R.; Kandlikar, S.G. Two-Phase Pressure Drop Characteristics During Low Temperature Transients in PEMFCs. In Proceedings of the ASME 2014 12th International Conference on Nanochannels, Microchannels and Minichannels (ASME), Chicago, IL, USA, 3–7 August 2014.
28. Banerjee, R.; Kandlikar, S.G. Two-phase pressure drop response during load transients in a PEMFC. *Int. J. Hydrog. Energy* **2014**, *39*, 19079–19086. [CrossRef]
29. US Fuel Cell Council. USFCC Single Cell Test Protocol # 05-014. Available online: http://www.members.fchea.org/core/import/PDFs/Technical%20Resources/MatComp%20Single%20Cell%20Test%20Protocol%2005-014RevB.2%20071306.pdf (accessed on 20 July 2018).
30. Cleghorn, S. TECHNICAL DATA–PRIMEA SERIES MEAs. Available online: https://www.dropbox.com/s/lzmrz9t4vuafhi1/2000%20Gore_Primea_NT-PEM_MEA_Einfahrprozedur.pdf (accessed on 20 July 2018).
31. Banerjee, R.; See, E.; Kandlikar, S.G. Pressure Drop and Voltage Response of PEMFC Operation under Transient Temperature and Loading Conditions. *ECS Trans.* **2013**, *58*, 1601–1611. [CrossRef]

Article

Investigation of the Fuel Utilization Factor in PEM Fuel Cell Considering the Effect of Relative Humidity at the Cathode

Mojtaba Baghban Yousefkhani [1], Hossein Ghadamian [1], Keyvan Daneshvar [2,*], Nima Alizadeh [3] and Brendy C. Rincon Troconis [2]

1 Department of Energy, Materials and Energy Research Center (MERC), Tehran 4777-14155, Iran; mjtb_baghban@yahoo.com (M.B.Y.); h.ghadamian@merc.ac.ir (H.G.)

2 Department of Mechanical Engineering, University of Texas at San Antonio, San Antonio, TX 78249, USA; brendy.rincon@utsa.edu

3 Department of Natural Resources and Environment, Islamic Azad University, Science and Research Branch (SRBIAU), Tehran 775-14515, Iran; nima.alizadeh.g@gmail.com

* Correspondence: keyvan.daneshvar@utsa.edu

Received: 22 October 2020; Accepted: 20 November 2020; Published: 22 November 2020

Abstract: This research consists of both theoretical and experimental sections presenting a novel scenario for the consumption of hydrogen in the polymer electrolyte membrane fuel cell (PEMFC). In the theory section, a new correction factor called parameter δ is used for the calculation of fuel utilization by introducing concepts of "useful water" and "non-useful water". The term of "useful water" refers to the state that consumed hydrogen leads to the production of liquid water and external electric current. In the experimental section, the effect of the relative humidity of the cathode side on the performance and power density is investigated by calculating the parameter δ and the modified fuel utilization at 50% and 80% relative humidity. Based on the experimental results, the maximum power density obtained at 50% and 80% relative humidity of the cathode side is about 645 mW/cm^2 and 700 mW/cm^2, respectively. On the other hand, the maximum value of parameter δ for a value of 50% relative humidity in the cathode side is about 0.88, while for 80% relative humidity it is about 0.72. This means that the modified fuel utilization for 50% relative humidity has a higher value than that for 80%, which is not aligned with previous literature. Therefore, it is necessary to find an optimal range for the relative humidity of the cathode side to achieve the best cell performance in terms of the power generation and fuel consumption as increasing the relative humidity of the cathode itself cannot produce the best result.

Keywords: PEM fuel cell; useful water; hydrogen consumption scenarios; modified fuel utilization

---

1. Introduction

Hydrogen as a fuel obtained through renewable and non-renewable energy sources can provide reliable solutions to the energy problem and environmental aspects caused by fossil fuels. Fuel cells can continuously generate electrical energy as long as they are fed with oxidants and fuels, unlike batteries, which are energy-storage systems. On the other hand, raising a few percent of the efficiency in fuel cells will save significant energy in the related industry.

A large number of empirical studies have been carried out to evaluate the performance of the polymer electrolyte membrane fuel cell (PEMFC) and the effects of various parameters on their efficiency [1–11]. Many models have also been proposed to examine the processes occurring in the PEMFC up to now. Most of these models were presented in the last two decades.

Jeon et al. [12] investigated the effect of cathode relative humidity on the performance of PEMFC. Their results indicate that the performance of the cell improves by increasing the relative humidity of the cathode side. Nevertheless, improvement in the cell performance was identified only taking into consideration the reduction of ohmic losses due to the increased water content in the membrane. These activation losses, on both sides at the anode and cathode, are insignificant in comparison to the effect of relative humidity of the cathode side. Therefore, by increasing the relative humidity, the location of the highest local current density is transferred from the end of the electrode towards the center. Abdullahzadeh et al. [13] analyzed the cathode of the PEMFC by a two-dimensional model. According to the results of the proposed model, in a two-phase mode, voltage over-potential is higher than single-phase mode, especially at higher voltages. The main reason for this phenomenon is flooding in the gas diffusion layer of the cathode side. Moreover, increasing the relative humidity in the inlet gas stream decreases the performance of the fuel cell due to the higher probability of forming liquid water.

Nishikawa et al. [14] presented a solution based on stack separation methods used to increase fuel utilization for a 30-kW fuel cell; their system consumes hydrogen with high purity, which can increase fuel utilization without requiring additional equipment. According to their results for the mode of humidification in the cathode, the probability of an imbalance in the currents and the voltage oscillations will decrease as the fuel utilization increases. Therefore, it will be possible to achieve higher fuel utilization and lower voltage oscillations simultaneously. Iranzo et al. [15] studied the effect of both relative humidity of the reactants and stoichiometric cathode, focusing on the distribution of liquid water in a commercial PEMFC with a cross-section of 50 cm^2. Their experimental results indicate that an increase in relative humidity of the cathode causes a decrease in the cell resistance, and consequently the cell voltage will increase substantially. Indeed, the relative humidity of the inlet oxygen to the cathode leads to a current equilibrium in both anode and cathode due to water transfer to the membrane through the back diffusion. It should be noted that this achievement is related to the cell performance at low current densities and the fact that increasing relative humidity does not result in flood conditions in the gas distributor and electrodes of the cathode. In other words, low humidity is required at high current densities. Muirhead et al. [16] investigated the effects of the relative humidity on liquid water accumulation and mass transport resistance at high current densities in PEMFC. Based on their results, high relative humidity leads to higher concentration polarization losses. The reason is that the excess liquid water in the carbon fiber of the gas diffusion layer (GDL), which is in contact with the flow channels, causes an increase in the total mass transfer resistance.

According to the previous researches, it can be concluded that the relative humidity in the cathode is very effective on the fuel cell utilization and the generated power density. On the other hand, the relative humidity of the cathode can change the amount of hydrogen in the electrochemical reaction of the fuel cell by influencing the concentration polarization losses. Since the fuel cell is a power generation system and converts the chemical energy of a fuel into electrical energy through an electrochemical reaction, it is very important to establish an optimal ratio between the overall system efficiency that is directly related to the fuel consumption and the amount of produced power. In a different approach, Baghban et al. [17] calculated the fuel utilization factor of the PEMFC through a transfer phenomenon; also, the effect of inlet pressure and mass flow rate of input hydrogen on the fuel utilization factor was studied via modeling. The results of their modeling indicated that increasing pressure is associated with a significant increase in fuel utilization. Furthermore, with increasing temperature, the effect of increasing pressure on fuel utilization is decreased. According to the other results of this study, fuel utilization is a differential property, and in order to calculate it accurately, the equations must be considered differentially.

In this study at first, a simple definition of the fuel utilization factor is presented. Subsequently, using the basic concepts of the electrochemistry, various scenarios for the hydrogen consumption within the fuel cell are analyzed. It should be noted that the main innovation of this study is the introduction of a novel scenario to define the various routes of consumption or conversion of input hydrogen into liquid water and the production of current in a PEMFC. Then, based on the electrochemical reactions, a direct relationship between the amount of hydrogen consumed and the amount of water produced at the cathode side of the PEMFC is demonstrated. Lastly, by using a different and novel approach in comparison to the previous studies, the effect of the relative humidity of the cathode side on the generated power density was experimentally investigated.

2. Theory Analysis of the Fuel Utilization

2.1. Definition of the Fuel Utilization Factor

Most current research on fuel cell technology are based on the use of hydrogen with high purity (about 99%) as a fuel. PEMFC is one of the most important possible solutions to achieve an environmentally friendly energy resource. In cases where a demand for hydrogen with a high purity content is needed, if the hydrogen does not return to the system, it will be wasted and energy will be dissipated from a systematic point of view.

In the overall form (according to Equation (1)), the ratio of the hydrogen consumed in a fuel cell to generate thermal and electric energy to the amount of the hydrogen fed is called the fuel utilization factor (U_F) and it is expressed in percentage terms as follows:

$$U_F = \frac{H_{2,con}}{H_{2,in}} = \frac{H_{2,in} - H_{2,out}}{H_{2,in}} \tag{1}$$

The fuel utilization of a PEMFC can be improved by applying an unused hydrogen recycling cycle or a suitable design to prevent the flood condition. Figure 1 shows a schematic of the current production in a PEMFC and the utilization of hydrogen.

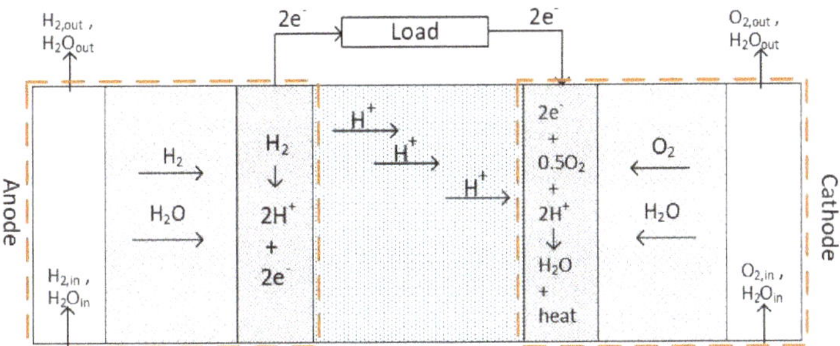

Figure 1. Schematic of a polymer electrolyte membrane fuel cell (PEMFC) showing the electrochemical reactions and the current production.

The main issue in determining the fuel utilization factor is the calculation of hydrogen consumption. In terms of the overall reaction within a fuel cell, the amount of hydrogen consumed is directly related to the amount of water produced:

$$H_2 \leftrightarrow 2H^+ + 2e^- \tag{2}$$

$$\frac{1}{2}O_2 + 2H^+ + 2e^- \leftrightarrow H_2O \tag{3}$$

$$H_2 + \frac{1}{2}O_2 \leftrightarrow H_2O + \text{heat}\ (\Delta E = -286kJ/mol\ @25\ ^\circ C) + \text{electricity} \tag{4}$$

Therefore, it seems that the most important weakness in the theoretical calculation of the fuel utilization is related to the calculation of hydrogen consumption. In the following, new scenario is outlined for the path of hydrogen within the cell and the circumstances taking place when hydrogen passes through the fuel cell.

2.2. A Novel Scenario for H_2 Consumption

According to the studies carried out in the field of the fuel utilization [18–20], it turns out that there is a significant difference between the theoretical and experimental values reported in the literature, which indicates a weakness in the formulation. This weakness in the formulation and the unreasonable intervals for calculating the fuel utilization is notable in the various formulations presented in previous studies. Therefore, different paths for hydrogen consumption in the fuel cell need to be carefully considered. In Figure 2, the different paths for hydrogen flow in a PEMFC are proposed.

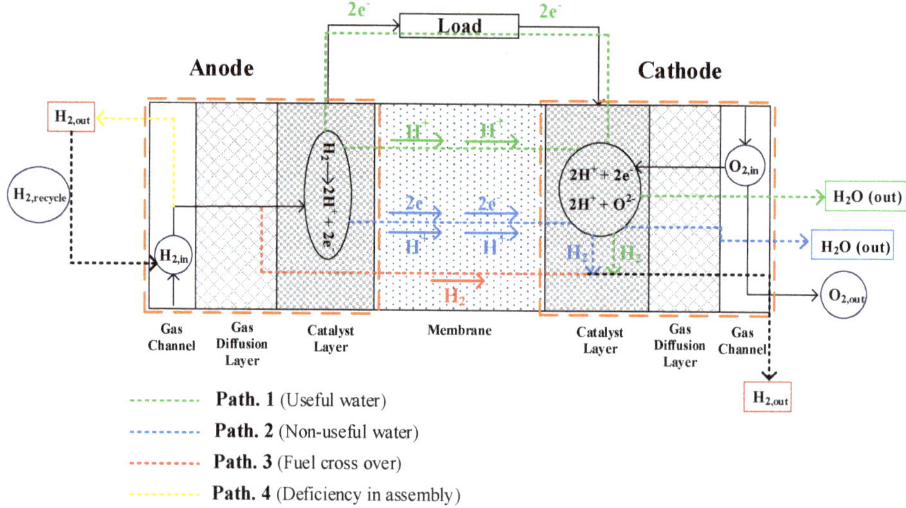

Figure 2. Different scenario for hydrogen flow paths in a PEMFC.

According to the diagram presented in Figure 2, the input hydrogen to the fuel cell's anode may pass through four different paths:

- Path No. 1: The input hydrogen in the catalyst layer is converted into H^+ and electron. Hydrogen ions pass through the membrane and the electrons move through the external circuit to the cathode, generating current. In this path, the hydrogen ion also passes through the membrane on the other side ("Cathode") and then can react either with the free electrons and re-generate hydrogen, or react with the oxygen ion due to the shown reduction reaction in the cathode side and generate water. The water generated in this state is called "useful water".
- Path No. 2: In this path, the input hydrogen in the catalytic layer is converted into H^+ ions and electrons, but the electrons produced along with the hydrogen ions pass through the membrane. In this state, electrons pass through the membrane using the momentum drag. That means the movement of hydrogen ions causes dragging of electrons. Hydrogen ions and electrons react in the cathode side after passing through the membrane and re-generate hydrogen. Although it may be possible that the hydrogen ion passing through the membrane reacts with the oxygen ion due to a reduction reaction in the cathode side and generates liquid water, this will be "non-useful water" since the generation of this water is not accompanied by the production of an electric current. In this state, part of the hydrogen is consumed and generates water, but it is not useful. It is obvious that most of the produced electrons in the catalytic layer move through the external circuit to the cathode, which results in the production of external current. However, a small number of electrons produced in the catalytic layer travel through the membrane to the cathode due to the polarity created. To further clarify "Path No. 2", it can be said that the faradic efficiency of a PEMFC is not completely 100% for two reasons:

(i) The short circuit current, which can arise because of the electrons transfer across the membrane. Such a current is negligible in normal operating conditions despite the membrane can be assumed as a dielectric. This last contribution can be very important as a consequence either of the membrane ageing or the same manufacturing flaws. In the worst case, such a conduction can also arise from the local contact between the two electrodes;

(ii) The gas crossover across the membrane, which in this case can be associated with "Path No. 3".

- Path No. 3: Hydrogen passes through the membrane without participating in the oxidation reaction, which is called fuel crossover. Hydrogen passing through the membrane leaves the cathode side of the fuel cell. The amount of hydrogen in this path is negligible, even at high temperatures and high currents.
- Path No. 4: In this path, the reason for not having an effective hydrogen passage is due to either the lack of proper design of the flow channels or the failure to assemble the components of the fuel cell correctly. This will cause the hydrogen to move out of the output channel in the anode side without having a chance to participate in the oxidation reaction.

Figure 2 illustrated that part of the input hydrogen to the fuel cell is exited from the cathode side (Paths 1–3), and only in Path 4 does unused hydrogen pass through the anode side's channel. Moreover, even though there are two paths where ionized hydrogen passes through the membrane and liquid water is generated (Path 1 and 2), only in path No. 1 is the water generated a measure of the production of the electric current in the fuel cell. If the value of the fuel utilization is calculated in accordance with Equation (1), it will certainly be a different value from the theoretical factor. In order to modify the fuel utilization formulation based on the results of the diagram presented in Figure 2, a new coefficient was determined as follows:

$$\delta = \frac{H_2O_{out} \text{ with charge transfer}}{H_2O_{out} \text{ with charge transfer} + H_2O_{out} \text{ without charge transfer}} \quad (5)$$

By the multiplication of the parameter δ in the main relationship of U_F (Equation (1)) more accurate results can be obtained by using the following equation,

$$U_{F,\ modified} = \frac{H_{2\ in} - H_{2\ out}}{H_{2\ in}} \times \delta \quad (6)$$

Therefore, it can be concluded that one of the most important and influential factors for determining the exact amount of the fuel utilization factor is the amount of liquid water on the cathode side. Hence, it has been proposed to experimentally analyze the effect of the relative humidity of the cathode side on the produced power density following the modified fuel utilization factor.

3. Experimental Measurement

In this research, a FCTS-125W fuel-cell testing machine manufactured by the Asian Hydrogen New Science Company, located in Isfahan, Iran was used to carry out the experiments. An overview of the FCTS-125W test pilot system is presented in Figure 3.

The characteristics of the tested PEMFC and the operating conditions are listed in Table 1. It should be noted that in this paper, as the study has been performed on fuel utilization and fuel cell performance, the amount of reactive gases is considered as additional (with the same value mentioned in Table 1) to prevent the undesirable activity of a limiting factor on performance. However, in studies focusing on flow gas channels and gas diffuser layer, it is necessary to observe stoichiometric ratios to properly analyze the effect of these two parts on the performance of the fuel cell, since these two parts of the fuel cell affect the rate of transfer of reactive gases to the reaction sites (catalytic layer).

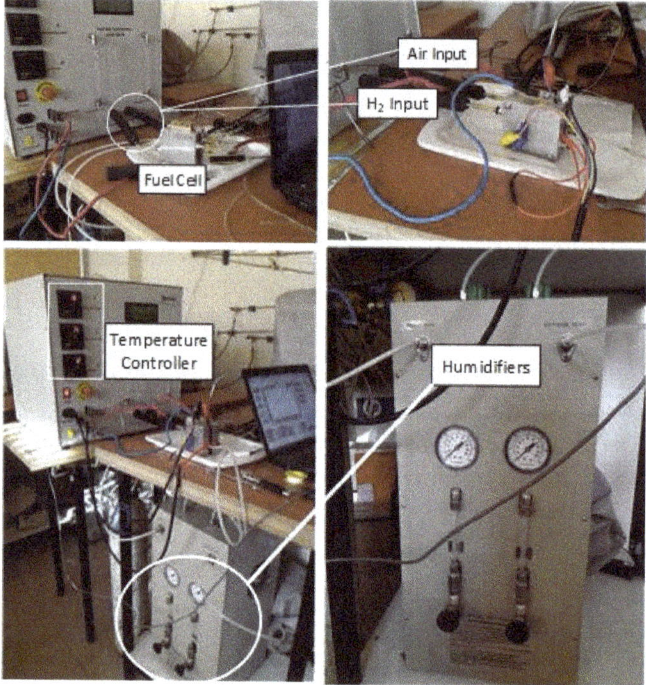

Figure 3. An overview of the FCTS-125W test pilot system.

Table 1. Technical and physical characteristics of the tested PEMFC and operating conditions.

Characteristic	Description/Value
Cell parameters	
Flow channel type	Single serpentine
Flow collectors material	Stainless steel
End plate material	Aluminum alloy
Maximum operating temperature (°C)	90
Maximum operating pressure (bar)	2
Active surface area (cm^2)	5
Operating conditions	
Anode pressure (bar)	1
Cathode pressure (bar)	1
Inlet flow rate of H_2 (lit/min)	0.3
Inlet flow rate of O_2 (lit/min)	0.3
Cell Temperature (°C)	60
Anode Temperature (°C)	60
Cathode Temperature (°C)	56
Anode relative humidity (%)	100

4. Results and Discussion

4.1. The Effect of Cathode Relative Humidity on PEMFC Power Density

Regarding the water production on the cathode side, the main issue of the relative humidity is its value on the cathode side. In this research, the relative humidity of the cathode side is considered 50% and 80%. The relative humidity of the anode side is always set to be 100%. In Figure 4, performance curves are presented for comparison between 50% and 80% relative humidity in the cathode side.

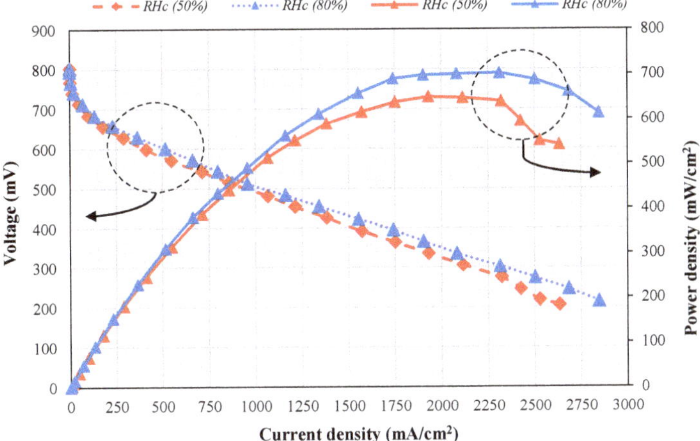

Figure 4. The effect of increasing the relative humidity of the cathode side on the performance curve.

According to Figure 4, the open circuit voltage (OCV) value for the tested fuel cell is about 0.8 V, which is less than the theoretical value (about 0.9 V). However, the produced current density is relatively high, which causes the final power density to be reasonable. In other words, the difference in the relative humidity of the cathode side has no significant effect on the OCV value, and only increases the range of fuel-cell performance in terms of produced current density. In terms of quantitative comparison, in cathode relative humidity of 50%, the maximum power density is approximately 645 mW/cm^2, which is achieved at 2000 mA/cm^2 current density. On the other hand, when the relative humidity of the cathode side is equal to 80%, the maximum power density is approximately 700 mW/cm^2, which is achieved at 2310 mA/cm^2 current density. This shows an increase of 8.5%, compared with the 50% relative humidity of the cathode side at the maximum power density.

The results displayed in Figure 4 can be analyzed in three current ranges:

1. At a current density lower than 500 mA/cm^2, increasing the relative humidity of the cathode side does not affect the polarization curve. In fact, in this current density range, the effective factor showing the catalytic effectiveness is the activation losses. The relative humidity change has no effect on the activation losses.
2. At current density values between 500 and 1750 mA/cm^2, an increase in relative humidity leads to a gradual increase in the fuel-cell performance curve. Considering the range of medium current densities, the most important losses are the ohmic losses; therefore, increasing the relative humidity of the cathode side causes the membrane to hydrate and the ohmic losses to decrease.
3. At current densities larger than 1750 mA/cm^2, the difference between the two performance curves is relatively constant up to a current density of about 2200 mA/cm^2 for 50% and 80% relative humidity. On the other hand, at current densities larger than 2200 mA/cm^2, the performance curve with a relative humidity of 50% shows a sudden decrease not seen for 80% relative humidity. As the current density depends on the concentration of the reactants at the catalyst layer, it can be concluded that in high current density, the most important voltage loss is due to concentration losses; the main cause of it is the high reaction rate inside the fuel cell and the exothermic nature of the half-reaction of the cathode side. Therefore, increasing the relative humidity can lead to enhancement of the cell's performance through the further transfer of water into the cell and pushing the membrane to the full hydrated state, despite the increase in the reaction rate inside

the fuel cell. It should be noted that excessive moisture might result in a flooding condition in the cathode.

In summary, increasing the relative humidity of the cathode side will increase the output power density of the fuel cell. On the other hand, due to the fact that the fuel cell is a power generation system and converts the fuel's chemical energy into electrical energy through an electrochemical reaction, it is also necessary to consider the effect of changing relative humidity of the cathode side from the fuel consumption point of view. For this purpose, the modified fuel utilization factor is calculated based on the scenario presented in Figure 2, which is discussed in the next section.

4.2. Calculation of the Modified Fuel Utilization Factor by Using the Parameter δ

In order to calculate the modified fuel utilization factor in accordance with Equation (6), initially, it is necessary to measure the flow rate of both consumable hydrogen and input hydrogen to the fuel cell. According to Faraday's law, the relationship between the consumed hydrogen and the produced current is as follows [21]:

$$I = N_{H_2} \cdot n \cdot F \tag{7}$$

where, I is produced current in A, N_{H_2} is the amount of consumable hydrogen in mol/s and $n \cdot F$ is the transferred flux in C/mol. Furthermore, the amount of the inlet flow rate of the hydrogen is calculated in terms of mol/s (Table 1) by using the input fuel density, which is a function of the temperature and pressure of the input flow.

Moreover, parameter δ is calculated according to its definition in the form of Equation (5). To determine the practical value of parameter δ with the appropriate justification, the denominator of Equation (5) is calculated as the sum of the input water to the fuel cell, which is due to the moisture content and the water generated by the electrical current produced. The reason for this is the impossibility of direct measuring of the non-useful water applied in the denominator of Equation (5), as well as performing the measuring under the stable condition of the fuel cell performance.

Under stable conditions, the amount of moisture entering the fuel cell acts as the regulator of the amount of water absorbed by the swollen membrane and the amount of the input fuel's flow rate to the fuel cell. By determining the consumed hydrogen, the input hydrogen and the parameter δ, the modified fuel utilization factor can be calculated in terms of the produced current. In Figure 5, the results of calculations for the numerical value of the parameter δ and the modified fuel utilization at the 50% and 80% relative humidity of the cathode side are shown. To cover the comparison, the experimental data of other studies have been used for this issue at the 35% relative humidity of the cathode side [22].

As shown in Figure 5, the parameter δ increases with increasing produced current density, which appears quite logical. This incremental ratio is more intense at lower current densities and the variation of the parameter δ decreases with increasing produced current density. Also, it was found that the maximum value achieved for the parameter δ at 50% relative humidity is approximately 0.88, while in the case of 80% relative humidity, the maximum obtained value is approximately 0.72. On the other hand, the modified fuel utilization factor at the 50% relative humidity is higher than that at 80% relative humidity in the cathode side. Furthermore, as shown in Figure 5, at a relative humidity of 35% compared to 50% and 80%, the higher fuel utilization is obtained, which is again aligned with the idea presented in this study. This result is the outcome of the new approach considered in Figure 2 for the consumption of hydrogen and because of the calculation of the modified fuel utilization factor based on the water generated by produced current.

Another important point is that in Figure 5, the value of the parameter δ at the 50% relative humidity is equal to 0.7 at 900 mA/cm^2 current density. While at 80% relative humidity, when the produced current density attains about 2500 mA/cm^2, the value of the parameter δ is still equal to 0.7. This means that despite the relative humidity of the cathode side being able to increase both produced current density and power generation of the fuel cell, at the same time, as shown in Figure 5, the modified fuel utilization factor which is under the direct effect of the parameter δ is reduced. Therefore, it should be noted that it is necessary to find an optimal range for the relative humidity of the cathode side to achieve the best performance of the fuel cell in terms of power generation and fuel consumption. Hence, increasing the relative humidity of the cathode side itself cannot provide the best result.

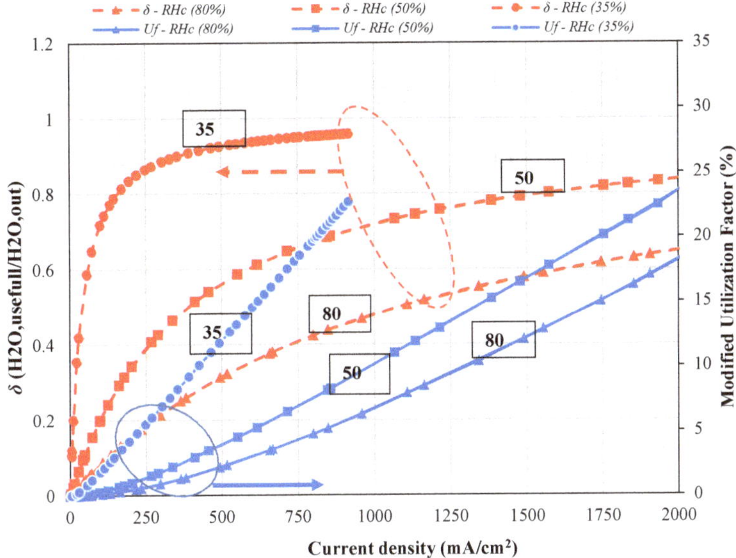

Figure 5. δ and the modified fuel utilization in terms of the total produced current density, at 35%, 50% and 80% relative humidity of the cathode side.

5. Conclusions

In this work, which contains both theoretical and experimental analysis, a PEMFC was studied in terms of the fuel utilization factor. In the theoretical part, a new scenario was presented providing a detailed discussion of the various paths that hydrogen can go through in a PEMFC and how it is consumed. The main reason for this theoretical discussion is the existence of an irrational interval in the various formulations presented in the literature to calculate the fuel utilization factor, which indicates a weakness in the formulation. Two modes for the water production were considered in the analysis: "useful water" and "non-useful water". The term of "usefulness" refers to the state that consumed hydrogen leads to the production of an external electrical current. Considering the fuel utilization factor is directly related to the hydrogen consumption, a correction coefficient (δ) was presented to calculate the theoretical fuel utilization factor, taking into account the two concepts of "useful water" and "non-useful water". That is the innovation of the theoretical part of this study leading to the calculation of the modified fuel utilization factor. According to the result of the theoretical discussion, one of the effective factors to determine the proper concluded value of the fuel utilization is the amount of liquid water generated on the cathode side, which contains the sum of useful and non-useful water.

An increase in relative humidity to 80% improved the fuel cell performance curve. At a relative humidity of 50% in the cathode side, the maximum power density is approximately 645 mW/cm^2, which is achieved at 2000 mA/cm^2 current density. At 80% relative humidity, the maximum power density is about 700 mW/cm^2, which results in 2310 mA/cm^2 current density and shows an increase of about 8.5% with respect to the maximum power density at 50% relative humidity.

An important factor to calculate the exact amount of the fuel utilization is the amount of liquid water at the cathode side. This is considered in the novel introduced parameter δ that was obtained by dividing the useful water to the sum of the useful and non-useful water. The parameter δ increases as the produced current density increases, but this incremental ratio is more intense at the lower current densities. With increasing the produced current density, the variation in the parameter δ decreases and approaches a certain amount of limit. The application of the parameter δ at a 50% and 80% relative humidity in the cathode side causes the modified fuel utilization factor at 50% relative humidity be higher than that at 80%. This conclusion is not aligned with the result achieved in the study of the effect of the relative humidity in the cathode side on the fuel consumption. Therefore, it should be noted that it is necessary to find an optimal range for the relative humidity in the cathode side to achieve the best performance of the fuel cell in terms of power generation and fuel consumption since increasing the relative humidity of the cathode side itself cannot ensure the best result.

Author Contributions: Conceptualization, M.B.Y., H.G. and K.D.; Data curation, M.B.Y., H.G. and N.A.; Formal analysis, H.G., K.D. and B.C.R.T.; Investigation, M.B.Y., H.G., K.D., N.A. and B.C.R.T.; Methodology, M.B.Y., H.G., K.D. and N.A.; Resources, H.G.; Supervision, H.G. and B.C.R.T.; Validation, M.B.Y. and K.D.; Writing—original draft, M.B.Y.; Writing—review & editing, M.B.Y., H.G., K.D., N.A. and B.C.R.T. All authors have read and agreed to the published version of the manuscript.

Funding: This research received no external funding.

Conflicts of Interest: The authors declare no conflict of interest.

Nomenclature

e^-	Electron
F	Faraday constant (96,487 $C{\cdot}mol^{-1}$)
GDL	Gas Diffusion Layer
H_2	Hydrogen
H_2O	Water
H^+	Hydrogen ion
I	Current (A)
n	Number of electrons in the electrochemical reaction
N_{H_2}	Amount of consumable hydrogen ($mol{\cdot}s^{-1}$)
O_2	Oxygen
OCV	Open circuit voltage (V)
PEMFC	Proton Exchange Membrane Fuel Cell
RH	Relative humidity (%)
U_F	Fuel utilization
$U_{F,\ modified}$	Modified fuel utilization
Greek Letters	
δ	Correction factor for fuel utilization
Superscripts	
a	anode
c	cathode
con	consumed
in	inlet
out	outlet

References

1. Ghadamian, H.; Saboohi, Y. Quantitative analysis of irreversibilities causes voltage drop in fuel cell (simulation & modeling). *Electrochim. Acta* **2004**, *50*, 699–704.
2. Ghadamian, H.; Bakhtary, K.; Namini, S.S. An algorithm for optimum design and macro-model development in PEMFC with exergy and cost considerations. *J. Power Sources* **2006**, *163*, 87–92. [CrossRef]
3. Kazeminasab, B.; Rowshanzamir, S.; Ghadamian, H. Multi-objective multivariable optimization of agglomerated cathode catalyst layer of a proton exchange membrane fuel cell. *Bulg. Chem. Commun.* **2015**, *47*, 38–48.

4. Kim, H.Y.; Kim, K. Numerical study on the effects of gas humidity on proton-exchange membrane fuel cell performance. *Int. J. Hydrog. Energy* **2016**, *41*, 11776–11783. [CrossRef]
5. Ozen, D.N.; Timurkutluk, B.; Altinisik, K. Effects of operation temperature and reactant gas humidity levels on performance of PEM fuel cells. *Renew. Sustain. Energy Rev.* **2016**, *59*, 1298–1306. [CrossRef]
6. Kazeminasab, B.; Rowshanzamir, S.; Ghadamian, H. Nitrogen doped graphene/cobalt-based catalyst layers of a PEM fuel cell: Performance evaluation and multi-objective optimization. *Korean J. Chem. Eng.* **2017**, *34*, 2978–2983. [CrossRef]
7. Ou, K.; Yuan, W.W.; Choi, M.; Yang, S.; Kim, Y.B. Performance increase for an open-cathode PEM fuel cell with humidity and temperature control. *Int. J. Hydrog. Energy* **2017**, *42*, 29852–29862. [CrossRef]
8. Muirhead, D.; Banerjee, R.; George, M.G.; Ge, N.; Shrestha, P.; Liu, H.; Lee, J.; Bazylak, A. Liquid water saturation and oxygen transport resistance in polymer electrolyte membrane fuel cell gas diffusion layers. *Electrochim. Acta* **2018**, *274*, 250–265. [CrossRef]
9. Hasheminasab, M.; Kermani, M.J.; Nourazar, S.S.; Khodsiani, M.H. A novel experimental based statistical study for water management in proton exchange membrane fuel cells. *Appl. Energy* **2020**, *264*, 114713. [CrossRef]
10. Wang, B.; Wu, K.; Xi, F.; Xuan, J.; Xie, X.; Wang, X.; Jiao, K. Numerical analysis of operating conditions effects on PEMFC with anode recirculation. *Energy* **2019**, *173*, 844–856. [CrossRef]
11. Rahman, M.A.; Mojica, F.; Sarker, M. Abel Chuang, P.Y. Development of 1-D multiphysics PEMFC model with dry limiting current experimental validation. *Electrochim. Acta* **2019**, *320*, 134601. [CrossRef]
12. Jeon, D.H.; Kim, K.N.; Baek, S.M.; Nam, J.H. The effect of relative humidity of the cathode on the performance and the uniformity of PEM fuel cells. *Int. J. Hydrog. Energy* **2011**, *36*, 12499–12511. [CrossRef]
13. Abdollahzadeha, M.; Pascoa, J.C.; Ranjbar, A.A.; Esmaili, Q. Analysis of PEM (Polymer Electrolyte Membrane) fuel cell cathode two-dimensional modeling. *Energy* **2014**, *68*, 478–494. [CrossRef]
14. Nishikawa, H.; Sasou, H.; Kurihara, R.; Nakamura, S.; Kano, A.; Tanaka, K.; Aoki, T.; Ogami, Y. High fuel utilization operation of pure hydrogen fuel cells. *Int. J. Hydrog. Energy* **2008**, *33*, 6262–6269. [CrossRef]
15. Iranzo, A.; Boillat, P.; Biesdorf, J.; Salva, A. Investigation of the liquid water distributions in a 50 cm^2 PEM fuel cell: Effects of reactants relative humidity, current density, and cathode stoichiometry. *Energy* **2015**, *82*, 914–921. [CrossRef]
16. Muirhead, D.; Banerjee, R.; Lee, J.; George, M.G.; Ge, N.; Liu, H.; Chevalier, S.; Hinebaugh, J.; Han, K.; Bazylak, A. Simultaneous characterization of oxygen transport resistance and spatially resolved liquid water saturation at high-current density of polymer electrolyte membrane fuel cells with varied cathode relative humidity. *Int. J. Hydrog. Energy* **2017**, *42*, 29472–29483. [CrossRef]
17. Yousefkhani, M.B.; Ghadamian, H.; Massoudi, A.; Aminy, M. Quantitative and qualitative investigation of the fuel utilization and introducing a novel calculation idea based on transfer phenomena in a polymer electrolyte membrane (PEM) fuel cell. *Energy Convers. Manag.* **2017**, *131*, 90–98. [CrossRef]
18. Farhat, Z.N. Modeling of catalyst layer microstructural refinement and catalyst utilization in a PEM fuel cell. *J. Power Sources* **2004**, *138*, 68–78. [CrossRef]
19. Hwang, J.W.; Lee, J.Y.; Jo, D.H.; Jung, H.W.; Kim, S.H. Polarization characteristics and fuel utilization in anode-supported solid oxide fuel cell using three-dimensional simulation. *Korean J. Chem. Eng.* **2011**, *28*, 143–148. [CrossRef]
20. Han, I.; Jeong, J.; Shin, H.K. PEM fuel-cell stack design for improved fuel utilization. *Int. J. Hydrogen Energy* **2013**, *38*, 11996–12006. [CrossRef]
21. Barbir, F. *PEM Fuel Cells: Theory and Practice*; University of Connecticut: Storrs, CT, USA; Elsevier: San Diego, CA, USA, 2005; pp. 33–34.
22. Arif, M.; Cheung, S.C.P.; Andrews, J. A systematic approach for matching simulated and experimental polarization curves for a PEM fuel cell. *Int. J. Hydrogen Energy* **2019**, *45*, 2206–2223. [CrossRef]

Article

Eulerian Two-Fluid Model of Alkaline Water Electrolysis for Hydrogen Production

Damien Le Bideau [1], Philippe Mandin [1,*], Mohamed Benbouzid [2], Myeongsub Kim [3], Mathieu Sellier [4], Fabrizio Ganci [5] and Rosalinda Inguanta [5]

1 Institut de Recherche Dupuy de Lôme (UMR CNRS 6027 IRDL), University Bretagne Sud, 56100 Lorient, France; damien.le-bideau@univ-ubs.fr
2 Institut de Recherche Dupuy de Lôme (UMR CNRS 6027 IRDL), University of Brest, 29238 Brest, France; Mohamed.Benbouzid@univ-brest.fr
3 Department of Ocean and Mechanical Engineering, Florida Atlantic University, Boca Raton, FL 33431, USA; kimm@fau.edu
4 Department of Mechanical Engineering, University of Canterbury, Christchurch 8140, New Zealand; mathieu.sellier@canterbury.ac.nz
5 Dipartimento di Ingegneria, University of Palermo, Viale delle Scienze, 90128 Palermo, Italy; fabrizio.ganci@unipa.it (F.G.); rosalinda.inguanta@unipa.it (R.I.)
* Correspondence: philippe.mandin@univ-ubs.fr

Received: 1 June 2020; Accepted: 30 June 2020; Published: 2 July 2020

Abstract: Hydrogen storage is a promising technology for storage of renewable energy resources. Despite its high energy density potential, the development of hydrogen storage has been impeded, mainly due to its significant cost. Although its cost is governed mainly by electrical energy expense, especially for hydrogen produced with alkaline water electrolysis, it is also driven by the value of the cell tension. The most common means of electrolyzer improvement is the use of an electrocatalyst, which reduces the energy required for electrochemical reaction to take place. Another efficient means of electrolyzer improvement is to use the Computational Fluid Dynamics (CFD)-assisted design that allows the comprehension of the phenomena occurring in the electrolyzer and also the improvement in the electrolyzer's efficiency. The designed two-phase hydrodynamics model of this study has been compared with the experimental results of velocity profiles measured using Laser Doppler Velocimetry (LDV) method. The simulated results were in good agreement with the experimental data in the literature. Under the good fit with experimental values, it is efficient to introduce a new physical bubble transfer phenomenon description called "bubble diffusion".

Keywords: hydrogen production; alkaline water electrolysis; two-phases flow; CFD; two-phase process

---

1. Introduction

A key challenge of the 21st century is to deal with climate change. The Intergovernmental Panel on Climate Change (IPCC) declares that, to stay below the 3 °C of global temperature increase from the pre-industrial era, zero net emission of greenhouse gases must be achieved at the mid-21 century. Thus, the consumption of fossil-based fuels must be reduced from the current level as much as possible. The present global energy demands can hardly be met simply by replacing conventional fossil-based energy sources (e.g., thermal or nuclear energy) by renewable ones. The integration of renewables into the national electrical grid brings the issue of energy storage because of the intermittent nature of renewable energy resources. Energy storage is achieved using several processes such as battery, Pumped Hydroelectric Energy Storage (PHES) or super capacitor. However, only a few processes allow interseasonal storage (synthetic natural gas and hydrogen (H_2)). There are several processes that do not emit greenhouse gases and produce H_2, such as thermochemical hydrogen process and electrolytic

H_2 process. Thermochemical processes for hydrogen production use heat to decompose water into hydrogen and oxygen [1], whereas electrolytic H_2 processes use electric energy to achieve water splitting. Electrolytic hydrogen is a promising storage technology due to its high specific potential energy and storage time, but its cost hampers its development. Another interesting advantage of hydrogen is the diversity of its use. Indeed, it can be used as a fuel for fuel cell for electricity production for buildings, as a reactant in chemical processes and as a fuel for powering vehicles. Thus, using hydrogen, especially for fueling heavy vehicles, could be one of the solutions to reduce the dependence on fossil fuels in the transport sector. There are three leading electrolysis technologies, but only two have been employed in industrial applications: they are the alkaline and Proton Exchange Membrane (PEM) electrolyzer. The latter is more efficient than the former but, due to the requirement of acidic electrolyte, rare materials must be used (such as platinum). Alkaline water electrolyzer has several advantages over PEM electrolyzer such as its robustness, cheaper costs, and system mature. One way of decreasing the cost is to increase the efficiency of the process by decreasing the cell voltage. The cell voltage is a sum of overvoltage Equation (1).

$$U_{cell}(j) = E_{rev}(P,T) + \eta(j)_{cath\ act} + \eta(j)_{an\ act} + R(T,\ Y_{KOH})\ j + \eta(j)_{cath\backslash an\ conc} \tag{1}$$

E_{rev} is the reversible voltage in V, η_{act} is the activation overvoltage in V, R the sum of the electrodes, membrane and electrolyte resistance in Ω cm^2 and η_{conc} the concentration overpotential in V, j the current density in A m^{-2}.

The E_{rev} voltage represents the minimum voltage where the water electrolysis occurs. This value is around 1.23 V in terms of atmospheric pressure and 25 °C. This minimal voltage can be decreased by increasing the temperature. The majority of previous studies have focused on finding a cheap, robust and electroactive electrode material to decrease the activation overpotential, which is the second term of the Equation (1) [2–4]. There is also another research trend that suggests that another way of increasing the efficiency is to decrease the ohmic resistance through the electrolyzer design using simulations or experiments. Another interesting research trend is to find another redox couple, such as those that exist in urea electrolysis [5] or through the use of an electrocatalytically driven cyclic process involving iron oxide species that allow a low OER overvoltage [6]. The latter research trend tries to decrease the cell voltage using material science. The present approach uses the chemical engineering approach. Even if it is theoretically possible to perform monophasic alkaline water electrolysis [7], this process is performed under two-phase flow configuration in the industry. Thus, the bubble impact on the efficiency must be properly modeled or simulated. The first well-known study about this subject was performed by Tobias et al. [8] and Hine et al. [9]. Both studies focused on the bubble effect on the electrolyte resistance. More recently, the simulation has been used to study this two-phase flow phenomenon. There are two main models: Euler–Euler and Euler–Lagrange models. The latter considers the electrolyte as a continuum phase and the bubbles as a discrete phase. Mandin et al. [10] used this model in their study, and the bubble dispersion is taken into account using a horizontal source term. Hreiz et al. [11] also used this model. In order to prevent the use of the source term, the bubble injection has been shifted away from electrode surfaces. Using this technique, it has been concluded that the drag force was sufficient to describe the bubble–liquid interaction. However, the presented comparison with the experiments in their study was mainly qualitative. For Euler–Euler models, two types of models have been used: the mixture model and the two-fluid model. The former has been employed by Dahlkild [12], Wedin et al. [13] and Schillings [14]. In those three studies, the model used is called semi-empirical because an empirical correlation is chosen. In Wedin et al. [13] and Schillings et al. [14], the calculations are compared to the experiments performed by Boissonneau et al. [15]. Those models prevent the use of turbulence closing model. However, Boissonneau et al. [15] observed a bubble-induced turbulent behavior in the upper part of the narrow and small electrochemical cell. In their study, Aldas et al. [16] used a laminar two-fluid model and found that this model underestimates the void fraction distribution compared to the experiment. They concluded that local weak turbulence must be taken into account. In another study, Mat [17] developed a two-fluid model that takes into account local turbulence. The computed results were compared with the experimental void fraction distribution results of Riegel [18] and good agreements were found. In this study, a two-fluid Euler–Euler is developed and will be compared with the experimental results of Boissonneau et al. [15] and the numerical results of Schillings et al. [14]. In this study, a new force is added to the traditional Fluent© commercial software Eulerian model. The pertinence of its use from a numerical and theoretical point of view will be discussed.

2. Numerical Methods

This section describes the geometry and the type of mesh used in the current study. The mathematical formulation, the boundary conditions and the numerical procedure are also introduced. Comparison of the numerical results with the experiments and discussion will be followed in the next section.

2.1. Geometry and Mesh

The geometry of the computational domain is identical to the experimental apparatus of Boissonneau et al. [15]. Figure 1 shows experimental apparatus of Boissonneau et al. [15]. It is composed of a channel with a height of 120 mm, a width of 3 mm and a thickness of 30 mm. The whole channel is submerged in the electrolyte. The electrodes are 40 mm high and are placed 40 mm away from the bottom. When the current is applied, bubbles are generated on electrode surfaces. The detached bubbles rise and trigger the electrolyte causing the bubble-induced spontaneous movement of the surrounding electrolyte. This phenomenon is called gas-lift configuration.

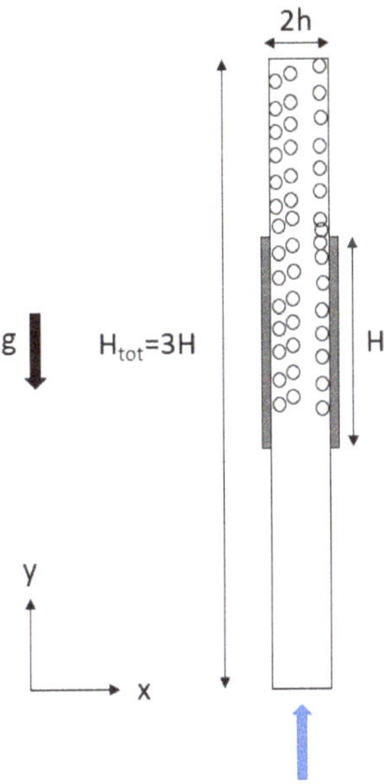

Figure 1. Geometry of the computation domain. h stands for the half-width of the electrolyte, H for the electrode height. The blue arrow symbolizes the pumping electrolyte induced by the electrogenerated bubbles.

In numerical calculations, a grid sensitivity study must be performed in order to obtain a grid-independent solution. In this study, the grid is refined until the difference between two solutions reaches zero. On the other hand, in the 2D two-fluid model, this is not always possible. Indeed, Picardi et al. [19] suggested that there is an ideal grid resolution for the 2D two-fluid model simulation. They determined that the grid resolution must correspond to the bubble-to-cell ratio $1/\sqrt{2}$. If a finer grid is chosen, the results become non-physical or the calculations diverge. The same results have been reproduced by Law et al. [20]. Finally, Picardi et al. [19] stated that this problem could be resolved when a 3D geometry was employed. However, their calculation took more time than in 2D (from 10 CPU hours to 94 days using a single processor on a quad-core 2.26 GHz Intel Nehalem processor). Panicker et al. [21] and Vaidheeswaran et al. [22] explained this phenomenon by the fact that the problem is ill-posed. An ill-posed problem means that the solution is elliptic and the solution of the Nth step depends on the N+1th step. They solved the problem by adding a collision or dispersion term. They also affirmed that although the virtual mass force has an insignificant influence on the results, it ensures a better stability on the calculation. Panicker et al. [19] declared that other authors solved this problem by artificially increasing the liquid viscosity or adding an interface pressure term to the model. Both articles used a linear stability analysis and, using the results of this study, Panicker et al. [19] added a dispersion term that uses the void fraction gradient. To ensure the hyperbolicity, a parameter depending on the void fraction was added to the dispersion term. As a result, the resulting model gives better results when compared with the model without the dispersion term. In this study, we choose to add the dispersion term used in studies by Marfaing et al. [23] and Davidson [24]. The Figure 2 shows the results of the grid sensitivity with four different meshes: 100, 120, 150 and 200 μm. It can be noted that there is a discrepancy of around 10% in the all domains with the grid size between a 200 and 150 μm. The maximum difference between the 150 μm and the other grids are located at the velocity and void fraction peaks. Even if it seems that 100 μm is sufficient to describe the void fraction and velocity distribution. a mesh of 60 μm has chosen in order to be sure that the results converge with no grid-dependency.

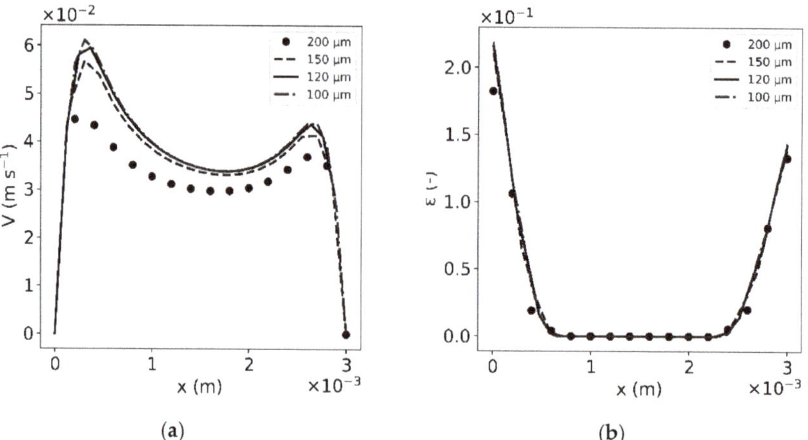

Figure 2. Results of the mesh sensitivity study. The graph (**a**) presents the mesh dependence of the velocity. The graph (**b**) presents the mesh dependence of the void fraction.

2.2. Mathematical Formulation

2.2.1. The Bubble Dispersion Problem

Lee et al. [25], Abdelouahed et al. [26] and Hreiz et al. [11] experimentally visualized and measured the void fraction distribution in the electrolyte in a cell under forced convection [25] and not net flow configuration cell (in this type of configuration, the gases escape at the air–liquid interface, but the electrolyte stay confined into the inter-electrode gap) [11,26]. They all reported that the bubbles were spread into the electrolyte. After simulating their experiments, Abdelouahed et al. [26] observed that classical models of lift (e.g., Saffman-Mei) and drag failed to reproduce their experimental results. They attributed this spreading to a lift force that has a negative coefficient. Although their model fits their experimental data, the model seems to rely on numerical diffusion triggered by a coarse mesh, so its use is questionable, at least in our present configuration, due to the fact that the electrochemical cell is narrow. This is because, to obtain their results, a coarse mesh must be used, but for our present geometry, this coarse mesh does not satisfy all the mesh independence conditions (the reader can refer to Section 2.1). Hreiz et al. [11] attributed this spreading to the numerical diffusion. Hreiz et al. [11] modeled the same experimental study using a Euler–Lagrange model (gas is modeled as a discrete phase). They succeeded in quantitative reproduction of their experimental data by using the drag force only, but the mesh independence condition was not satisfied. The model of Mat [17] is a Euler–Euler model that succeeds in simulated results that fit experimental data, but the results seem odd. They measured and calculated the diphasic boundary layer and void fraction increase with increasing electrolyte flow. Those results have been invalidated experimentally by Lee et al. [25] and numerically by Schillings et al. [14]. In addition, an increase in void fraction value triggers an increase in the cell voltage and it has been measured that increasing the electrolyte flow decreases the cell voltage. Schillings et al. [14] and Wedin et al. [13] used a mixture model developed by Ishii that simulated results close to the experimental data from Boissonneau et al. [15]. In addition to the drag force and Saffman lift force, their closure term is composed of three terms of bubble–bubble interactions. Thus, in this study, we decided to introduce a bubble dispersion term. Indeed, a high concentration of bubbles increases the bubble collision and hence trigger of bubble diffusion from high concentration to low concentration. This phenomenon has been observed by Ham et al. [27]. The additional force used in this study is presented in the Equation (2). As described earlier, the additional force is inspired by the force used in Marfaing et al. [23] and Davidson [24]. However, in their model, the present term was multiplied by the drag coefficient. The void fraction gradient is used to mathematically traduce the diffusive nature of this force.

$$\vec{F}_{BD} = -\underbrace{\varepsilon_g \rho \frac{K_g}{d_b} |U_r| \vec{\nabla} \varepsilon_g}_{Bubble\ dispersion\ force} \tag{2}$$

ε is the gas or liquid fraction, the subscript k can be either g (O_2, H_2) or liq (for liquid), ρ is the density in kg m^{-3}, d_b is the bubble diameter in m, U_r is the gas phase velocity minus the liquid velocity in m s^{-1}.

Using this term is expected to reproduce the turbulence-like behavior of the electrolyte flow observed by Boissonneau et al. [15].

2.2.2. Model

The model has been designed using the following hypotheses:

- The flow is Newtonian, viscous and incompressible;
- Due to the high heat transfer induced by the bubbles and the relatively low surface and volume inducing a low heat injection by ohmic and activation losses, the flow is considered isothermal
- At the same time, numerical simulations were carried out in order to highlight the influence of ions on the velocity and void fraction distribution. There were only very little differences when the ions distribution was taking into account. Thus, the electrolyte is considered as extremely well mixed. This hypothesis has been made also by Abdelouahed et al. [26] and Schillings et al. [14];
- Oxygen, hydrogen and electrolytes are three continuum media;
- Because the Reynolds number calculated via the data from Boissonneau et al. [15] is between 240 and 480, the flow is considered as laminar;
- The effect of the surface tension is neglected;
- Nagai et al. [28] observed a dependency of the diameter with respect with the current density and in addition to this dependency, Boissonneau et al. [15] observed that the bubble diameter increases with height. Schillings et al. [14] chose to make the diameter increase with the height and current density. In the current study and in other studies [12,13], it was decided that a constant bubble diameter is taken for a given current density. Although this hypothesis is made in a lot of studies,

it does not reflect the reality. There is a distribution of the bubble diameter, but since it is not yet possible to model this distribution, we are forced to make this assumption;

- The current density distribution does not affect the flow distribution [29,30]. Thus, the current density distribution is taken as uniform for the validation.

The accuracy of the two-dimensional (*x*,*y*) flow hypothesis needs to be discussed. Therefore, for each phase, the equation set can be written

$$\frac{\partial \varepsilon_k \rho_k}{\partial t} + \vec{\nabla} \cdot \left(\varepsilon_k \rho_k \vec{V}_k \right) = S_g \tag{3}$$

ε is the gas or liquid fraction, the subscript k can be either g (O_2, H_2) or liq, ρ is the density in kg m^{-3}, V the velocity in m s^{-1}, and S_g is the term source in kg m^{-3} s^{-1}.

$$\frac{\partial}{\partial t}\left(\varepsilon_k \rho_k \vec{V}_k \right) + \vec{\nabla} \cdot \left(\varepsilon_k \rho_k \vec{V}_k \vec{V}_k \right) = -\varepsilon_k \vec{\nabla} p + \vec{\nabla} \cdot (\varepsilon_k \tau) + \varepsilon_k \rho \vec{g} + \vec{F}_k \tag{4}$$

p is the pressure in Pa, τ is the stress tensor in Pa, g the gravitational acceleration in m s^{-2}, and $\vec{F}_k$ is the exchange term in N m^{-3}.

The stress tensor is written as follows

$$\tau = \mu_k \left[\left(\nabla \vec{V}_k + \nabla \vec{V}_k^{\,T} \right) - \frac{2}{3} \vec{\nabla} \cdot \vec{V}_k I \right] \tag{5}$$

with μ_k the viscosity of the phase k in Pa s and I the unit tensor.

$$\vec{F}_k = \vec{F}_D + \vec{F}_L + \vec{F}_{BD} \tag{6}$$

$$\vec{F}_k = \underbrace{-\frac{3}{4} \varepsilon_g \rho \frac{C_D}{d_b} |U_r| U_r}_{Drag\ force} - \underbrace{\varepsilon_g \rho C_L |U_r| rot\left(\vec{V}_l \right)}_{Lift\ force} - \underbrace{\varepsilon_g \rho \frac{K_g}{d_b} |U_r| \vec{\nabla} \varepsilon_g}_{Bubble\ dispersion\ force} \tag{7}$$

2.3. Boundary Conditions

The water electrolysis performed in Boissonneau et al. [15] is neither acidic or alkaline. This electrolysis is called as aqueous in this study. The supporting electrolyte is Na_2SO_4 concentrated at 50 g L^{-1}. Therefore, the reaction occurring at the anode and the cathode are Equations (8) and (9), respectively.

$$H_2O_{(liq)} \rightarrow 2H^+{}_{(aq)} + \frac{1}{2}\, O_{2(g)} + 2e^- \tag{8}$$

$$2H_2O_{(liq)} + 2e^- \rightarrow 2OH^-{}_{(aq)} + H_{2(g)} \tag{9}$$

If the reader wants to simulate an alkaline electrolysis, he must change the thermophysical inputs from Na_2SO_4 to KOH or NaOH. The reader can refer to Le Bideau et al. [31], where the thermophysical inputs are given with their sensitivity with respect to temperature and electrolyte mass fraction.

The quantity of produced gases is directly correlated to the current density through the Faraday's law Equation (10)

$$q_m = \mathrm{n} \frac{j\,S}{F} M_g \tag{10}$$

q_m is the mass flow of produced gas in kg s^{-1}, M_g is the molar mass of the gas kg mol^{-1}, S is the electrode surface in m^2, F is the Faraday constant 96,500 C mol^{-1}, and n is the ratio of the stoichiometric number of the gas and the number of electrons exchanged during the reaction.

In the two-fluid equation, in most studies [9–11,22,25,26], the input parameters for the boundary conditions are the velocity and the void fraction. The value of the void fraction is fixed arbitrarily.

According to Alexiadis [32,33], this value does not influence the hydrodynamic characteristic of the flow. However, better results have been obtained using a source term that produces gas in the cell in the vicinity of the electrodes. This method has been used by Charton et al. [34]. This source term is written as Equation (11)

$$S_g = \mathrm{n}\frac{j}{F \times \Delta x} M_g \quad (11)$$

Δx is the width of the first cell next to the electrode in m. For the bottom and top boundary condition, a pressure inlet (P_{Tot} = 0) and pressure outlet condition (P = 0) is fixed. For the other wall, a no-slip condition is fixed meaning that the velocity is set to 0 m s^{-1}. The Table 1 summarizes the boundary conditions for the present problem.

Table 1. The boundary conditions to solve the problem.

Position	Boundary Conditions
$x = 0\ H_{elec} < y < 2\ H_{elec}$	$\vec{V_l} = \vec{V_{H2}}0$
$x = 2\text{h}\ 0 < y < 3\ H_{elec}$	$\vec{V_l} = \vec{V_{O2}} = 0$
$0 < x < 2\text{h}\ y = 0$	$P_{Tot} = 0$
$0 < x < 2\text{h}\ y = 3\ H_{elec}$	$P = 0$

2.4. Numerical Procedure

Equations (3) and (4) are solved using the commercial code, Ansys Fluent. This commercial code solves equations using the finite volume method by discretizing the geometry in volume and subsequently integrating the governing equation over the volume. The governing equation is expressed as the following algebraic Equation (12)

$$a_p \varphi = \sum_i^N a_i \varphi + b \quad (12)$$

It is considered that the convergence is met when the residuals remain stable and when the average gas and liquid velocity, as well as the average gas void fraction, reach the value of 10^{-3}. In order to reach this convergence, for 2D simulation, the flow is initialized with a forced convection by imposing pressure at the bottom. When the convergence is reached, the obtained results are used as initial guess for the bubble-driven flow. The calculation time can vary, depending on the numerical procedure, between 20 and 30 min with an Intel Core i7-6700HQ CPU @2.60 GHz.

Table 2 gives the input data for the validation of the numerical model. It can be seen that the current density varies between 500 and 2000 A m^{-2}. The most interesting datapoints are those obtained at 2000 A m^{-2}. Indeed, industrial alkaline electrolyzers generally use a current density between 2000 and 5000 A m^{-2} [35]. The input data for the electrolyte were taken from the work of Isono et al. [36]. As the bubble diameter is taken as a constant, an average value has been calculated from the correlation of Schillings et al. [14]. As aforementioned, the term K in the Equation (5) is used to fit the experimental data of Boissonneau et al. [15]. A sensitivity study has been performed to choose the parameter. The results of this study are presented in Table 3. The parameter K is always bigger for oxygen than hydrogen.

Table 2. The input values for the problem.

Name	Value
Geometry inputs	
H_{elec} (mm)	40
L (mm)	30
h (mm)	1.5
H_{Tot} (mm)	120
Physical Inputs	
ρ_l (kg m^{-3})	1040
ρ_{O2} (kg m^{-3})	1.3
ρ_{H2} (kg m^{-3})	0.08
υ_l (m^2 s^{-1})	9.97×10^{-7}
Two-Phase Inputs	
d_b (μm)	for 500 A m^{-2} d_b = 50 μm for 1000 A m^{-2} d_b = 58 μm for 2000 A m^{-2} d_b = 78 μm

Table 3. Sensitivity study results for the parameter *K*.

500 A m^{-2}	1000 A m^{-2}	2000 A m^{-2}
K_{O2}/d = 10.5 K_{H2}/d = 5	K_{O2}/d = 9 K_{H2}/d = 4	K_{O2}/d = 5 K_{H2}/d = 2.5

In order to use the current model to other design, the Vaschy–Buckingham has been used and the sensitivity of the *K* parameter to dimensionless groups has been calculated in Equations (13)–(16).

$$Re_G = \frac{\rho_l \, V_G \, H_{elec}}{\mu_L} \tag{13}$$

$$Fr_G = \frac{g \, H_{elec}}{V_G{}^2} \tag{14}$$

$$r^* = \frac{d}{2\,H} \tag{15}$$

$$h^* = \frac{h}{H} \tag{16}$$

Therefore, the following correlation Equation (17) was used to calculate the *K* parameter. However, this correlation was used for height in the order of 10 cm and for Fr_G numbers higher than 10^5.

$$\frac{K}{V_G\,H} = 0.197\,Re_G{}^{0.108} + 0.5\,r^{*0.124} + 0.668\,h^{*-0.408} + 0.000323\,Fr_G{}^{0.661} + 0.375 \tag{17}$$

3. Results

The experimental data of Boissonneau et al. [15] are the liquid velocity at three locations: 5 mm before the electrodes (y = 35 mm), at the mid-section of the electrodes (y = 60 mm) and 5 mm before the ends of the electrodes (y = 75 mm). Figures 2–4 present the results of the current study and the experimental data. At the entrance of the channel, a Poiseuille liquid velocity distribution is observed, and this distribution is flattened at the center (also called bulk). The exchange of momentum between the gas phases and the liquid phase is easily observed next to the anode and cathode. It can be noted that the peak induced by hydrogen is bigger than the oxygen peak, because the injected volume of hydrogen is two times bigger than oxygen one. The fact is that in the bulk, a plateau is observed due to bubble induced turbulence [15]. Those figures allow for comparison of the current numerical results with experimental data. The "Abdelouahed-like" model is a model that uses the same method as presented in Abdelouahed et al. [26]. In their study [26], the authors used the lift coefficient to fit their simulated data with their experimental values. They showed that a negative coefficient permits a good agreement between their measurements and predicted values. In the present study, a sensitivity study was performed to have the coefficient that fits the most the experimental data from Boissonneau et al. [15]. First of all, the numerical results for all the models show some discrepancies with the experimental data. Schilling's model and Abdelouahed-like model accurately predict the velocity distribution on the cathode side, but the accuracy of the simulated data drops on the anode side. The Figure 3 shows that the "Abdelouahed-like" model describes the anode side and cathode side velocity distribution with accuracy, but the bulk velocity distribution shows larger errors. However, as shown in Figures 4 and 5, the more the current density increases, the less precise the model is.

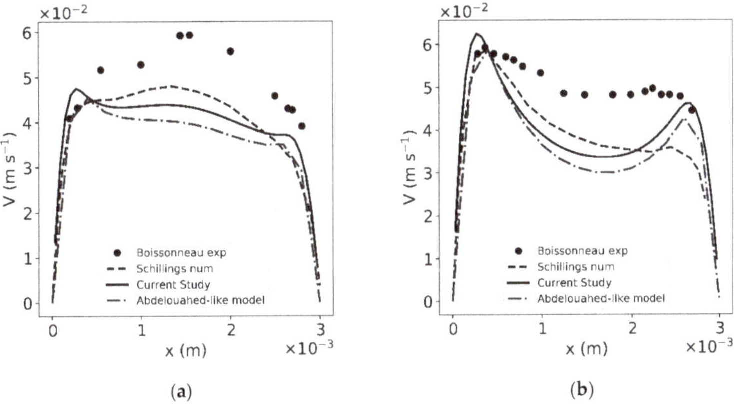

Figure 3. Result of the liquid velocity in different models at j = 500 A m^{-2} (**a**) at y = 60 mm and (**b**) at y = 75 mm. The dot are the data from Boissonneau et al. [15], the grey dashed line is the numerical results of the Schilling et al. model [14], the dotted and dashed line is a model using the method of Abdelouahed et al. [26]. Finally, the black solid line the results from the current study with 2D approximation.

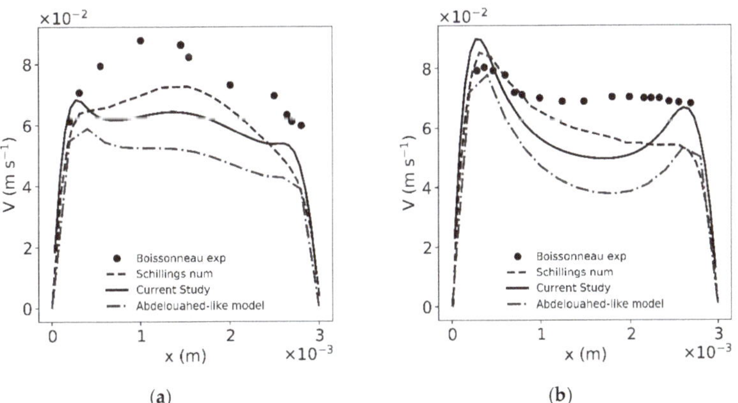

Figure 4. Results of the liquid velocity in different models at j = 1000 A m^{-2} (**a**) at y = 60 mm and (**b**) at y = 75 mm. The dot are the data from Boissonneau et al. [15], the grey dashed line is the numerical results of the Schilling et al. model [14], the dotted and dashed line is a model using the method of Abdelouahed et al. [26]. Finally, the black solid line the results from the current study with 2D approximation.

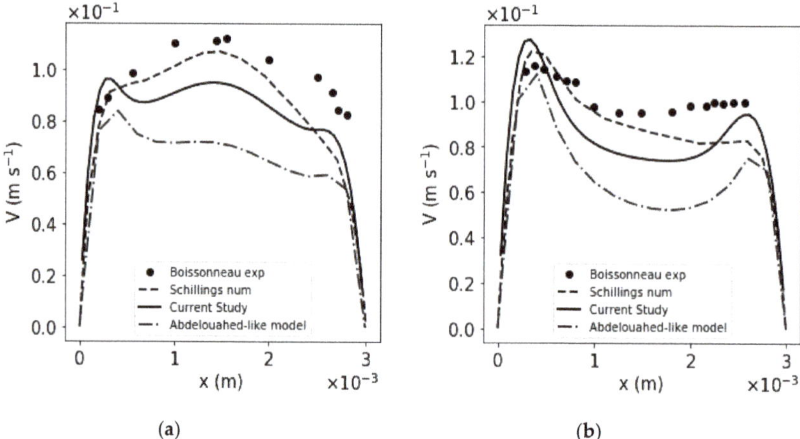

Figure 5. Result of the liquid velocity in different models at j = 2000 A m^{-2} (**a**) at y = 60 mm and (**b**) at y = 75 mm. The dot are the data from Boissonneau et al. [15], the grey dashed line is the numerical results of the Schilling et al. model [14], the dotted and dashed line is a model using the method of Abdelouahed et al. [26]. Finally, the black solid line the results from the current study with 2D approximation.

Figure 6 shows the void fraction evolution predicted by the current model. With increasing current density, the diphasic boundary layer of oxygen and hydrogen and the maximum oxygen and hydrogen void fraction increase. However, there are two parameters (the current density and the bubble diameter) that change between the two cases. Therefore, the calculated evolution cannot be clearly attributed to one parameter. The prediction of this evolution, depending on electrode height, is very important to predict the current density distribution. The prediction of the diphasic boundary layer thickness is also an essential output parameter because the electrolyte conductivity decreases with the void fraction.

Table 4 shows the average relative different of the different model with the Boissonneau et al. [15] at three current density (500, 1000 and 2000 A m^{-2}) at three horizontal zones and two vertical locations (y = 60 mm and y = 75 mm). The main drawback of the current model is that it does not describe the velocity at the bulk zone (between x = 0.5 and x = 2.5 mm) with high accuracy. However, for all current density and location, the current model has more accuracy than the Schillings et al. [14], model the velocity in O_2 zone (between x = 2.5 and x = 3 mm). The present model is less accurate than the model of Schillings et al. However, the model of Schillings et al. uses more semi-empiric term than the model of this study. Moreover, the mathematical formalism of the current study contains less equation than the model of Schillings et al. Thus, this model is easier to program.

Table 4. Average relative difference of the different model compared to Boissonneau et al. [15] at two vertical positions 60 and 75 mm and three horizontal zones for the three current density. H_2 is the zone between [0, 500] μm and O_2 is zone between [2.5, 3] mm.

j (A m^{-2})	Model Comparaison	Zone	60 mm	75 mm
500 A m^{-2}	Boissonneau vs. Abdelouahed	H_2	6.95×10^{-2}	3.96×10^{-2}
		O_2	2.48×10^{-1}	1.47×10^{-1}
		Bulk	2.91×10^{-1}	2.91×10^{-1}
	Boissonneau vs. Schillings	H_2	6.00×10^{-2}	1.60×10^{-2}
		O_2	2.85×10^{-1}	2.65×10^{-1}
		Bulk	1.90×10^{-1}	2.06×10^{-1}
	Boissonneau vs. Current Study	H_2	1.76×10^{-1}	4.55×10^{-2}
		O_2	2.01×10^{-1}	7.40×10^{-2}
		Bulk	2.48×10^{-1}	2.36×10^{-1}
1000 A m^{-2}	Boissonneau vs. Abdelouahed	H_2	1.04×10^{-1}	6.13×10^{-2}
		O_2	3.48×10^{-1}	2.69×10^{-1}
		Bulk	3.83×10^{-1}	3.51×10^{-1}
	Boissonneau vs. Schillings	H_2	1.21×10^{-1}	4.76×10^{-2}
		O_2	2.99×10^{-1}	2.27×10^{-1}
		Bulk	1.69×10^{-1}	1.31×10^{-1}
	Boissonneau vs. Current Study	H_2	1.04×10^{-1}	7.95×10^{-2}
		O_2	2.35×10^{-1}	6.37×10^{-2}
		Bulk	2.50×10^{-1}	1.99×10^{-1}
2000 A m^{-2}	Boissonneau vs. Abdelouahed	H_2	1.43×10^{-1}	7.09×10^{-2}
		O_2	2.69×10^{-1}	3.58×10^{-1}
		Bulk	3.51×10^{-1}	3.70×10^{-1}
	Boissonneau vs. Schillings	H_2	4.76×10^{-2}	4.31×10^{-2}
		O_2	2.27×10^{-1}	2.90×10^{-1}
		Bulk	1.31×10^{-1}	6.27×10^{-2}
	Boissonneau vs. Current Study	H_2	7.95×10^{-2}	7.44×10^{-2}
		O_2	6.37×10^{-2}	1.71×10^{-1}
		Bulk	1.99×10^{-1}	2.57×10^{-1}

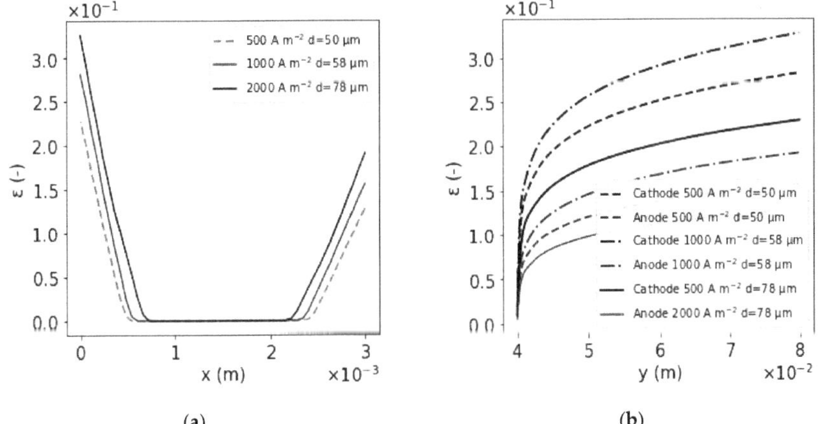

Figure 6. Void fraction evolution depending on the electrolyte width (**a**) and electrode height (**b**) for the three current density 500, 1000 and 2000 A m^{-2}.

Table 5 summarizes the results of the study. It must be noticed that the diphasic boundary layer of hydrogen is thinner than the oxygen one. These results are compared with the those of Schillings et al. [14]. In Schillings' study, the authors performed a dimensional study and found that the diphasic boundary layer depends on one dimensionless group, which is presented in Equations (18) and (19) (called Rayleigh-like number). The current results correspond well to the Schillings prediction. Thus, when the radius increases, the diphasic boundary layer increases, and when the equivalent injection gas velocity increases, the boundary layer decreases.

$$\log\left(\frac{\delta}{h}\right) = -0.25\log(Ra) + Cst \tag{18}$$

$$Ra = \frac{\nu\, U_g h^5}{g\, r^6\, H_{elec}} \tag{19}$$

The conductivity evolution depending on electrolyte gas content is well described by the Bruggeman correlation (Equation (20)). This correlation has been compared to experimental results in Hine et al. [9].

$$\frac{\sigma(\varepsilon)}{\sigma_0} = (1-\varepsilon)^{1.5} \tag{20}$$

σ_0 is the gas-free electrolyte conductivity in S m^{-1}.

Thus, the ohmic resistance presented in Equation (1) becomes Equation (21) [37]:

$$R(\varepsilon) = \frac{\delta_{H2}}{\sigma_0\,(1-\varepsilon)^{1.5}} + \frac{2h-\delta_{O2}-\delta_{H2}}{\sigma_0} + \frac{\delta_{O2}}{\sigma_0\,(1-\varepsilon)^{1.5}} \tag{21}$$

In order to compare biphasic system with a gas-free system, the previous resistance can be divided by the hypothetical gas-free resistance.

$$\frac{R(\varepsilon)}{R} = \frac{\frac{\delta_{H2}}{\sigma_0\,(1-\varepsilon)^{1.5}} + \frac{2h-\delta_{O2}-\delta_{H2}}{\sigma_0} + \frac{\delta_{O2}}{\sigma_0\,(1-\varepsilon)^{1.5}}}{\frac{2h}{\sigma_0}} \tag{22}$$

Table 5 shows that with an increasing current density and bubble radius, the resistance increases. This statement shows that the bubble management is an important issue.

Table 5. Maximum liquid velocity at cathode and anode sides, maximum oxygen and hydrogen void fraction and hydrogen and oxygen diphasic boundary layer for the three cases.

j (A m^{-2})	$V_{max\ liq\ cath}$ (m s^{-1})	$V_{max\ liq\ an}$ (m s^{-1})	$\varepsilon_{max\ cath}$	$\varepsilon_{max\ an}$	δ_{cath} (µm)	δ_{an} (µm)	$R(\varepsilon)/R$
500 A m^{-2}	7.8×10^{-2}	4.8×10^{-2}	0.23	0.13	515	600	1.032
1000 A m^{-2}	9×10^{-2}	6.8×10^{-2}	0.28	0.16	566	677	1.043
2000 A m^{-2}	1.25×10^{-1}	9.1×10^{-2}	0.33	0.19	690	800	1.057

4. Conclusions

In this study, a two-fluid multi-physics model with a new bubble transfer description has been proposed. This new description allows a good agreement with the results of Boissonneau et al.'s [15] experimental velocity profiles. It has been found out a bubble dispersion force allows a good agreement with experimental data and a better numerical convergence than the one obtained without this additional force. That observation is correlated with the studies of other studies such as Picardi et al. [19]. From the numerical point of view, the grid resolution remains a problem because even if the dispersion bubbles force improves the results, there is still a maximum grid resolution. This maximum grid resolution could lead to some inaccuracy if the fluid viscosity is too small or if the electrolyte width is too thin.

Further calculations must be performed to suppress these limitations by using a 3D geometry [19]. From the physical point of view, it can be concluded that the bubble radius and current density are two important parameters that influence the hydrodynamic. Further, calculations better properly characterize the output parameters sensitivities (velocity, void fraction etc.) to the current density and bubble diameter.

Author Contributions: Conceptualization, D.L.B. and P.M.; Methodology, D.L.B. and P.M.; Software, D.L.B. and P.M.; Validation, D.L.B. and P.M.; Investigation, D.L.B. and P.M.; Data curation, D.L.B. and P.M.; Writing—original draft preparation, D.L.B. and P.M.; writing—review and editing, P.M., M.B., M.K., M.S., F.G., and R.I.; Visualization, P.M. and M.B.; Supervision, P.M. and M.B.; Project administration, P.M.; Funding acquisition, P.M. All authors have read and agreed to the published version of the manuscript.

Funding: This research was funded by the national association ADEME and Bretagne Region (France).

Acknowledgments: We would like to deeply thank the national association ADEME for funding our research and the Bretagne region to support our research.

Conflicts of Interest: The authors declare no conflict of interest.

References

1. Castagnola, L.; Lomonaco, G.; Marotta, R. Nuclear Systems for Hydrogen Production: State of Art and Perspectives in Transport Sector. *Glob. J. Energy Technol. Res. Updates* **2014**, *1*, 4–18.
2. Schäfer, H.; Chatenet, M. Steel: The Resurrection of a Forgotten Water-Splitting Catalyst. *ACS Energy Lett.* **2018**, *3*, 574–591. [CrossRef]
3. Lavorante, M.J.; Franco, J.I. Performance of stainless steel 316L electrodes with modified surface to be use in alkaline water electrolyzers. *Int. J. Hydrogen Energy* **2016**, *41*, 9731–9737. [CrossRef]
4. Bocca, C.; Cerisola, G.; Barbucci, A.; Magnone, E. Oxygen evolution on Co2O3 and Li-doped Co_2O_3 coated electrodes in an alkaline solution. *Int. J. Hydrogen Energy* **1999**, *24*, 699–707. [CrossRef]
5. Boggs, B.K.; King, R.L.; Botte, G.G. Urea electrolysis: Direct hydrogen production from urine. *Chem. Commun.* **2009**, *32*, 4859–4861. [CrossRef] [PubMed]
6. Huck, M.; Ring, L.; Küpper, K.; Klare, J.P.; Daum, D.; Schäfer, H. Water Splitting Mediated by an Electrocatalytically Driven Cyclic Process Involving Iron Oxide Species; Huck, Marten, Lisa Ring, Karsten Küpper, Johann Klare, Diemo Daum, and Helmut Schäfer. *J. Mater. Chem. A* **2020**, *8*, 9896–9910. [CrossRef]
7. Babu, R.; Das, M.K. Experimental studies of natural convective mass transfer in a water-splitting system. *Int. J. Hydrogon Energy* **2019**, *44*, 14467–14480. [CrossRef]
8. Tobias, C.W. Effect of Gas Evolution on Current Distribution and Ohmic Resistance in Electrolyzers. *J. Electrochem. Soc.* **1959**, *106*, 6. [CrossRef]
9. Hine, F. Bubble Effects on the Solution IR Drop in a Vertical Electrolyzer Under Free and Forced Convection. *J. Electrochem. Soc.* **1980**, *127*, 292. [CrossRef]
10. Mandin, P.; Hamburger, J., Bessou, S., Picard, G. Modelling and calculation of the current density distribution evolution at vertical gas-evolving electrodes. *Electrochim. Acta* **2005**, *51*, 1140–1156. [CrossRef]
11. Hreiz, R.; Abdelouahed, L.; Fünfschilling, D.; Lapicque, F. Electrogenerated bubbles induced convection in narrow vertical cells: PIV measurements and Euler–Lagrange CFD simulation. *Chem. Eng. Sci.* **2015**, *134*, 138–152. [CrossRef]
12. Dahlkild, A.A. Modelling the two-phase flow and current distribution along a vertical gas-evolving electrode. *J. Fluid Mech.* **2001**, *428*, 249–272. [CrossRef]
13. Wedin, R.; Dahlkild, A.A. On the Transport of Small Bubbles under Developing Channel Flow in a Buoyant Gas-Evolving Electrochemical Cell. *Ind. Eng. Chem. Res.* **2001**, *40*, 5228–5233. [CrossRef]
14. Schillings, J., Doche, O., Deseure, J. Modeling of electrochemically generated bubbly flow under buoyancy-driven and forced convection. *Int. J. Heat Mass Transf.* **2015**, *85*, 292–299. [CrossRef]
15. Boissonneau, P.; Byrne, P. An experimental investigation of bubble-induced free convection in a small electrochemical cell. *J. Appl. Electrochem.* **2000**, *30*, 767–775. [CrossRef]
16. Aldas, K.; Pehlivanoglu, N.; Mat, M.D. Numerical and experimental investigation of two-phase flow in an electrochemical cell. *Int. J. Hydrogen Energy* **2008**, *33*, 3668–3675. [CrossRef]

17. Mat, M. A two-phase flow model for hydrogen evolution in an electrochemical cell. *Int. J. Hydrogen Energy* **2004**, *29*, 1015–1023. [CrossRef]
18. Riegel, H.; Mitrovic, J.; Stephan, K. Role of mass transfer on hydrogen evolution in aqueous media. *J. Appl. Phys.* **1998**, *28*, 10–17.
19. Picardi, R.; Zhao, L.; Battaglia, F. On the Ideal Grid Resolution for Two-Dimensional Eulerian Modeling of Gas–Liquid Flows. *J. Fluids Eng.* **2016**, *138*, 114503. [CrossRef]
20. Law, D.; Battaglia, F.; Heindel, T.J. Model validation for low and high superficial gas velocity bubble column flows. *Chem. Eng. Sci.* **2008**, *63*, 4605–4616. [CrossRef]
21. Panicker, N.; Passalacqua, A.; Fox, R.O. On the hyperbolicity of the two-fluid model for gas—Liquid bubbly flows. *Appl. Math. Model.* **2018**, *57*, 432–447. [CrossRef]
22. Vaidheeswaran, A.; de Bertodano, M.L. Stability and convergence of computational eulerian two-fluid model for a bubble plume. *Chem. Eng. Sci.* **2017**, *160*, 210–226. [CrossRef]
23. Marfaing, O.; Guingo, M.; Laviéville, J.; Bois, G.; Méchitoua, N.; Mérigoux, N.; Mimouni, S. An analytical relation for the void fraction distribution in a fully developed bubbly flow in a vertical pipe. *Chem. Eng. Sci.* **2016**, *152*, 579–585. [CrossRef]
24. Davidson, M.R. Numerical calculations of two-phase flow in a liquid bath with bottom gas injection: The central plume. *Appl. Math. Model.* **1990**, *14*, 67–76. [CrossRef]
25. Lee, J.W.; Sohn, D.K.; Ko, H.S. Study on bubble visualization of gas-evolving electrolysis in forced convective electrolyte. *Exp. Fluids* **2019**, *60*, 156. [CrossRef]
26. Abdelouahed, L.; Hreiz, R.; Poncin, S.; Valentin, G.; Lapicque, F. Hydrodynamics of gas bubbles in the gap of lantern blade electrodes without forced flow of electrolyte: Experiments and CFD modelling. *Chem. Eng. Sci.* **2014**, *111*, 255–265. [CrossRef]
27. Ham, J.; Homsy, G. Hindered settling and hydrodynamic dispersion in quiescent sedimenting suspensions. *Int. J. Multiph. Flow* **1988**, *14*, 533–546. [CrossRef]
28. Nagai, N.; Takeuchi, M.; Nakao, M. Influences of Bubbles between Electrodes onto Efficiency of Alkaline Water Electrolysis. *Pac. Cent. Therm. Fluids Eng.* **2003**, *10*, F4008.
29. Abdelouahed, L. Gestion Optimale du Gaz Électrogénéré Dans un Réacteur D'électroréduction de Minerai de fer. Ph.D. Thesis, Université de Lorraine, Nancy, France, 2013.
30. Schillings, J. Etude Numérique et Expérimentale D'écoulements Diphasiques: Application aux Écoulements à Bulles Générées par Voie Électrochimique. Ph.D. Thesis, Université Grenoble Alpes, Grenoble, France, 2018.
31. Le Bideau, D.; Mandin, P.; Sellier, M.; Benbouzid, M.; Myeongsub, K. Review of necessary thermophysical properties and their sensivities with local temperature and electrolyte mass fraction for alkaline water electrolysis multiphysics modelling. *Int. J. Hydrogon Energy* **2019**, *44*, 4553–4569. [CrossRef]
32. Alexiadis, A.; Dudukovic, M.P.; Ramachandran, P.; Cornell, A.; Wanngård, J.; Bokkers, A. Liquid–gas flow patterns in a narrow electrochemical channel. *Chem. Eng. Sci.* **2011**, *66*, 2252–2260. [CrossRef]
33. Alexiadis, A.; Dudukovic, M.P.; Ramachandran, P.; Cornell, A.; Wanngård, J.; Bokkers, A. On the electrode boundary conditions in the simulation of two phase flow in electrochemical cells. *Int. J. Hydrogen Energy* **2011**, *36*, 8557–8559. [CrossRef]
34. Charton, S.; Rivalier, P.; Ode, D.; Morandini, J.; Caire, J.P. *Hydrogen Production by the Westinghouse Cycle: Modelling and Optimization of the Two-Phase Electrolysis Cell*; WIT Press: Bologna, Italy, 2009; pp. 11–22.
35. Grigoriev, S.A.; Fateev, V.N.; Bessarabov, D.G.; Millet, P. Current status, research trends, and challenges in water electrolysis science and technology. *Int. J. Hydrogon Energy* **2020**. [CrossRef]
36. Isono, T. Density, viscosity, and electrolytic conductivity of concentrated aqueous electrolyte solutions at several temperatures. Alkaline-earth chlorides, lanthanum chloride, sodium chloride, sodium nitrate, sodium bromide, potassium nitrate, potassium bromide, and cadmium nitrate. *J. Chem. Eng. Data* **1984**, *29*, 45–52. [CrossRef]
37. Mandin, P.; Derhoumi, Z.; Roustan, H.; Rolf, W. Bubble Over-Potential during Two-Phase Alkaline Water Electrolysis. *Electrochim. Acta* **2014**, *128*, 248–258. [CrossRef]

Article

A Comprehensive Study of Custom-Made Ceramic Separators for Microbial Fuel Cells: Towards "Living" Bricks

Jiseon You [1,*], Lauren Wallis [1], Nevena Radisavljevic [2], Grzegorz Pasternak [2,3], Vincenzo M. Sglavo [4], Martin M Hanczyc [2], John Greenman [1] and Ioannis Ieropoulos [1,*]

1 Bristol BioEnergy Centre, University of the West of England, Bristol BS16 1QY, UK; lauren.wallis@uwe.ac.uk (L.W.); john.greenman@uwe.ac.uk (J.G.)

2 Laboratory for Artificial Biology, Department of Cellular, Computational and Integrative Biology (CIBIO), University of Trento, Via Sommarive 9, 38123 Povo TN, Italy; nevena.radisavljevic@unitn.it (N.R.); grzegorz.pasternak@pwr.edu.pl (G.P.); martin.hanczyc@unitn.it (M.M.H.)

3 Laboratory of Microbial Electrochemical Systems, Department of Polymer and Carbonaceous Materials, Wroclaw University of Science and Technology, Wyb. Wyspiańskiego 27, 50-370 Wrocław, Poland

4 Department of Industrial Engineering, University of Trento, via Sommarive 9, 38123 Trento, Italy; vincenzo.sglavo@unitn.it

* Correspondence: jiseon.you@uwe.ac.uk (J.Y.); ioannis.ieropoulos@brl.ac.uk (I.I.); Tel.: +44-117-328-6318 (I.I.)

Received: 16 September 2019; Accepted: 21 October 2019; Published: 24 October 2019

Abstract: Towards the commercialisation of microbial fuel cell (MFC) technology, well-performing, cost-effective, and sustainable separators are being developed. Ceramic is one of the promising materials for this purpose. In this study, ceramic separators made of three different clay types were tested to investigate the effect of ceramic material properties on their performance. The best-performing ceramic separators were white ceramic-based spotty membranes, which produced maximum power outputs of 717.7 ± 29.9 μW (white ceramic-based with brown spots, 71.8 $W \cdot m^{-3}$) and 715.3 ± 73.0 μW (white ceramic-based with red spots, 71.5 $W \cdot m^{-3}$). For single material ceramic types, red ceramic separator generated the highest power output of 670.5 ± 64. 8 μW (67.1 $W \cdot m^{-3}$). Porosity investigation revealed that white and red ceramics are more porous and have smaller pores compared to brown ceramic. Brown ceramic separators underperformed initially but seem more favourable for long-term operation due to bigger pores and thus less tendency of membrane fouling. This study presents ways to enhance the function of ceramic separators in MFCs such as the novel spotty design as well as fine-tuning of porosity and pore size.

Keywords: microbial fuel cell; low-cost ceramics; separator; membrane; porosity; pore size; water absorption; mercury intrusion

1. Introduction

Although a large body of work regarding microbial fuel cell (MFC) technology is still at the laboratory stage, in the recent years, various attempts to evolve the technology towards commercialization have been made. Successful case studies of large-scale systems have been reported [1–4] (total reactor volume of over 90 L) and an online biosensor based on MFC technology has been released to the market [5]. Recently Trapero et al. claimed that MFC technology is economically feasible and sufficiently competitive when compared to conventional wastewater treatment processes such as activated sludge [6]. As with every other technology, more effort needs to be made to improve the technology to be more cost-effective, reliable, and efficient. One of the major challenges for MFC commercialization is the cost of materials, especially separators [6–8]. In MFCs, separators mediate ion transport and physically separate the anodes and cathodes. Separator-less systems have shown higher power outputs in some cases due to lower internal resistance [9], thus being an attractive option to lower system costs. However, they have limitations in terms of low coulombic efficiency due to the occurrence of ionic species crossover and undesirable side reactions [10]. Moreover, for the systems requiring anolyte/catholyte separation, separators are essential.

For efficient MFC operation, separators should have high proton transfer rate and low internal resistance. Chemical and mechanical strength is also an important requirement for long-term operation. One of the common issues related to MFC separators is membrane fouling, especially biofouling, which increases internal resistance of the system and leads to system failure. Koók et al. elucidated membrane biofouling mechanism and related membrane properties in their recent review paper [11]. Separators used in MFCs can be classified in two groups, namely porous and non-porous separators. In-depth reviews on separators tested in MFCs can be found in [12–14]. As an effort into seeking well-performing, cost-effective, and sustainable materials for the MFC separator, ceramics have drawn attention. The use of ceramic separators in MFCs has proven its comparable performance to that of selective ion exchange membranes with much less cost. Jana et al. built MFCs with cylindrical earthen ceramic separators which generated a maximum power output (P_{MAX}) of 14.6 $W \cdot m^{-3}$ [15]. Another study using much smaller cylindrical terracotta separators (internal volume of 10 mL) reported a P_{MAX} of 44.8 $W \cdot m^{-3}$ [16]. Besides the aforementioned advantages, ceramics are also thought to be suitable materials for MFC scaling up because of its structural durability and plasticity [17,18]. Moreover, ceramic making capabilities exist around the world including ODA countries. Combining this with a ubiquitous fuel such as urine, could increase accessibility of MFC technology.

Ceramic separators are considered to be porous separators whose pores facilitate ion transport while separating anolyte and catholyte. Considering the pore size of common ceramic materials, they can be classified as ultra-filtration (UF, pore diameter of 10–100 nm) or micro-filtration (MF, pore diameter of 100–10^4 nm) [13,19]. Not only the porosity affects MFC performance, but also clay composition, wall thickness, pore size distribution, and density may well play important roles in MFC performance, which in turn these parameters can be optimized for a specific target application [17].

Winfield et al. compared earthenware and terracotta in cylindrical MFCs and reported that earthenware generated a 75% higher power than terracotta, which was a similar level of power produced from a cation exchange membrane (CEM) [20]. Another study by Pasternak et al. also suggested earthenware is compatible to conventional ion exchange membranes for MFCs in terms of performance and cost after comparing four different types of ceramic: mullite, earthenware, pyrophyllite and alumina in cylindrical single-chamber MFCs [21]. In that study, the highest performance was observed in ceramics with porosities ranging between 2 and 14%. Jimenez et al. investigated the effect of ceramic separator thickness on MFC performance [22]. They tested cylindrical fine fire clay ceramic separators with different thicknesses (2.5, 5 and 10 mm) and reported the higher power and catholyte production was obtained from the thinnest separator of 2.5 mm. In a different study looking into the effect of thickness and porosity of ceramic separators, optimum ceramic separator thickness for MFC power generation depends on the porosity [23]. For highly porous ceramics (porosity of 30.5%), power output was proportional to the thickness (highest power output of 321 $mW \cdot m^2$ obtained from 9 mm thick senators), whereas a thinner separator (thickness of 3 mm) was more favourable for less porous ceramics (porosity of 11.0%). If a separator is extremely porous such as tissue paper, then there is insufficient separation between the electrodes and a short circuit can occur [24]. In less extreme cases, high porosity can result in anolyte crossover to the cathode chamber, both reducing the amount of available substrate for electrochemically active biofilms and promoting heterotrophic bacterial growth on the cathode. Oxygen transfer from the cathode to the anode can also be an issue in this case.

For successful implementation of ceramic separators in MFCs, understanding the relations between their properties and performance as MFC separators is crucial. Also testing long-term operation is required for practical applications. Nevertheless, there is still much need for comprehensive studies looking into these aspects in the field. The main objective of the present study was to investigate the effect of ceramic material properties on their performance as MFC separators. Three different clay-based ceramics with the same thickness of 3 mm were compared in terms of power output both in short-term and long-term operations. Not only porosity but also pore size of each test ceramic material was measured and its influence on the performance was analysed.

This piece of work is part of a bigger study into "Living Architecture" [25], which is a field of work that investigates how smart homes and smart cities of the future could be developed. In particular, living technology such as the MFC can be integrated into bricks for households or other structures, thereby allowing on-site treatment and electricity generation in real time. This is the "living brick" context under which the ceramic samples were developed and tested.

2. Materials and Methods

2.1. Custom-Made Ceramics

The ceramic clays used in this study are a mixture of a plastic clay and 25% chamotte, which has a diameter ranging from 0 to 0.5 mm (Georg & Schneider, Siershahn, Germany). Three types of clay were used and named "brown" (product no.: 366), "red" (product no.: 364) and "white" (product no.: 264); the names were based on the resultant colour after kilning. Technical data of all three clay types are available from the manufacturer and are presented in Table S1. All ceramics were fired at 960 °C for 20 min, at a rate of 150 $°C \cdot h^{-1}$. This temperature is lower than the firing temperature recommended by the manufacturer, and was chosen to achieve a higher porosity by preventing the matrix from vitrifying and thus closing the majority of the interstices [26]. The recommended firing temperatures are 1070–1120 °C for the brown, 1070–1240 °C for the red and white clay.

Ceramic separators were made in a cylindrical shape with one end sealed. Bottom sealed cylinders were 55 mm long, with inner diameter of 18 mm and thickens of 3 mm. The ceramic separators used for the study are shown in Figure 1. As these were all hand made, and then fired, the dimensions vary by a few millimetres between samples.

Figure 1. Images of ceramic cylinders used in the study. Spotty ceramics are exemplars of red ceramic-based with brown spots, named RB.

As well as making separators from a single clay type (brown, red or white ceramic), combinations of the three raw materials were used to produce novel "spotty" separators. This type of ceramic separators was expected to offer additional benefits to the MFCs, resulting from the combination of the two clay types.

2.2. Microbial Fuel Cell Designs, Inoculation and Operation

Cylindrical MFCs were built to be tested as ceramic separators. For the anode, plain carbon fibre veil (20 $g \cdot m^{-2}$ carbon loading; PRF Composite Materials Poole, Dorset, UK) was cut into 270 cm^2 (30 × 9 cm) and then folded to fit into the anode chambers. A hot-pressed activated carbon cathode, prepared as previously described [27] with a total surface area of 35 cm^2, was placed onto the separator; this cathode was open to air. MFCs were inserted in 50 mL plastic containers, which was partially filled with water to keep the moisture of the cathodes. This allowed wetting of the air-cathodes, and improved the cathode performance. The schematic diagram of MFC design used in this study is shown in Figure 2 and the actual tested design is shown in Figure S2.

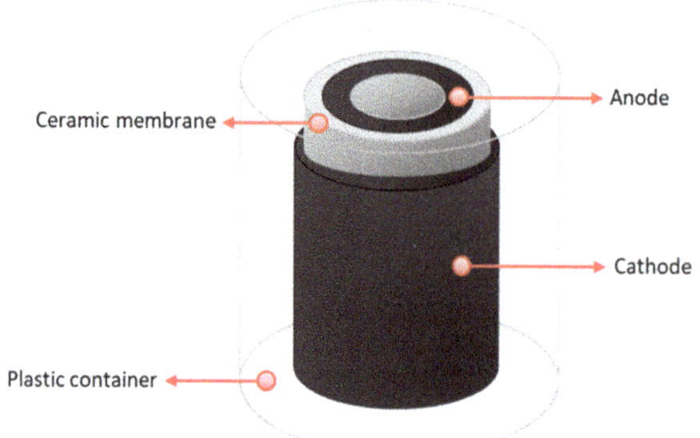

Figure 2. Schematic diagram of an MFC reactor used in the study.

Anaerobic sewage sludge from a local wastewater treatment plant (Wessex Water, Saltford, UK) was used to inoculate the MFCs, after being enriched with 1% tryptone and 0.5% yeast extract. MFCs were fed in batch (once every 1–3 days) with neat human urine, donated from consenting adults. Typically, the urine had a pH of 9.2–9.3, and conductivity of 28–30 mS·cm^{-1}. In each feed, 5 mL of anolyte was replaced with fresh feedstock. The total volume of anolyte was 10 mL.

Throughout the work, variable external loads were connected to each MFC, which were determined based on polarisation runs that were carried out periodically. All tests were carried out in triplicates.

2.3. Polarisation Test and Data Logging

For polarisation experiments, various external resistances ranged from 4.8 kΩ to 4 Ω were loaded every 5 min and the potential between the anode and cathode was recorded every 30 s. Power output of the MFCs was monitored in real time in volts (V) against time using a multi-channel Agilent 34972A DAQ unit (Agilent Technologies, California, USA) every 5 min. All experiments were carried out in a temperature-controlled environment, at 22 ± 2 °C.

2.4. Physiochemical Property Analysis of Ceramic Separators

2.4.1. Porosity Investigation

Porosity investigation was carried out using both the ASTM (American Society for Testing and Materials) C373 procedure [28] and mercury intrusion porosimetry (PASCAL 140, CE Instruments, Wigan, UK and CARLO ERBA Poro 2000, Science Exchange, Cobham, UK) methods. Prior to the analysis, single compound separator samples were thermally treated at 500 °C to remove even the very limited presence of organics and then subjected to thermal gravimetric analysis (TGA) to detect the presence of residues. The TGA results showed no evident weight loss, except for a very low amount at low temperature, probably corresponding to the evolution of humidity. The overall residue was below the sensitivity of the instrument (0.1%) and no thermal effects were noticed. Mercury intrusion tests were performed once for each sample to find surface area, porosity, pore size, and pore distribution data. ASTM tests were repeated on five different specimens for each sample, and the structural properties of a material such as bulk density, apparent porosity, and specific gravity to the precision of approx. ±0.1% were determined.

Scanning electron microscopy (SEM) analysis of the ceramic surface and cross-section was performed using Quanta FEG 650 (Thermo Fisher Scientific, Massachusetts, USA) and JSM-5500 (JEOL, Tokyo, Japan). The cross-section images were processed to find percentage porosity using MATLAB (MathWorks, Massachusetts, USA).

2.4.2. Composition Analysis

Energy dispersive X-ray Spectroscopy (EDXS) analysis was carried out to determine chemical compositions of the ceramics (Quanta FEG 650, Thermo Fisher Scientific, Massachusetts, USA).

2.5. Statistical Analysis

To evaluate if the differences between test ceramics in terms of power generation performance are statistically meaningful, analysis of variance (ANOVA: ordinary one-way) was performed using GraphPad Prism 8 (GraphPad Software, California, USA). Power produced in joules from 4 feeding cycles in days between 44 and 52 were calculated. This resulted in 108 data points for all 9 types of ceramic separator.

3. Results and Discussion

3.1. Effect of Ceramic Type on MFC Performance

During the electroactive biofilm maturing period, beginning with inoculation, initially a 1 kΩ external resistor was connected. Then, it was replaced by a heavier load (lower resistance value) as the cell voltage levels reached 300 mV, which took between 2 and 10 days, each cell differed. Figure 3 shows power outputs of the first 13 days of MFC operation. The different types of ceramic separators tested are illustrated in Figure S1. While most of the MFCs showed similar power generating patterns, which are typical for batch-fed MFCs in initial stage, no voltage could be measured from MFCs with brown ceramic separators for the first 3 batch feedings (until day 4). This was thought to be due to its relatively low permeability. All MFCs had air breathing cathodes which use atmospheric oxygen as a terminal electron acceptor thus open to air. This type of cathode can be run in dry condition where external addition of catholyte is not required. In fact, for this type of cathode, flooding should be prevented for efficient MFC operation. However, keeping the cathode in wet conditions (but not flooded) can improve its performance by facilitating the continuous flow of cations from the anode, especially for cathodes with no electrochemically efficient catalysts such as Pt or Pt alloys. Water formation as a "circuit-closing" reaction of MFC electricity generation is well studied [29]. Also, catholyte production mainly by electro-osmotic drag is common for MFCs with ceramic separators that are non-ion-selective and relatively highly porous [22,30]. In this study, therefore MFC operation started without any initial catholyte present. However, after this finding, it was decided to add water to the cathode chambers to maintain the wet condition of both cathode and separator.

Once catholyte was present, peak values of power output for all MFCs increased in each batch feeding cycle then stabilised, implying the anodic biofilms matured. Figure 4 presents power output profiles of all test ceramics between day 43 and day 48, when MFCs were fully matured. Single composition ceramic types (brown, red, and white) produced similar levels of peak power in each feed, whereas difference in peak power was more distinct in spotty ceramics. The best-performing ceramic was RB (red ceramic-based with brown spots) producing 491.1 ± 1.7 µW of peak power. Although all single composition ceramic types showed similar peak power, power output of brown ceramic decreased more rapidly than the other two ceramics, red and white. This was the case for brown ceramic-based spotty types of BW (brown ceramic-based with white spots) and BR (brown ceramic-based with red spots). This could be related to the low permeability of brown ceramic, resulting in poor ion transfer through the ceramic matrix, which was also the reason both the separator and cathode electrode required hydration. When comparing the area under curve (AUC) of each ceramic tested, red and WB (white ceramic-based with brown spots) generated the highest output in two groups of single composition and spotty ceramic type. Overall, spotty types outperformed single composition separators both in terms of peak power and AUC, suggesting a great potential of this ceramic design type as an MFC separator.

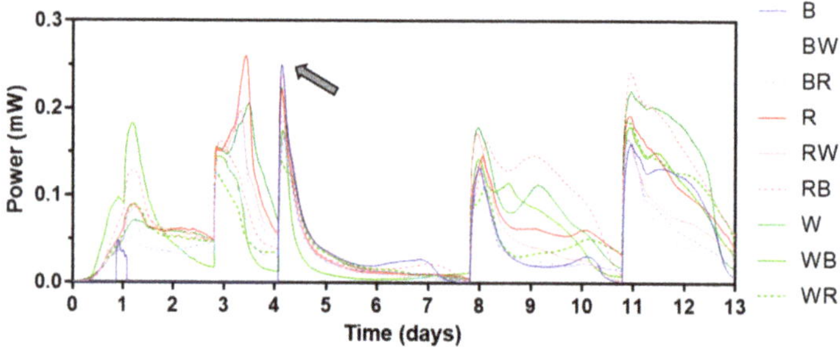

Figure 3. Power output profile of the first 13 days. The grey arrow indicates when continuous reading of power from brown ceramic separators began. Letters "B", "R", and "W" stand for "brown", "red", and "white" respectively. Data represent average values from triplicates of each ceramic type. See Figure S1 for the different types of ceramic separators.

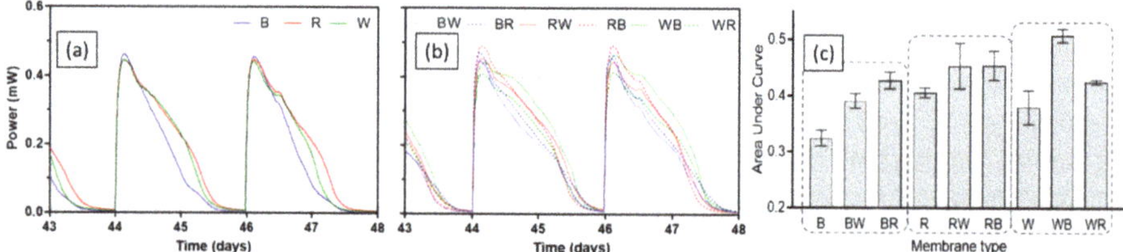

Figure 4. Power outputs of MFCs with test ceramics between day 43 and day 48; single composition ceramics (**a**), spotty ceramics (**b**), and area under curve of each batch feed (**c**). Data represent average values from triplicates of each ceramic type.

The difference in performance of test ceramics was also observed from the power curves. Figure 5 shows power curves of all ceramics from polarisation tests that were carried out in week 9. For single material ceramic types, the best-performing ceramic was red with a P_{MAX} of 670.5 ± 64.8 μW. This was 14.9% and 22.5% higher than white (570.5 ± 48.0 μW) and brown (519.9 ± 53.6 μW) respectively. On the other hand, all spotty ceramics showed similar levels of P_{MAX}, although white-based ceramics produced the highest power outputs (717.7 ± 29.9 μW for WB and 715.3 ± 73.0 μW for WR) and currents. Again, better performance of spotty ceramics in comparison to single composition ceramics (except red and red-based ceramics) was confirmed (Figure 5c). The statistical results support the findings. The ordinary one-way ANOVA test results indicate that there were significant differences (p-value less than 0.05) between nine test ceramics for power production of each feeding cycle (p-value: < 0.001).

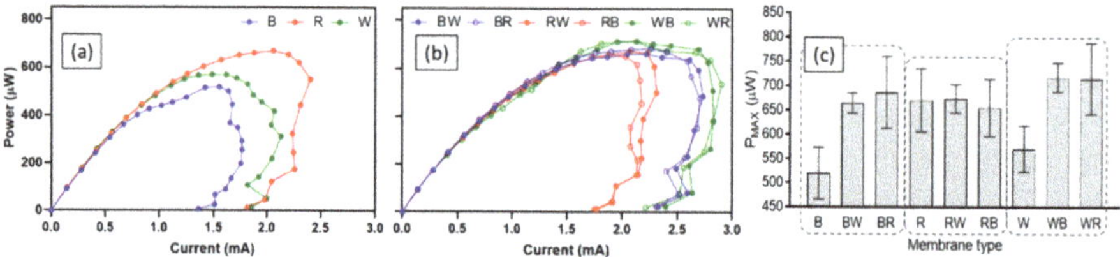

Figure 5. Power curves and P_{MAX} of MFCs with test ceramics; single composition ceramics (**a**), spotty ceramics (**b**), and P_{MAX} from each power curve (**c**). Data represent average values from triplicates of each ceramic type, and error bars of power curves are not shown for clarity.

3.2. Long-Term Operation

For the practical implementation of MFC technology, long-term performance should be verified. Figure 6 depicts power generating performance of all test ceramics between day 106 and day 110. Average peak power of brown ceramic during this period was 338.1 ± 22.8 μW, which was 26.4% lower compared to the peak power of the same ceramic during the mid-term. Brown-based spotty ceramics (BW and BR) showed similar extent of performance decrease of 30.7% and 28.1% respectively. This is interesting since the brown ceramic was the least performing separator during the mid-term, in terms of both peak power and AUC. On the other hand, peak power of red and white that performed better in the mid-term decreased by 55.4% and 76.9% respectively. Peak power of spotty ceramics based on both red and white also reduced by around 66.8–88.2%. The trend of change in peak power of each ceramic over the whole experimental period can be seen in Figure S3.

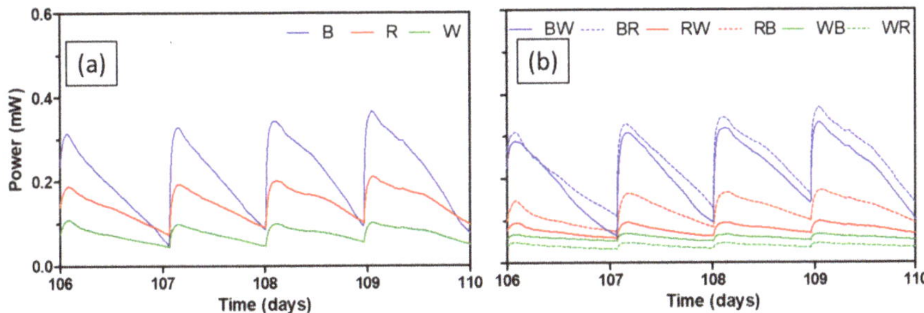

Figure 6. Power output levels of MFCs with test ceramics between day 106 and day 110; single composition ceramics (**a**), spotty ceramics (**b**). Data represent average values from triplicates of each ceramic type.

Performance decrease in MFC power generation is common due to an increase of internal resistance as a result of increasing thickness of the anodic biofilm, bio- and abiotic fouling of porous materials including electrodes and separators, or microbially influenced corrosion [31–33]. Although fully hydrolysed urine was used in this study as a feedstock, a small degree of precipitation of uric salts was expected since neat urine was used without pre-treatment such as filtration. This could aggravate the performance of ceramic separators by blocking pores (abiotic fouling). Interestingly, during MFC operation, a significant amount of precipitation on the cathode was observed in the most of MFCs, which could hinder the proton transfer and subsequently slow down the oxygen reduction reaction. This is mainly driven by electro-osmotic drag and the number of ions diffused is strongly related to MFC performance [22,34]. Therefore, it is thought that the initially better performing ceramics such as red and white which also had smaller pore sizes (Table 1 and Figure 7) had more propensity towards membrane fouling in the long run. To tackle this type of membrane fouling, using alkaline catholyte, or periodic washing of both the cathode and separator can be implemented.

Table 1. Results of porosity and pore size measurements.

	Density [1] ($g \cdot cm^{-3}$)	Apparent Porosity [1] (%)	Total Porosity [2] (%)	Average Pore Diameter [2] (nm)	Maximum Power (μW)	Maximum Power Density [3] ($W \cdot m^{-3}$)
Brown	2.7	23.3	22.0	188	519.9	52.0
Red	2.7	25.6	23.7	95	670.5	67.1
White	2.6	26.8	23.9	81	570.5	57.1

Measured by [1] ASTM C373 method, [2] Hg intrusion method, [3] power density normalized by liquid volume of the anode compartment.

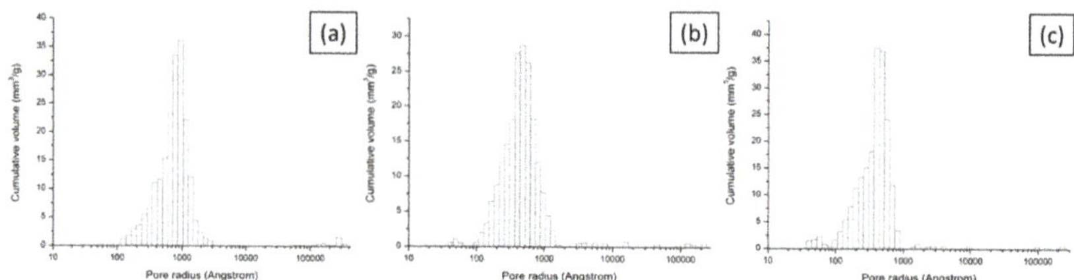

Figure 7. Pore size distribution for brown (**a**), red (**b**) and white (**c**) ceramic samples.

3.3. *Physiochemical Properties of Test Ceramics*

Mercury intrusion porosimetry and water absorption (ASTM C373) methods were applied since they are appropriate for a mesoporous (pore diameters between 2 nm and 50 nm) and macroporous (pore diameters of greater than 50 nm) materials such as this. The porosity and pore size distribution of single composition ceramics are shown in Table 1 and Figure 7. Although there are some differences between the two measurement methods, the porosity in all materials ranges between 22% and 27%, which is relatively high in comparison to other MFC ceramic separators previously tested [20,21,23]. Increased porosity by lowering firing temperature was confirmed by water absorption measurements (data not shown). Overall, brown ceramic is the least porous as speculated and this explains why addition of catholyte for brown ceramic was needed initially (Figure 3). Porosities of the other two ceramics are similar although porosity of white ceramic is slightly higher. As shown in Figure 7, the distribution of pore radii for all ceramics follows a log-Weibull distribution, peaking at less than 100 nm. The average pore size (diameter) is highest for the brown at 188 nm which is almost twice the size of the other two, making them all macroporous. This is of interest as pores over 1 µm may allow the passage of bacteria across the chambers. However, not all the pores will be open-ended and the likelihood of a bacterium finding an open-ended pore in the entire 3 mm thickness is low. The percentage of pores which are larger than 1 µm in radius is 1.92%, 1.32% and 0.56% for brown, red, and white ceramic respectively (Table S2), which is sufficiently low to assume that bacterial cross-contamination between MFC compartments is unlikely.

The SEM images (Figures 8 and 9) show that the surfaces of all ceramics are heterogeneous, with a range of features including smooth regions, cracks, vacancies, pores, and flakes. It is possible to see the inclusions of chamotte and other materials within the ceramic matrix, particularly in the cross-section images (Figure 9). From the surface images at lower magnification, in Figure 8, it is possible to see some inclusions on the surface, and in the case of the red ceramic, a micro-size crack is visible around one of these inclusions. At the higher magnification, some pores of roughly 1 µm are visible.

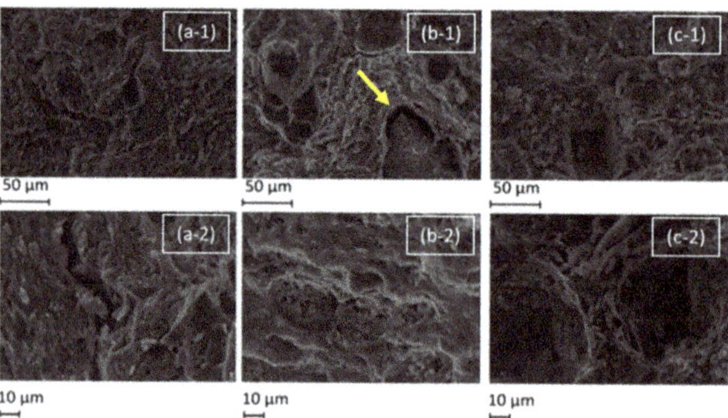

Figure 8. SEM micrographs of the surface of test ceramics: brown (**a**), red (**b**) and white (**c**) under magnifications of 500 (-1) and 1000 (-2) times. The yellow arrow points to a visible crack.

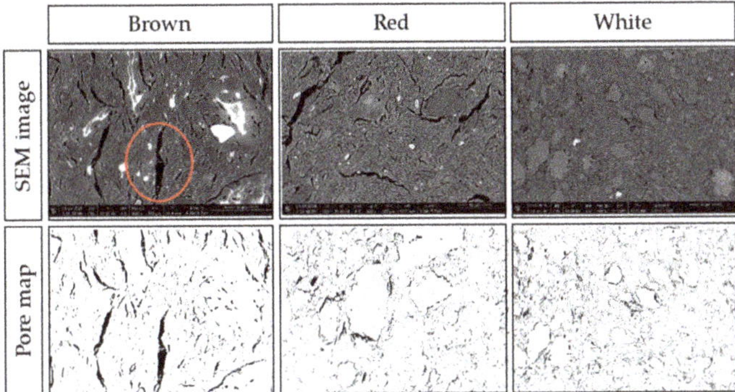

Figure 9. Cross-section SEM micrographs at 600x magnification of brown, white, and red ceramics with the associated pore maps.

Figure 9 shows the micro-scale cracks within the matrix, and the inclusions of different compositions and dimensions. Both brown and red ceramics have cracks of more than 10 μm in width and 100 μm in length. The largest crack in brown ceramic, circled in the bottom central part of the image, is 15 μm wide and 125 μm long. Some inclusions show as white in the SEM, and others have white regions, which is an artefact of the surface of the specimen. White ceramic, while also exhibiting cracks around inclusions, has instead much smaller pores. As the micro-scale cracks are a result of the matrix differential shrinkage during cooling, and the white ceramic has much smaller cracks, this suggests that the white ceramic suffered less differential shrinkage and thus less thermal stress during firing. The calculated percentage porosity from these images are 8.5%, 12.2% and 10.7% for brown, red, and white, respectively.

In summary, the brown ceramic has denser matrix with bigger pores in comparison to the red and white ceramics. This could be its intrinsic nature due to its composition or the adjusted firing temperature, perhaps both. Again, this can explain its initial underperformance but less propensity towards membrane fouling.

3.4. Composition and Structure of Ceramic Separators

The chemical composition of ceramics is shown in Table 2, with values from both the manufacturer data sheet and from EDXS analysis. The primary elements of all the ceramics were silicon and aluminium. The brown ceramic interestingly contains 4.8% MnO, which may be acting as a catalyst for the oxygen reduction reaction (ORR). Manganese oxides have attracted attention as a cost-effective alternative to platinum catalysts at the cathode in electrochemical energy conversion systems. While many stoichiometries have been investigated and each has different effects, in general manganese has been demonstrated to improve ORR kinetics by promoting the more efficient four-electron ORR pathway over the two-electron pathway [35]. Several papers use MnOx with carbon nanotubes as a catalyst in MFCs and find that in comparison to Pt/C catalyst it performs better [36], or similar [37–39]. However, in all of these papers, the manganese oxides were always MnO_2 or MnO_3, and were applied at the cathode surface, so without further investigation of our system, it is not clear whether the manganese is indeed acting as a catalyst or if other properties of the brown ceramic are dominant.

Table 2. Chemical composition of ceramics used.

Chemical Composition [1] (wt.%)		CO_2	SiO_2	Al_2O_3	TiO_2	Fe_2O_3	CaO	MgO	K_2O	Na_2O	MnO
Brown	MSDS	-	64.5	20.0	1.3	6.5	0.3	0.4	2.2	0.1	4.8
	EDXS	13.9	57.0	17.7	1.4	4.2	0.3	0.4	2.7	0.3	1.9
Red	MSDS		68.9	20.5	1.3	6.0	0.3	0.4	2.5	0.1	0.0
	EDXS	10.7	57.0	21.5	1.6	10.3	0.4	0.6	3.3	0.3	-
White	MSDS		72.0	22.0	1.8	1.0	0.3	0.2	2.3	0.3	0.0
	EDXS	18.1	51.3	21.5	2.2	2.7	-	0.5	3.2	0.3	-

[1] Data taken from the manufacturer's (Georg und Schneider) data sheets and from EDXS analysis on a Quanta 650 FEG scanning electron microscope, which only reveals elements, and the oxides have been assumed.

4. Conclusions

Commercially available ceramic materials with different compositions have been tested as MFC separators and as a (potential) chassis or vessel. Intensive physiochemical property analysis revealed that brown ceramic has lower porosity and larger pores in comparison to red and white ceramics. Single composition brown ceramic required cathode hydration due to its low porosity, which resulted in relatively poor performance initially. However, the brown ceramic separator is a more suitable choice for long-term operation, by outperforming other single composition ceramic types. This study confirms that pore size as well as porosity plays an important role in ceramic performance in MFCs.

Another important finding is the potential of spotty type ceramics. Although some of the spotty type ceramics had a leakage issue due to the nature of handmade ceramics and difference in extent of shrinkage of two materials during firing, in most cases spotty ceramics outperformed single composition types. Therefore, composite design could also be a way of enhancing the function of ceramic separators, as well as fine-tuning porosity and pore size. This may be particularly important in cases where the cathode is inside the cylinder (and the anode outside), which is an MFC topology that allows for catholyte synthesis [22]; in this way, it is possible to combine high power performance and high catholyte quality.

Ceramic separators for MFCs are still very much a lab making and not subject to rigorous manufacturing and quality control procedures; this is by far the main consideration that anyone working in this field should take seriously into account. Fab labs or facilities, whose core competency is the fabrication of new structures from new materials, could be the first option for scientists, since ceramics manufacturers, whose (successful) business model is distant from research, will be unable to engage, unless such proposition made sense. In the case when scientific labs can employ a rigorous fabrication process, then physical parameters (thickness, density/porosity and pore size) should take priority before physicochemical parameters can become the focus.

Lastly, the use of low-cost, locally sourced ceramic separators of customisable shape for high performance MFCs would allow for the design and implementation of this technology into the built environment, integrated into a living architecture [25].

Supplementary Materials: The following are available online at http://www.mdpi.com/1996-1073/12/21/4071/s1, Figure S1: All ceramic cylinder membranes tested in the study, Figure S2: Photos of experimental set-up (left) and individual MFCs (right), Figure S3: Temporal change of peak power of test ceramics, Table S1: Chemical compositions and physical properties of tested ceramic materials, Table S2: Percentage relative volume of pores for test ceramics.

Author Contributions: J.Y. and L.W. performed the MFC experiments. J.Y., L.W., N.R. and G.P. analysed materials and data. J.Y., L.W. and N.R. wrote the draft manuscript. V.M.S. supervised the EDXS and porosity measurements. I.I., J.G., M.M.H. and V.M.S reviewed and edited the manuscript. I.I. and J.G. also acquired the funding and provided an idea of the "spotty ceramic membranes". All authors have approved the submitted manuscript.

Funding: This research was funded by the European Commission Horizon 2020 FET-OPEN Living Architecture Project (grant no. 686585).

Acknowledgments: The authors would like to thank Barbara Imhof and Waltraut Hoheneder (LIQUIFER Systems Group GmbH) for providing ceramic materials used for experiments.

Conflicts of Interest: The authors declare no conflict of interest.

References

1. Dong, Y.; Qu, Y.; He, W.; Du, Y.; Liu, J.; Han, X.; Feng, Y. A 90-liter stackable baffled microbial fuel cell for brewery wastewater treatment based on energy self-sufficient mode. *Bioresour. Technol.* **2015**, *195*, 66–72. [CrossRef] [PubMed]
2. Goto, Y.; Yoshida, N. Scaling up Microbial Fuel Cells for Treating Swine Wastewater. *Water* **2019**, *11*, 1803. [CrossRef]
3. Ge, Z.; He, Z. Long-term performance of a 200 liter modularized microbial fuel cell system treating municipal wastewater: Treatment, energy, and cost. *Environ. Sci. Water Res. Technol.* **2016**, *2*, 274–281. [CrossRef]
4. Liang, P.; Duan, R.; Jiang, Y.; Zhang, X.; Qiu, Y.; Huang, X. One-year operation of 1000-L modularized microbial fuel cell for municipal wastewater treatment. *Water Res.* **2018**, *141*, 1–8. [CrossRef] [PubMed]
5. Online TOX/BOD System—KORBI. Available online: http://www.korbi.com/eng/products_type/habt-3000/?pageds=1&k=BOD (accessed on 6 September 2019).
6. Trapero, J.R.; Horcajada, L.; Linares, J.J.; Lobato, J. Is microbial fuel cell technology ready? An economic answer towards industrial commercialization. *Appl. Energy* **2017**, *185*, 698–707. [CrossRef]
7. Santoro, C.; Arbizzani, C.; Erable, B.; Ieropoulos, I. Microbial fuel cells: From fundamentals to applications. A review. *J. Power Sources* **2017**, *356*, 225–244. [CrossRef]
8. Kondaveeti, S.; Kakarla, R.; Kim, H.S.; Kim, B.G.; Min, B. The performance and long-term stability of low-cost separators in single-chamber bottle-type microbial fuel cells. *Environ. Technol.* **2018**, *39*, 288–297. [CrossRef]
9. Liu, H.; Logan, B.E. Electricity generation using an air-cathode single chamber microbial fuel cell in the presence and absence of a proton exchange membrane. *Environ. Sci. Technol.* **2004**, *38*, 4040–4046. [CrossRef]
10. Palanisamy, G.; Jung, H.Y.; Sadhasivam, T.; Kurkuri, M.D.; Kim, S.C.; Roh, S.H. A comprehensive review on microbial fuel cell technologies: Processes, utilization, and advanced developments in electrodes and membranes. *J. Clean. Prod.* **2019**, *221*, 598–621. [CrossRef]
11. Koók, L.; Bakonyi, P.; Harnisch, F.; Kretzschmar, J.; Chae, K.J.; Zhen, G.; Kumar, G.; Rózsenberszki, T.; Tóth, G.; Nemestóthy, N.; et al. Biofouling of membranes in microbial electrochemical technologies: Causes, characterization methods and mitigation strategies. *Bioresour. Technol.* **2019**, *279*, 327–338. [CrossRef]
12. Bajracharya, S.; Sharma, M.; Mohanakrishna, G.; Dominguez Benneton, X.; Strik, D.P.; Sarma, P.M.; Pant, D. An overview on emerging bioelectrochemical systems (BESs): Technology for sustainable electricity, waste remediation, resource recovery, chemical production and beyond. *Renew. Energy* **2016**, *98*, 153–170. [CrossRef]
13. Oliot, M.; Galier, S.; Roux de Balmann, H.; Bergel, A. Ion transport in microbial fuel cells: Key roles, theory and critical review. *Appl. Energy* **2016**, *183*, 1682–1704. [CrossRef]
14. Dhar, B.R.; Lee, H.S. Membranes for bioelectrochemical systems: Challenges and research advances. *Environ. Technol.* **2013**, *34*, 1751–1764. [CrossRef] [PubMed]
15. Jana, P.S.; Behera, M.; Ghangrekar, M.M. Performance comparison of up-flow microbial fuel cells fabricated using proton exchange membrane and earthen cylinder. *Int. J. Hydrog. Energy* **2010**, *35*, 5681–5686. [CrossRef]
16. Gajda, I.; Greenman, J.; Santoro, C.; Serov, A.; Atanassov, P.; Melhuish, C.; Ieropoulos, I.A. Multi-functional microbial fuel cells for power, treatment and electro-osmotic purification of urine. *J. Chem. Technol. Biotechnol.* **2019**, *94*, 2098–2106. [CrossRef]
17. Winfield, J.; Gajda, I.; Greenman, J.; Ieropoulos, I. A review into the use of ceramics in microbial fuel cells. *Bioresour. Technol.* **2016**, *215*, 296–303. [CrossRef]
18. Santoro, C.; Flores-Cadengo, C.; Soavi, F.; Kodali, M.; Merino-Jimenez, I.; Gajda, I.; Greenman, J.; Ieropoulos, I.; Atanassov, P. Ceramic microbial fuel cells stack: Power generation in standard and supercapacitive mode. *Sci. Rep.* **2018**, *8*, 3281. [CrossRef]

19. Baker, R.W. Membrane Separation. In *Encyclopedia of Separation Science*; Elsevier: Cambridge, MA, USA, 2000; pp. 189–210.
20. Winfield, J.; Greenman, J.; Huson, D.; Ieropoulos, I. Comparing terracotta and earthenware for multiple functionalities in microbial fuel cells. *Bioprocess Biosyst. Eng.* **2013**, *36*, 1913–1921. [CrossRef]
21. Pasternak, G.; Greenman, J.; Ieropoulos, I. Comprehensive study on ceramic membranes for low-cost microbial fuel cells. *ChemSusChem* **2016**, *9*, 88–96. [CrossRef]
22. Merino Jimenez, I.; Greenman, J.; Ieropoulos, I. Electricity and catholyte production from ceramic MFCs treating urine. *Int. J. Hydrog. Energy* **2017**, *42*, 1791–1799. [CrossRef]
23. Khalili, H.B.; Mohebbi-Kalhori, D.; Afarani, M.S. Microbial fuel cell (MFC) using commercially available unglazed ceramic wares: Low-cost ceramic separators suitable for scale-up. *Int. J. Hydrog. Energy* **2017**, *42*, 8233–8241. [CrossRef]
24. Winfield, J.; Chambers, L.D.; Rossiter, J.; Greenman, J.; Ieropoulos, I. Urine-activated origami microbial fuel cells to signal proof of life. *J. Mater. Chem. A* **2015**, *3*, 7058–7065. [CrossRef]
25. Living Architecture—Transform Our Habitats from Inert Spaces into Programmable Sites. Available online: https://livingarchitecture-h2020.eu/ (accessed on 10 September 2019).
26. Djangang, C.N.; Elimbi, A.; Melo, U.C.; Lecomte, G.L.; Nkoumbou, C.; Soro, J.; Bonnet, J.P.; Blanchart, P.; Njopwouo, D. Sintering of clay-chamotte ceramic composites for refractory bricks. *Ceram. Int.* **2008**, *34*, 1207–1213. [CrossRef]
27. You, J.; Walter, X.A.; Greenman, J.; Melhuish, C.; Ieropoulos, I. Stability and reliability of anodic biofilms under different feedstock conditions: Towards microbial fuel cell sensors. *Sens. Bio-Sens. Res.* **2015**, *6*, 43–50. [CrossRef]
28. *ASTM C373-18, Standard Test Methods for Determination of Water Absorption and Associated Properties by Vacuum Method for Pressed Ceramic Tiles and Glass Tiles and Boil Method for Extruded Ceramic Tiles and Non-Tile Fired Ceramic Whiteware Products*; ASTM International: West Conshohocken, PA, USA, 2018.
29. Logan, B.E.; Hamelers, B.; Rozendal, R.; Schröder, U.; Keller, J.; Freguia, S.; Aelterman, P.; Verstraete, W.; Rabaey, K. Microbial fuel cells: Methodology and technology. *Environ. Sci. Technol.* **2006**, *40*, 5181–5192. [CrossRef] [PubMed]
30. Gajda, I.; Greenman, J.; Melhuish, C.; Ieropoulos, I. Simultaneous electricity generation and microbially-assisted electrosynthesis in ceramic MFCs. *Bioelectrochemistry* **2015**, *104*, 58–64. [CrossRef] [PubMed]
31. Malvankar, N.S.; Tuominen, M.T.; Lovley, D.R. Biofilm conductivity is a decisive variable for high-current-density Geobacter sulfurreducens microbial fuel cells. *Energy Environ. Sci.* **2012**, *5*, 5790. [CrossRef]
32. Kundu, P.P.; Dutta, K.; Kumar, P.; Bharti, R.P.; Kumar, V.; Kundu, P.P. Polymer Electrolyte Membranes for Microbial Fuel Cells: Part A. Nafion-Based Membranes. In *Progress and Recent Trends in Microbial Fuel Cells*; Elsevier: Amsterdam, The Netherlands, 2018; pp. 47–72, ISBN 978-0-444-64017-8.
33. Natarajan, K.A. Biofouling and microbially influenced corrosion. In *Biotechnology of Metals*, 1st ed.; Elsevier: Amsterdam, The Netherlands, 2018; pp. 355–393, ISBN 978-0-12-804022-5.
34. Gajda, I.; Greenman, J.; Melhuish, C.; Santoro, C.; Li, B.; Cristiani, P.; Ieropoulos, I. Electro-osmotic-based catholyte production by Microbial Fuel Cells for carbon capture. *Water Res.* **2015**, *86*, 108–115. [CrossRef]
35. Stoerzinger, K.A.; Risch, M.; Han, B.; Shao-Horn, Y. Recent Insights into Manganese Oxides in Catalyzing Oxygen Reduction Kinetics. *ACS Catal.* **2015**, *5*, 6021–6031. [CrossRef]
36. Zhang, Y.; Hu, Y.; Li, S.; Sun, J.; Hou, B. Manganese dioxide-coated carbon nanotubes as an improved cathodic catalyst for oxygen reduction in a microbial fuel cell. *J. Power Sources* **2011**, *196*, 9284–9289. [CrossRef]
37. Rout, S.; Nayak, A.K.; Varanasi, J.L.; Pradhan, D.; Das, D. Enhanced energy recovery by manganese oxide/reduced graphene oxide nanocomposite as an air-cathode electrode in the single-chambered microbial fuel cell. *J. Electroanal. Chem.* **2018**, *815*, 1–7. [CrossRef]

38. Shahbazi Farahani, F.; Mecheri, B.; Reza Majidi, M.; Costa de Oliveira, M.A.; D'Epifanio, A.; Zurlo, F.; Placidi, E.; Arciprete, F.; Licoccia, S. MnOx-based electrocatalysts for enhanced oxygen reduction in microbial fuel cell air cathodes. *J. Power Sources* **2018**, *390*, 45–53. [CrossRef]
39. Zhang, L.; Liu, C.; Zhuang, L.; Li, W.; Zhou, S.; Zhang, J. Manganese dioxide as an alternative cathodic catalyst to platinum in microbial fuel cells. *Biosens. Bioelectron.* **2009**, *24*, 2825–2829. [CrossRef] [PubMed]

Article

A Paper-Based Microfluidic Fuel Cell Using Soft Drinks as a Renewable Energy Source

Jaime Hernández Rivera [1,2], David Ortega Díaz [3], Diana María Amaya Cruz [4], Juvenal Rodríguez-Reséndiz [5,*], Juan Manuel Olivares Ramírez [2,*], Andrés Dector [6,*], Diana Dector [1], Rosario Galindo [7] and Hilda Esperanza Esparza Ponce [1]

1 Centro de Investigación en Materiales Avanzados, Chihuahua 31136, Mexico; jhernandezr@utsjr.edu.mx (J.H.R.); diana.dector@cimav.edu.mx (D.D.); hilda.esparza@cimav.edu.mx (H.E.E.P.)
2 Renewable Energy Department, Universidad Tecnológica de San Juan del Río, Querétaro 76800, Mexico
3 Instituto Tecnológico de San Juan del Río, Querétaro 76800, Mexico; ortegadiaz15@gmail.com
4 Facultad de Ingeniería, Universidad Autónoma de Querétaro, Campus Amealco, Querétaro 76010, Mexico; diana.amaya@uaq.mx
5 Facultad de Ingeniería, Universidad Autónoma de Querétaro, Querétaro 76010, Mexico
6 Renewable Energy Department, CONACYT–Universidad Tecnológica de San Juan del Río, Querétaro 76800, Mexico
7 CONACYT–División de Ciencias Naturales y Exactas, Universidad de Guanajuato, Guanajuato 36050, Mexico; mr.galindo@ugto.mx
* Correspondence: juvenal@uaq.edu.mx (J.R.-R.); jmolivaresr@utsjr.edu.mx (J.M.O.R.); adector@conacyt.mx (A.D.); Tel.: +52-427-122-4463 (J.M.O.R.)

Received: 9 April 2020; Accepted: 11 May 2020; Published: 13 May 2020

Abstract: The research aims were to construct an air-breathing paper-based microfluidic fuel cell (paper-based μFC) and to evaluated it with different soft drinks to provide energy for their prospective use in portable devices as an emergency power source. First, in a half-cell configuration, cyclic voltammetry showed that glucose, maltose, and fructose had specific oxidation zones in the presence of platinum-ruthenium on carbon (PtRu/C) when they were individual. Still, when they were mixed, glucose was observed to be oxidized to a greater extent than fructose and maltose. After, when a paper-based μFC was constructed, PtRu/C and platinum on carbon (Pt/C) were used as anode and cathode, the performance of this μFC was mostly influenced by the concentration of glucose present in each soft drink, obtaining maximum power densities at room temperature of 0.061, 0.063, 0.060, and 0.073 mW cm^{-2} for Coca Cola, Pepsi, Dr. Pepper, and 7up, respectively. Interestingly, when the soft drinks were cooled, the performance was increased up to 85%. Furthermore, a four-cell stack μFC was constructed to demonstrate its usefulness as a possible power supply, obtaining a power density of 0.4 mW cm^{-2}, using Coca Cola as fuel and air as oxidant. Together, the results of the present study indicate an alternative application of an μFC using soft drinks as a backup source of energy in emergencies.

Keywords: fuel cell application; microfluidic fuel cell; power supply; soft drinks

1. Introduction

Actually, the performance of low-consumption electronic devices such as wireless sensors, digital clocks, and small medical devices, among others, is limited, due to the use a rechargeable lithium-ion battery [1]. In this way, the intensive research to obtain small energy sources as a possible application for low-consumption electronic devices has been directed to the use of micro fuel cells [2–6].

The paper based microfluidic fuel cells (paper based μFC) have been proposed as an external power source for low-consumption electronic devices [7–9]. The paper-based μFCs have the following advantages:

- The paper used in the paper-based μFCs construction is a very low-cost material, and consequently, the fuel cells are disposable.
- Due to the movement of reactants through capillary action, the use of external pumps is unnecessary [7].
- The ion transport is carried out through the electrolyte contained in the reactants, and as a consequence, the use of a proton or anion exchange membrane is unnecessary [10].
- These paper-based μFCs may use different liquid organic compounds as fuel, being reported to date ethanol [11], methanol [7,11], formate[12], urea [13,14], and glucose [8,15].

Research into paper-based μFCs using glucose as fuel has experienced a growing interest in the past few years and showing potential in electricity generation for portable devices [8,15,16]. The main advantages of using glucose had been non-toxicity and handling since it had been obtained from reactive grade glucose to body fluids such as human blood [9,17]. The above assumes that the use of other economic sources of glucose that does not require a previous treatment would constitute a significant achievement. In this context, soft drinks are an excellent alternative, since they are a cheap and highly available source; in addition, they contain other saccharides such as fructose, galactose, sucrose, lactose, and maltose, which could be used as fuel in a paper-based μFC.

Some relevant works evaluating paper-based μFCs under soft drinks are listed in the following lines:

The first research has evaluated different kinds of soft drinks (iced red tea, vegetable juice, fruit juice, and aerated water) in a miniature biofuel cell using glucose dehydrogenase and bilirubin oxidase as anode and cathode, respectively [18]. The μFCs in the best performance recollected 140 mW cm^{-2}, and 0.51 V of power density and open circuit potential, using glucose dehydrogenase as the biocatalyst on single-walled carbon nanohorn. The present investigation collocated the perspective of using the biofuel cell as a portable power source.

Other researchers used soft drinks such as Nutri-Express, Coca Cola, and Minute Maid grape juice, as fuels in a miniature origami biofuel cell employed glucose dehydrogenase as bio-anode, obtaining power in the order of microwatts [19]. The system was described as a new approach for effective green energy systems.

In a third work [20], was reported a 3-cell stack paper-based biofuel cell powered by Gatorade + NAD^{+1} mM biofuel solutions at pH 7.3 (physiological pH). The anode (4.5 cm^2) and the cathode (2.5 cm^2) employed in this 3-cell stack were constructed employed Glucose Dehydrogenase (GDH) and Bilirubin oxidase (Box), respectively. The performance obtained through the stack system was from 1.8 V and 0.18 mWmg^{-1} GDH (0.216 mW cm^{-2}, calculated in this work from 3 mg GDH reported by the authors and cathode area).

Finally, in more recent research a miniature self-pumping paper-based enzymatic biofuel cell was constructed using glucose oxidase and laccase as anode and cathode, respectively, and soft drinks as fuels. The authors reported that power density obtained was attributed to the glucose contained in the soft drinks, which was fresh watermelon juice > 7up > Mountain Dew > Pepsi (14.5, 13.5, 12, and 6.15 μWcm^{-2}) [21].

The above reports have used enzymatic electrodes for the oxidation of saccharides present in the soft drinks. As another alternative, the use of inorganic catalysts that oxidize saccharides in ideal and real conditions has been directed at the use of gold-based materials in many jobs [17,22–25].

In this sense, another material that has shown activity in the electro-oxidation of saccharides in fuel cells but has not had much application in real conditions has been the PtRu. This work has proposed a paper-based μFCs that use a catalyst anode such as PtRu/C, which is used due to a minimal poisoning rate and lower cost than Pt/C [26]. In addition, to assess their behavior, the individual voltammetry assessment of glucose, fructose, and maltose is reported, as well as the simulated Coca Cola, Pepsi, Dr. Pepper, and 7up, evaluation, this means the mixture of these saccharides according to their reported content. Finally, the real soft drinks were employed, for the first time, like fuel in a non-enzymatic paper-based μFC; and the effect of two temperatures (4 °C, as consumption temperature and 25 °C, as room temperature) on the performance of the fuel cell was evaluated. The novelty of this work is the use of inorganic catalyst material in paper-based μFC for power generation using soft drinks as fuels. This approach has not been widely studied in this type of material. In addition, a new concept of emergency energy through this paper-based μFC was created. Furthermore, due to the easy access to sugary soft drinks that we have today, these paper microfluidic fuel cells could actually be applied as possible backup energy sources for low-consumption electronic devices. This article is outlined as follows: Section 2 depicts methodology, begins with the catalytic material used; Pt/C and Pt-Ru/C. After describing the electrochemical characterization to saccharides oxidation, in this section, describe so the design of the paper microfluidic fuel cell. Section 3 displays the results, and discussion about the characterization of Ru distribution on carbon paper, a cyclic voltammogram of PtRu/C in the presence of fructose, glucose, and maltose, after in presence on the 7up, Coca Cola, Dr. Pepper, and Pepsi. Finally presents the polarization and power density curve.

2. Development of μFC

2.1. Materials

All chemicals were reagent grade and were used without further modification. PtRu/C (20 wt.%) and Pt/C (30%) were from E-TEK. Nafion 5%, glucose, fructose, and maltose acquired from Sigma-Aldrich. Carbon paper Toray was obtained from Technoquip Co Inc TGPH-120. Soft drinks were purchased from a local retail market. Isopropyl alcohol and KOH acquired from J.T. Baker. Whatman filter paper, grade Fusion 5 was used for paper-based μFC construction.

2.2. PtRu/C Characterization

The experimental methodology started with the electrochemical characterization [23], of the catalytic properties of Pt/Ru for the saccharides (glucose, fructose, and maltose) oxidation. This electrochemical characterization consisted of cyclic voltammetry in a standard three-electrode glass cell using a saturated calomel electrode (SCE) and a Pt wire as reference electrode and counter electrode, respectively, and a glassy carbon electrode (3 mm diameter) modified with the PtRu/C ink as working electrode. All cyclic voltammetry experiments were conducted at 20 mVs^{-1} and performed by potentiostat/galvanostat (Zahner Zennium). Half-cell studies consisted of four different conditions shown in Table 1. The conditions 1, 2, and 3 correspond to each concentration of the reported alone saccharides present in each soft drink [27] (glucose, fructose, and maltose) mixed in 20 mL of 0.3 M KOH. The condition 4 consisted of a mixture of all the previous saccharides inside each soft drink in 20 mL of 0.3 M KOH, obtaining a near simulated term. In all circumstances, the solutions were deoxygenated by N_2-saturation (Infra, 99.999% pure) for 20 min at 25 °C.

To obtain a mapping of Ru was used as a HITACHI, Scanning Electron Microscope, Model SU 3500 & QUANTAX: Electron Dispersive Scattering for a Micro-analysis.

2.3. Paper-Based Microfluidic Fuel Cell Assembly Process

After the evaluation of the electrocatalytic properties of PtRu, was proceeded to assembly the paper-based μFC. The electrodes consisted of a piece microporous carbon paper (Toray Technoquip Co Inc TGPH-120) from 1.0 × 0.5 cm covered with a catalytic ink prepared using Nafion 5% (Sigma-Aldrich), isopropyl alcohol (J.T. Baker) and PtRu/C (20 wt.% from E-TEK) or Pt/C (30 wt.% on Vulcan XC -72 from E-TEK) [9]. The ink over the electrodes was deposited using an airbrush, and 1 mg cm^{-2} metal loading was obtained. A foil of conducting aluminum was used to envelop the carbon paper electrodes as a contact.

Table 1. Half-cell conditions for soft drinks electro-oxidation on PtRu/C.

Condition	Saccharide	Coca Cola /mol L^{-1}	Pepsi /mol L^{-1}	Dr. Pepper /mol L^{-1}	7up /mol L^{-1}
1	Glucose	0.23	0.24	0.22	0.29
2	Fructose	0.35	0.36	0.34	0.25
3	Maltose	0.003		0.003	0.003
4	Glucose + Fructose + Maltose	Simulated Coca Cola	Simulated Pepsi	Simulated Dr. Pepper	Simulated 7up

The paper-based μFC was made of an absorbent paper strip (whatman, grade Fusion 5) composed with a fuel inlet of 1.0 × 0.3 cm and a reaction zone of 3.0 × 0.7 cm and delimited using a cutter plotter (Grapthec America Inc.) [9]. The schematic design of the paper-based μFC is presented in Figure 1. First, a rectangular size of 5.0×0.7 cm of one layer of glass slide adhesive was cut with the objective of providing mechanical support to the paper-based μFC. On this adhesive layer was assembled the anode electrode; subsequently, the paper strip and cathode were collocated to form a sandwich structure. Placing the cathode on top facilitated oxygen access from the atmosphere.

On the other hand, for the construction of the 2-cell and 4-cell stacks, continuous pairs of electrodes were collocated (anode and cathode in the bottom and top, respectively) and rotated on the paper strip. Finally were connected following a serial connection using aluminum foil.

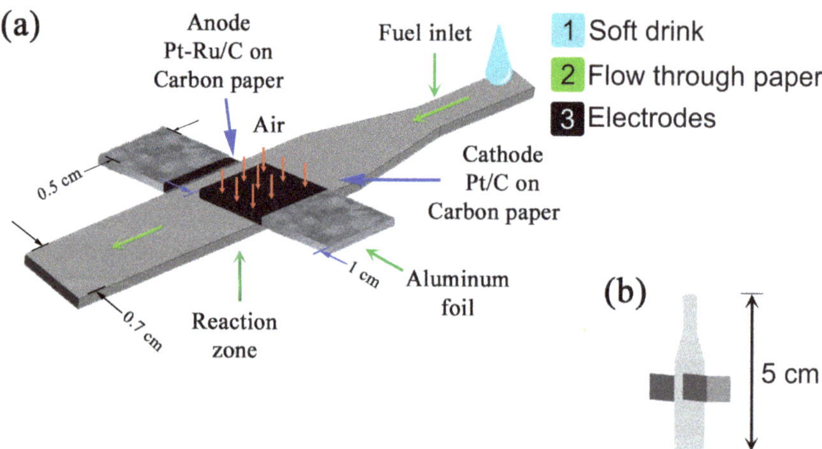

Figure 1. Components of fuel cell: (**a**) Schematic design of the paper-based microfluidic fuel cell and (**b**) dimension of the paper-based microfluidic fuel cell (μFC).

2.4. Performance of the Paper-Based Microfluidic Fuel Cell

In these experiments, the soft drinks were neither deoxygenated nor degassed. This condition could cause a bubble in the catalytic surface, causing an area in this surface without fuel, so the current and voltage could have a variation. This variation could cause unstable in the polarization curve. In all paper-based μFCs performance cases, the soft drinks were proved in two temperatures: at room temperature (27 °C) and cold (4 °C), this last temperature represents an adequate approximation of operation in a possible real application. Even though the soft drinks are available at a low cost in large quantities, the paper-based μFCs present in this work is intended to be fed with a drop of soft drinks from the straw or with a small amount dropped of soft drink while the μFCs remains connected to some device, which limits the available volume of fuel significantly. For this reason, we standardized the volume of the experiments to 15 μL, which is equivalent to the amount that could be present in a drop. Although the supply of soft drinks as fuel is not limited to this, for a possible real application. The resulting polarization and power density curves were recorded at 20 mVs^{-1} using a Zahner Zennium potentiostat (PP211).

3. Results and Discussion

3.1. PtRu/C Characterization

The cyclic voltammograms (CVs) corresponding to the saccharides (glucose, fructose, and maltose) founded in all soft drinks used in this study (7up, Coca Cola, Dr. Pepper, and Pepsi) showed similar behavior in all cases. In this sense, Figure 2 shows only the CVs in 0.3 M KOH on PtRu/C at a scan rate of 20 mVs^{-1} in a potential range of −0.9 to 0.7 V for the concentration of the saccharide in 7up.

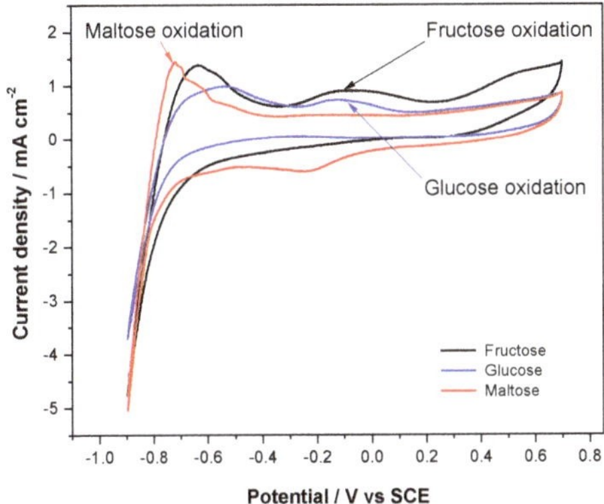

Figure 2. Cyclic voltammograms of PtRu/C on a glassy carbon electrode in the presence of fructose, glucose, and maltose in 0.3 M KOH, the scan rate of 20 mVs^{-1} at room temperature at the concentration corresponding to 7up.

Three peaks at −0.63, −0.10, and 0.57 V in the forward scan were observed for the glucose oxidation test (blue line in Figure 2). The peak at −0.63 V originated due to dehydrogenation of glucose (Figure 3a) in the hydrogen wave region; this peak was observed for fructose too at −0.55 V (black line in Figures 2 and 3b). The peaks at −0.10 and −0.14 V for glucose oxidation and fructose oxidation, respectively emerged at the potential region where the electrode surface is partially covered by adsorbed OH^-. In this region, the adsorbed OH^- on catalyst surface can oxidize glucose and fructose directly. The peak at 0.57 V was observed due to oxidation of the platinum surface [26,28–32]. For maltose oxidation, peak at −0.70 V in a forward scan was observed due to the oxidation of maltose to D-maltbionic acid (red line in Figures 2 and 3c) [33].

On the other hand, Figure 4 shows the cyclic voltammograms corresponding to simulated soft drinks, which are composed of the same concentrations of saccharides present in each soft drink as, shown in Table 1, but in this case, all are homogeneously dissolved in a 0.3 M KOH solution. In all cases, saccharide oxidation reaction curves are observed in PtRu/C at 20 mVs^{-1} in a potential range of −0.9 to 0.7 V.

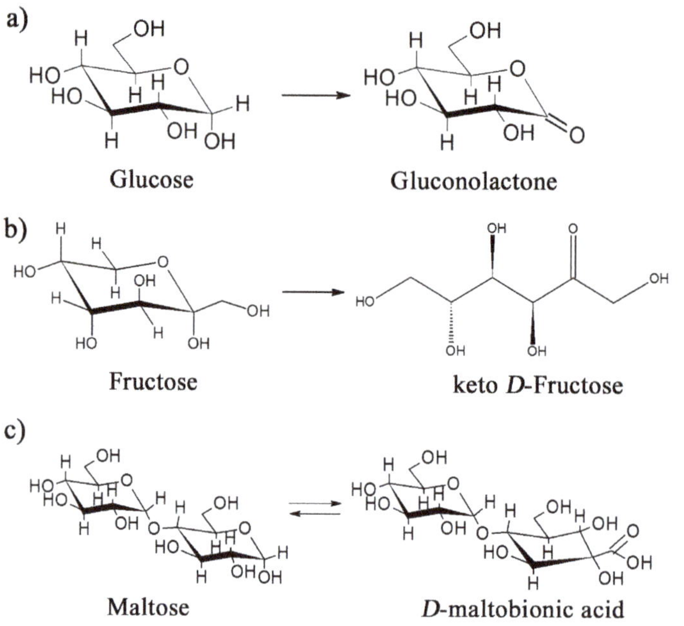

a)
Glucose
Gluconolactone

b)
Fructose
keto *D*-Fructose

c)
Maltose
D-maltobionic acid

Figure 3. Oxidation reactions of saccharides: (**a**) glucose, (**b**) fructose, and (**c**) maltose.

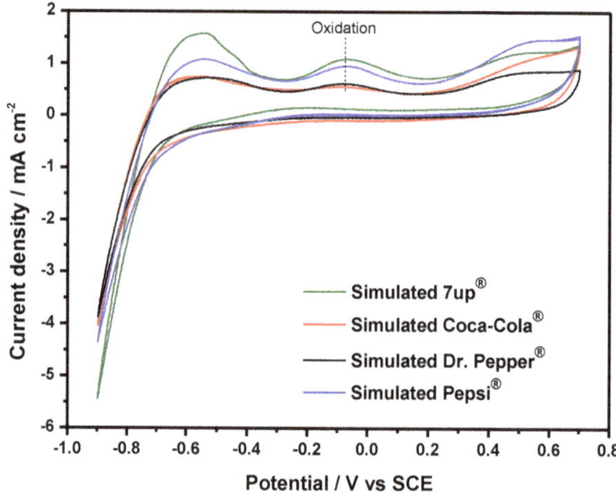

Figure 4. Cyclic voltammograms of PtRu/C on glassy carbon electrode in presence of simulated 7up, Coca Cola, Dr. Pepper, and Pepsi at a scan rate of 20 mVs^{-1} at room temperature.

The forms of the voltammograms were very similar between them and are characterized by the presence of three peaks in the forward scans. It is interesting that the position of the peaks for the simulated drinks, they very close to the position of the glucose oxidation peaks in PtRu/C found previously (Figure 2). Therefore the distribution of Ru it shows in Figure 5, and glucose oxidation is predominant on PtRu/C despite the presence of the other saccharides [31], and [34,35] used in this study.

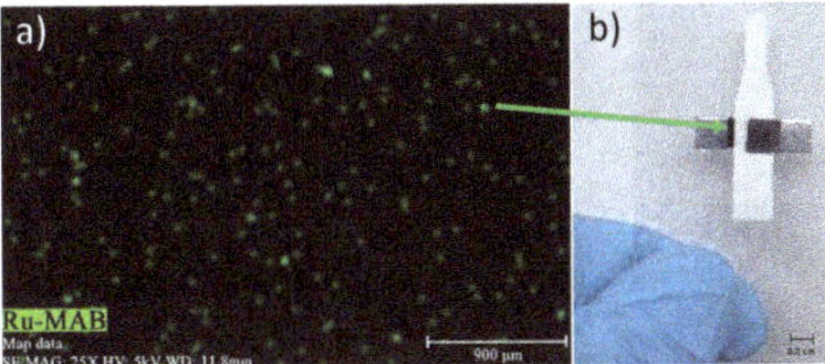

Figure 5. Fuel cell and material catalytic: (**a**) mapping of Ru, (**b**) micro-fuel cell photography.

These tests show an approach to check the oxidation of saccharides via electrochemistry, leaving as a basis for the possible use of soft drinks as a source of saccharides.

3.2. Soft Drinks Paper-Based Microfluidic Fuel Cell Performance

Figure 6 shows the polarization and power density curves for each paper-based microfluidic fuel cell operated with 7up, Coca Cola, Dr. Pepper, and Pepsi as an energy source, respectively, at room temperature (27 °C) and cold (4 °C). Table 2 shows the experimental parameters obtained from the overall performance of the paper-based μFCs with the different soft drinks as fuels. From Figure 6, can be seen that the fuels that offer the best performance are 7up, Coca Cola, Dr. Pepper in that order which yields maximum power densities of 0.073, 0.063, 0.061, and 0.060 mW cm^{-2} and 0.120, 0.113, 0.112, and 0.111 mW cm^{-2}, for room and cold temperature, respectively. These results have been resumed in Table 2, and confirm that glucose oxidation is predominant on PtRu/C in the presence of fructose and maltose since the order of the performance of the devices coincides with the order of the concentration of glucose in each soft drink, which is 7up > Pepsi > Coca Cola > Dr. Pepper.

Table 2. Experimental parameters obtained from the overall performance of the paper-based μFCs with different soft drinks as fuels. Results are expressed as mean values (n = 3) ± standard deviation.

Soft Drink	OCP /V	J /mA cm^{-2}	P MAX /mW cm^{-2}
Coca Cola at room temperature	0.63 ± 0.09	0.43 ± 0.03	0.061 ± 0.004
Cold Coca Cola	0.60 ± 0.08	0.75 ± 0.06	0.112 ± 0.008
Pepsi at room temperature	0.55 ± 0.06	0.34 ± 0.03	0.063 ± 0.003
Cold Pepsi	0.62 ± 0.07	0.64 ± 0.05	0.113 ± 0.007
Dr. Pepper at room temperature	0.64 ± 0.08	0.45 ± 0.04	0.060 ± 0.004
Cold Dr. Pepper	0.67 ± 0.07	0.67 ± 0.05	0.111 ± 0.007
7up at room temperature	0.69 ± 0.06	0.37 ± 0.03	0.073 ± 0.005
Cold 7up	0.64 ± 0.05	0.62 ± 0.05	0.120 ± 0.007

In all cases, the power density values show an increase for each cold soft drink in comparison with room temperature, suggesting that the temperature plays an essential role for the improved electro-catalytic performance of the device, as is known, the operation of these devices is limited by the O_2 reduction reaction that exists in the cathode according to Equation (1), despite the use of Pt, which is the best catalyst for this reaction and taking into account that the way in which the reaction occurs is highly influenced by the surface on which it is given, in addition to the conditions of the reaction medium [36–38].

$$O_2 + 4H^+ + 4^{e^-} \rightarrow 2H_2O \qquad E^0 = 1.229 \text{ V} \tag{1}$$

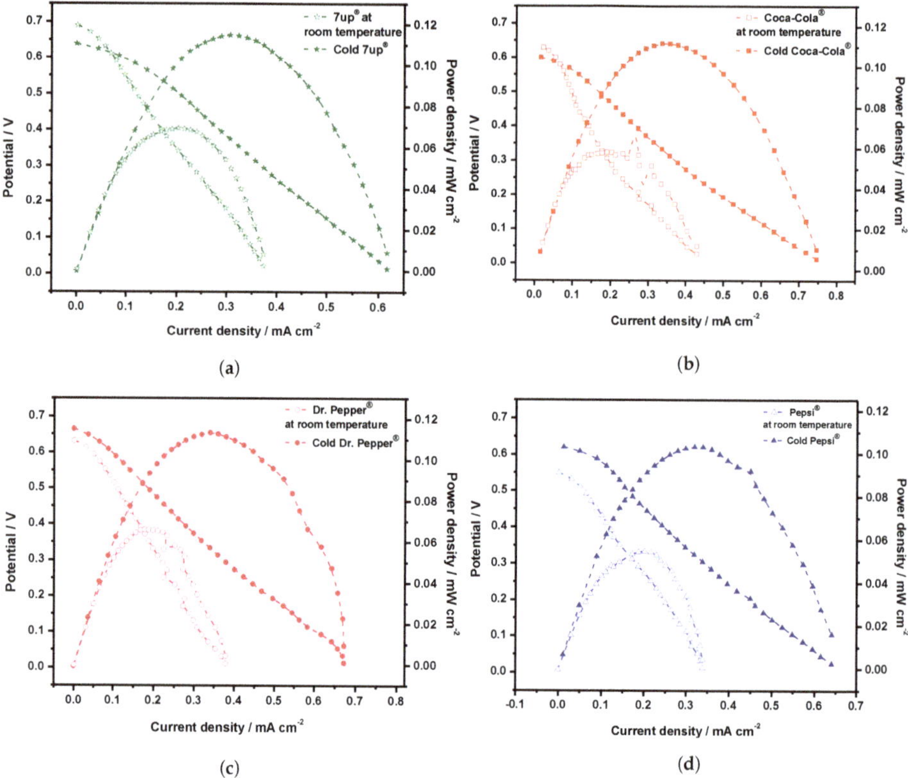

Figure 6. Polarization and power density curves of PtRu/C and Pt/C as anode and cathode, respectively that uses real soft drink cold (solid Figures) and soft drink at room temperature (open Figures) as fuel and air as oxidant. (**a**) 7up, (**b**) Coca Cola, (**c**) Dr. Pepper, and (**d**) Pepsi.

Despite its high spontaneity, the problem with oxygen reduction lies in a large amount of energy that must be supplied to the molecule for the activation of the double bond (498 kJ mol^{-1}), which makes this reaction kinetically impeded [38]. Kinetically, it is ideal that the reaction occurs as close as possible to the thermodynamic reaction potential [39], with a high reaction speed, or at least sufficient for the needs of each case. The current-overpotential relationship is given by the Butler–Volmer equation (Equation (2)).

$$i_c = i^0(e^{\frac{n\alpha F\eta_c}{RT}}) - e^{\frac{n(\alpha-1)F\eta_c}{RT}} \quad (2)$$

where i_c is the current density of the oxygen reduction reaction, i^0 is the current exchange density, n is the number of electrons transferred in the step that determines the reaction rate, α is the transfer coefficient, η_c is the overpotential for the oxygen reduction reaction, F is Faraday's constant, R the constant of the ideal gases, and T the temperature in Kelvin degrees. For the current density of the reaction to be high to on low overpotentials, it must be fulfilled that, the exchange current (i^0) be high, or the factor $\alpha F(RT)^{-1}$ be big. A decrease in temperature would cause an increase in this factor, considering that the other parameters do not change because the same surface is used as a catalyst and the same medium, increasing the contribution of the oxygen reduction reaction current and, therefore, the performance of the cell.

The power density obtained when each soft drink is used as fuel is a maximum advance from the economic and environmental standpoints; it is possible to obtain power densities using a cheaper, common, and easily available soft drink as fuel. It is important to note that these soft drinks are less expensive than other fuel commonly used in paper-based μFCs, because the soft drink does not need preparation in the laboratory. Figure 7, shows the development of the open circuit potential (OCP), for a paper-based μFCs for cases 7up, Coca Cola, Dr. Pepper, and Pepsi.

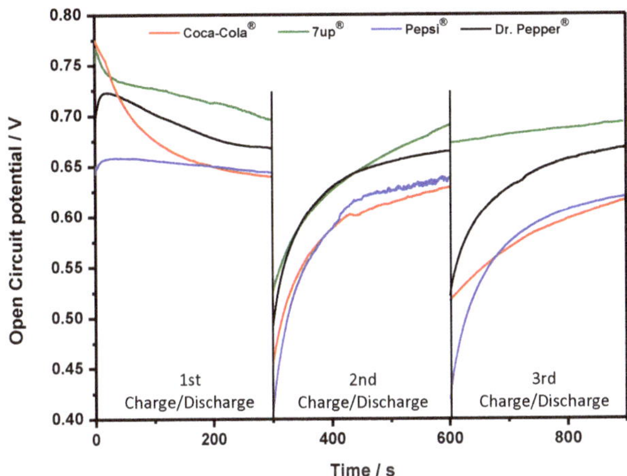

Figure 7. Open circuit potential stability, for a paper-based μFCs for cases 7up, Coca Cola, Dr. Pepper, and Pepsi.

These tests (Figure 7) evaluated the capabilities of the paper-based μFC to recover after three cycles of charging/discharging its open circuit potential. The time required for this test was determined based on the drying of 15 μL of soft drink (10 to 15 min). The charge–discharge experiments using OCP refer to the following: first, the OCP is measured vs. time until it reaches its maximum value (stability), then a discharge curve is realized (voltage vs. current) until the voltage reaches zero. After that, the recovery of the OCP vs. time is measured again; this process is done three times. In this way, we analyze the OCP obtain after each cycle of charge/discharge. It can observe an increase in potential as soon as the fuel comes into contact with the electrodes showing that the oxidation reaction of the glucose is taking place. In addition, the potential is stabilized at 700, 670, 650, and 640 mV for 7up, Dr. Pepper, Pepsi, and Coca Cola, respectively. During the charging period, the potential tends to return to the initial value at which it stabilized and remains constant as time passes and with each cycle. These results indicate that the devices are quite stable after charging/discharging cycles so that only one of them could be occupied in a successive manner for the generation of electricity using soft drinks as fuel.

For comparative purposes, Table 3 shows the results obtained in different reports in which they have used soft drinks as fuel and this work. It is seen that the obtained maximum values are within the range of those already reported for these fuels. However, this work offers the advantage of the use of devices that do not use enzymatic anodes, so the lifetime of the paper-based μFCs is bigger, and the cost of generating electricity would be much lower.

3.3. *Soft-Drinks Paper-Based Microfluidic Fuel Cell Performance Stack as a Possible Backup Power Supply*

The 2-cell and 4-cell stacks were developed and demonstrated the usefulness of the paper-based μFCs as a backup possible power supply. The polarization and power density curves for each fuel cell-stack operated with Coca Cola as fuel are showed in Figure 8. Coca Cola was chosen as fuel for this test because it is one of the most consumed soft drinks in the world, so it would have a higher chance of being used in a real application.

Table 3. Different reports where soft drinks have been used as a fuel.

Soft Drink	Fuel Cell Type	OCP /V	PMAX /mW cm^{-2}	Reference
Vegetable juice	Miniature glucose/air biofuel cell	0.71	0.245	[18]
Nutri-Express	Miniature origami biofuel cell	0.3	≈0.008	[19]
Coca Cola	Miniature origami biofuel cell	≈0.09	≈0.0006	[19]
Gatorade	3-cell stack paper based biofuel cell	1.8	0.216	[20]
7up	Miniature self-pumping paper-based enzymatic biofuel cell	0.31	0.0135	[21]
Mountain Dew	Miniature self-pumping paper-based enzymatic biofuel cell	0.39	0.012	[21]
Pepsi	Miniature self-pumping paper-based enzymatic biofuel cell	0.32	0.00615	[21]
Coca Cola at room temperature	Paper-based microfluidic fuel cell	0.6	0.061	This work
Cold 7up	Paper-based microfluidic fuel cell	0.64	0.12	This work
Cold Coca-Cola	2 and 4-cell stack paper-based microfluidic fuel cell	1.07 1.6	0.27 0.39	This work This work

An increase in open circuit potential and maximum power density was found when a serial connection for 2 and 4 paper-based μFC (1.07, 1.6 V and 0.27, 0.4 mW cm^{-2}, respectively) was realized respect to a single a paper-based μFC (0.6 V). However, according to the Kirchhoff's voltage law, an optimal doubling and quadrupling of the 2-cell and 4-cell performance were expected; nevertheless, this was not found due losses associated with the resistance imposed by the electric circuit used in the stack and some contact deficiencies of the electrodes with the conductor wires because its small size makes it difficult to handle.

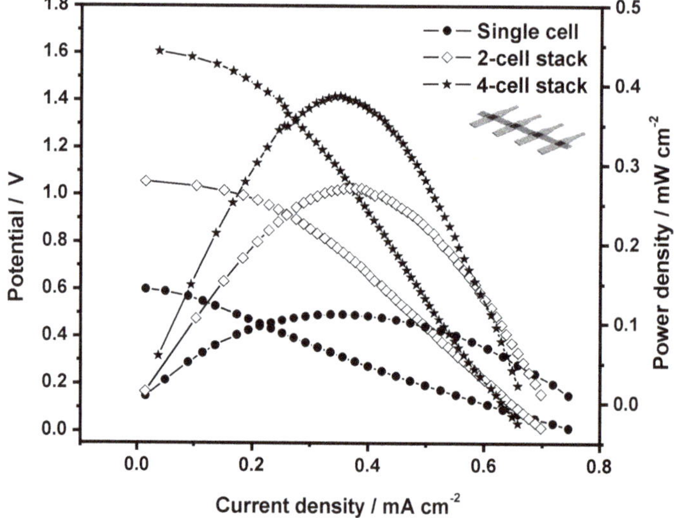

Figure 8. Polarization and power density curves of PtRu/C and Pt/C as anode and cathode, respectively in 2- and 4-cell paper-based microfluidic stacks using Coca Cola as fuel and air as oxidant.

4. Conclusions

In this work, the soft drinks have been used as an energy source in paper-based μFCs. In addition, the use of cold soft drinks showed a better performance than soft drinks at room temperature. However, in a possible real application, the temperature would increase as a function of time, causing a decrease in the performance of the fuel cell. The study of this decrease in performance with respect to temperature could be of interest for future research. On the other hand, the material used (commercial PtRu/C) as a catalyst for the oxidation of the saccharides showed that glucose oxidation is predominant in the presence of maltose and fructose so that the performance of the devices was directly affected by the glucose concentration. We foresee that combining the advantages in the materials field to use other catalysts on the paper-based μFCs would lead to better performances. The design of the presented paper-based μFC has been inspired in a single flow. In this sense, the study of the flow velocity induced by capillary action respect to the configuration of the fuel cell could simulate through software such as ANSYS. The use of different soft drinks as energy sources represents a great contribution to the field of fuel cells due to the price, availability, and environmentally friendly of these beverages. This suggests that the use of soft drinks as fuels in paper-based μFCs constitutes a hopeful alternative for small portable electronic devices. In addition, the use of other easily accessible fuels could be considered.

Author Contributions: Conceptualization, J.H.R. and D.D.; methodology, H.E.E.P.; validation, J.M.O.R., A.D.; and J.R.-R.; formal analysis, D.O.D.; investigation, R.G.; resources, D.M.A.C. All authors have read and agreed to the published version of the manuscript

Funding: The authors appreciate the founding of CONACyT for "Cátedras CONACyT" project 513 and PRODEP.

Acknowledgments: The authors gratefully acknowledge to UAQ, UTSJR, SEP and CONACyT.

Conflicts of Interest: The authors declare no conflict of interest.

References

1. Xie, L.; Li, J.; Cai, S.; Li, X. Design and experiments of a self-charged power bank by harvesting sustainable human motion. *Adv. Mech. Eng.* **2016**, *8*, 1687814016651371. [CrossRef]
2. Alanís-Navarro, J.; Reyes-Betanzo, C.; Moreira, J.; Sebastian, P. Fabrication and characterization of a micro-fuel cell made of metallized PMMA. *J. Power Sources* **2013**, *242*, 1–6. [CrossRef]
3. Esquivel, J.; Senn, T.; Hernández-Fernández, P.; Santander, J.; Lörgen, M.; Rojas, S.; Löchel, B.; Cané, C.; Sabaté, N. Towards a compact SU-8 micro-direct methanol fuel cell. *J. Power Sources* **2010**, *195*, 8110–8115. [CrossRef]
4. Moreno-Zuria, A.; Dector, A.; Cuevas-Muñiz, F.; Esquivel, J.; Sabaté, N.; Ledesma-García, J.; Arriaga, L.; Chávez-Ramírez, A. Direct formic acid microfluidic fuel cell design and performance evolution. *J. Power Sources* **2014**, *269*, 783–788. [CrossRef]
5. Hashemi, S.; Neuenschwander, M.; Hadikhani, P.; Modestino, M.; Psaltis, D. Membrane-less micro fuel cell based on two-phase flow. *J. Power Sources* **2017**, *348*, 212–218. [CrossRef]
6. Morales-Acosta, D.; Godinez, L.A.; Arriaga, L. Performance increase of microfluidic formic acid fuel cell using Pd/MWCNTs as catalyst. *J. Power Sources* **2010**, *195*, 1862–1865. [CrossRef]
7. Esquivel, J.; Del Campo, F.; De La Fuente, J.G.; Rojas, S.; Sabate, N. Microfluidic fuel cells on paper: meeting the power needs of next generation lateral flow devices. *Energy Environ. Sci.* **2014**, *7*, 1744–1749. [CrossRef]
8. González-Guerrero, M.J.; Del Campo, F.J.; Esquivel, J.P.; Leech, D.; Sabaté, N. based microfluidic biofuel cell operating under glucose concentrations within physiological range. *Biosens. Bioelectron.* **2017**, *90*, 475–480. [CrossRef]
9. Dector, A.; Galindo-De-La-Rosa, J.; Amaya-Cruz, D.; Ortíz-Verdín, A.; Guerra-Balcázar, M.; Olivares-Ramírez, J.; Arriaga, L.; Ledesma-García, J. Towards autonomous lateral flow assays: Paper-based microfluidic fuel cell inside an HIV-test using a blood sample as fuel. *Int. J. Hydrogen Energy* **2017**, *42*, 27979–27986. [CrossRef]
10. Kjeang, E.; Djilali, N.; Sinton, D. Microfluidic fuel cells: A review. *J. Power Sources* **2009**, *186*, 353–369. [CrossRef]
11. Lau, C.; Moehlenbrock, M.J.; Arechederra, R.L.; Falase, A.; Garcia, K.; Rincon, R.; Minteer, S.D.; Banta, S.; Gupta, G.; Babanova, S.; et al. Paper based biofuel cells: Incorporating enzymatic cascades for ethanol and methanol oxidation. *Int. J. Hydrogen Energy* **2015**, *40*, 14661–14666. [CrossRef]
12. Copenhaver, T.S.; Purohit, K.H.; Domalaon, K.; Pham, L.; Burgess, B.J.; Manorothkul, N.; Galvan, V.; Sotez, S.; Gomez, F.A.; Haan, J.L. A microfluidic direct formate fuel cell on paper. *Electrophoresis* **2015**, *36*, 1825–1829. [CrossRef]
13. Chino, I.; Muneeb, O.; Do, E.; Ho, V.; Haan, J.L. A paper microfluidic fuel cell powered by urea. *J. Power Sources* **2018**, *396*, 710–714. [CrossRef]
14. Castillo-Martínez, L.; Amaya-Cruz, D.; Gachuz, J.; Ortega-Díaz, D.; Olivares-Ramírez, J.; Dector, D.; Duarte-Moller, A.; Villa, A.; Dector, A. Urea oxidation in a paper-based microfluidic fuel cell using Escherichia coli anode electrode. *J. Phys. Conf. Ser.* **2018**, *1119*, 012004. [CrossRef]

15. González-Guerrero, M.J.; del Campo, F.J.; Esquivel, J.P.; Giroud, F.; Minteer, S.D.; Sabaté, N. based enzymatic microfluidic fuel cell: From a two-stream flow device to a single-stream lateral flow strip. *J. Power Sources* **2016**, *326*, 410–416. [CrossRef]
16. Shyu, J.C.; Wang, P.Y.; Lee, C.L.; Chang, S.C.; Sheu, T.S.; Kuo, C.H.; Huang, K.L.; Yang, Z.Y. Fabrication and test of an air-breathing microfluidic fuel cell. *Energies* **2015**, *8*, 2082–2096. [CrossRef]
17. Dector, D.; Olivares-Ramírez, J.; Ovando-Medina, V.; Dominguez, A.S.; Villa, A.; Duarte-Moller, A.; Sabaté, N.; Esquivel, J.; Dector, A. Fabrication and evaluation of a passive SU8-based micro direct glucose fuel cell. *Microsyst. Technol.* **2019**, *25*, 211–216. [CrossRef]
18. Wen, D.; Xu, X.; Dong, S. A single-walled carbon nanohorn-based miniature glucose/air biofuel cell for harvesting energy from soft drinks. *Energy Environ. Sci.* **2011**, *4*, 1358–1363. [CrossRef]
19. Yu, Y.; Han, Y.; Lou, B.; Zhang, L.; Han, L.; Dong, S. A miniature origami biofuel cell based on a consumed cathode. *Chem. Commun.* **2016**, *52*, 13499–13502. [CrossRef]
20. Villarrubia, C.W.N.; Lau, C.; Ciniciato, G.P.; Garcia, S.O.; Sibbett, S.S.; Petsev, D.N.; Babanova, S.; Gupta, G.; Atanassov, P. Practical electricity generation from a paper based biofuel cell powered by glucose in ubiquitous liquids. *Electrochem. Commun.* **2014**, *45*, 44–47. [CrossRef]
21. Rewatkar, P.; Goel, S. Microfluidic paper based membraneless biofuel cell to harvest energy from various beverages. *J. Electrochem. Sci. Eng.* **2020**, *10*, 49–54. [CrossRef]
22. Dector, A.; Arjona, N.; Guerra-Balcázar, M.; Esquivel, J.; Campo, F.D.; Sabaté, N.; Ledesma-García, J.; Arriaga, L. Non-Conventional Electrochemical Techniques for Assembly of Electrodes on Glassy Carbon-Like PPF Materials and Their Use in a Glucose Microfluidic Fuel-Cell. *Fuel Cells* **2014**, *14*, 810–817. [CrossRef]
23. López Zavala, M.Á.; González Peña, O.I.; Cabral Ruelas, H.; Delgado Mena, C.; Guizani, M. Use of Cyclic Voltammetry to Describe the Electrochemical Behavior of a Dual-Chamber Microbial Fuel Cell. *Energies* **2019**, *12*, 3532. [CrossRef]
24. Olivares-Ramírez, J.; Ovando-Medina, V.; Ortíz-Verdín, A.; Amaya-Cruz, D.; Coronel-Hernandez, J.; Marroquín, A.; Dector, A. Lateral flow assay HIV-based microfluidic blood fuel cell. *J. Phys. Conf. Ser.* **2018**, *1119*, 012022. [CrossRef]
25. Ovando-Medina, V.; Dector, A.; Antonio-Carmona, I.; Romero-Galarza, A.; Martínez-Gutiérrez, H.; Olivares-Ramírez, J. A new type of air-breathing photo-microfluidic fuel cell based on ZnO/Au using human blood as energy source. *Int. J. Hydrogen Energy* **2019**, *44*, 31423–31433. [CrossRef]
26. Basu, D.; Basu, S. Synthesis, characterization and application of platinum based bi-metallic catalysts for direct glucose alkaline fuel cell. *Electrochim. Acta* **2011**, *56*, 6106–6113. [CrossRef]
27. Walker, R.W.; Dumke, K.A.; Goran, M.I. Fructose content in popular beverages made with and without high-fructose corn syrup. *Nutrition* **2014**, *30*, 928–935. [CrossRef]
28. Haynes, T.; Dubois, V.; Hermans, S. Particle size effect in glucose oxidation with Pd/CB catalysts. *Appl. Catal. A Gen.* **2017**, *542*, 47–54. [CrossRef]
29. Amani-Beni, Z.; Nezamzadeh-Ejhieh, A. NiO nanoparticles modified carbon paste electrode as a novel sulfasalazine sensor. *Anal. Chim. Acta* **2018**, *1031*, 47–59. [CrossRef]
30. Ibrahim, A.A.; Umar, A.; Amine, A.; Kumar, R.; Al-Assiri, M.; Al-Salami, A.; Baskoutas, S. Highly sensitive and selective non-enzymatic monosaccharide and disaccharide sugar sensing based on carbon paste electrodes modified with perforated NiO nanosheets. *New J. Chem.* **2018**, *42*, 964–973. [CrossRef]
31. Mahshid, S.S.; Mahshid, S.; Dolati, A.; Ghorbani, M.; Yang, L.; Luo, S.; Cai, Q. Template-based electrodeposition of Pt/Ni nanowires and its catalytic activity towards glucose oxidation. *Electrochim. Acta* **2011**, *58*, 551–555. [CrossRef]
32. Basu, D.; Basu, S. A study on direct glucose and fructose alkaline fuel cell. *Electrochim. Acta* **2010**, *55*, 5775–5779. [CrossRef]
33. Morrison, R.; Boyd, R. *Organic Chemistry*, 5th ed.; Allyn and Bacon: Boston, MA, USA, 1987.
34. Tao, B.; Zhao, K.; Miao, F.; Jin, Z.; Yu, J.; Chu, P.K. Fabrication of ordered porous silicon nanowires electrode modified with palladium-nickel nanoparticles and electrochemical characteristics in direct alkaline fuel cell of carbohydrates. *Ionics* **2016**, *22*, 1891–1898. [CrossRef]
35. Liu, X.; Long, L.; Yang, W.; Chen, L.; Jia, J. Facilely electrodeposited coral-like copper micro-/nano-structure arrays with excellent performance in glucose sensing. *Sens. Actuators B Chem.* **2018**, *266*, 853–860. [CrossRef]
36. Perez, J.; Villullas, H.M.; Gonzalez, E.R. Structure sensitivity of oxygen reduction on platinum single crystal electrodes in acid solutions. *J. Electroanal. Chem.* **1997**, *435*, 179–187. [CrossRef]

37. Su, F.; Poh, C.K.; Tian, Z.; Xu, G.; Koh, G.; Wang, Z.; Liu, Z.; Lin, J. Electrochemical behavior of Pt nanoparticles supported on meso-and microporous carbons for fuel cells. *Energy Fuels* **2010**, *24*, 3727–3732. [CrossRef]
38. Hsueh, K.; Gonzalez, E.; Srinivasan, S. Electrolyte effects on oxygen reduction kinetics at platinum: A rotating ring-disc electrode analysis. *Electrochim. Acta* **1983**, *28*, 691–697. [CrossRef]
39. Gewirth, A.A.; Thorum, M.S. Electroreduction of dioxygen for fuel-cell applications: Materials and challenges. *Inorg. Chem.* **2010**, *49*, 3557–3566. [CrossRef]

MDPI
St. Alban-Anlage 66
4052 Basel
Switzerland
Tel. +41 61 683 77 34
Fax +41 61 302 89 18
www.mdpi.com

Energies Editorial Office
E-mail: energies@mdpi.com
www.mdpi.com/journal/energies

www.ingramcontent.com/pod-product-compliance
Lightning Source LLC
LaVergne TN
LVHW070950170726
843515LV00005B/956